Calculus Co

Connecting Middle School
and College Mathematics
(CM)²

Calculus Connections
Mathematics for Middle School Teachers

Asma Harcharras
Department of Mathematics
University of Missouri-Columbia

Dorina Mitrea
Department of Mathematics
University of Missouri-Columbia

PEARSON
Prentice
Hall

Upper Saddle River, New Jersey

Library of Congress Cataloging-in-Publication Data

Harcharras, Asma.

    Calculus connections : connecting middle school and college mathematics (CM)^2 / Asma Harcharras, Dorina Mitrea.

      p. cm—(Prentice Hall series in mathematics for middle school teachers)

    Includes bibliographical references and index.

    ISBN 0-13-144923-0

    1. Calculus—Study and teaching (Middle School) 2. Calculus—Study and teaching (Higher) 3. Mathematical readiness—Study and teaching (Middle school) I. Title: Connecting middle school and college mathematics (CM)^2 II. Mitrea, Dorina, 1965- III. Title. IV. Series.

QA303.3.H37 2007
515'.071'2—dc22

2005052348

Editor in Chief: *Sally Yagan*
Executive Acquisitions Editor: *Petra Recter*
Project Manager: *Michael Bell*
Production Management: *Progressive Publishing Alternatives*
Assistant Managing Editor: *Bayani Mendoza de Leon*
Senior Managing Editor: *Linda Mihatov Behrens*
Executive Managing Editor: *Kathleen Schiaparelli*
Manufacturing Manager: *Alexis Heydt-Long*
Manufacturing Buyer: *Maura Zaldivar*
Marketing Assistant: *Jennifer de Leeuwerk*
Art Director: *Jayne Conte*
Cover Designer: *Bruce Kenselaar*
Art Studio/Formatter: *Laserwords*
Editorial Assistant/Supplement Editor: *Joanne Wendelken*
Cover Image: *Chad Ehlers/Stock Connection*

 ©2007 Pearson Education, Inc.
Pearson Prentice Hall
Pearson Education, Inc.
Upper Saddle River, New Jersey 07458

Pearson Prentice Hall™ is a trademark of Pearson Education, Inc.
Development of these materials was supported by a grant from the National Science Foundation (ESI 0101822).

Printed in the United States of America

ISBN: 0-13-144923-0

Pearson Education LTD., *London*
Pearson Education Australia PTY, Limited, *Sydney*
Pearson Education Singapore, Pte. Ltd.
Pearson Education North Asia Ltd., *Hong Kong*
Pearson Education Canada, Ltd., *Toronto*
Pearson Educación de Mexico, S.A. de C.V.
Pearson Education—Japan, *Tokyo*
Pearson Education Malaysia, Pte. Ltd.

To our husbands, Bill and Marius.

Contents

Preface

THE $(CM)^2$ PROJECT

Improving the quality of mathematics education for middle school students is of critical importance at this time, and creating new opportunities for students to learn important mathematical skills under the leadership of well-prepared and dedicated teachers is essential. New standards-based curricula and instructional models, coupled with on-going professional development and teacher preparation, are fundamental to this effort.

These sentiments are eloquently articulated in the Glenn Commission Report: *Before It's Too Late: A Report to the Nation from the National Commission on Mathematics and Science Teaching for the 21st Century* (U.S. Department of Education, 2000). In fact, the principal message of the Glenn Commission Report is that America's students must improve their mathematics and science performance if they are to be successful in our rapidly changing technological world. To this end, the Report recommends that we greatly intensify our focus on improving the quality of mathematics and science teaching in grades K–12 by bettering the quality of teacher preparation, and it also stresses the necessity of developing creative plans to attract and retain substantial numbers of future mathematics and science teachers.

Some fifteen years ago, mathematics teachers, mathematics educators, and mathematicians collaborated to develop the architecture for standards-based reform, and their recommendations for the improvement of school mathematics, instruction, and assessment have been articulated in three seminal documents published by the National Council of Teachers of Mathematics: *Curriculum and Evaluation Standards for School Mathematics* (1989), *Professional Standards for School Mathematics* (1991), and *Assessment Standards in School Mathematics* (1995). More recently, these three documents have been updated and combined into a single book, *NCTM Principles and Standards for School Mathematics* (2000).

The vision of school mathematics laid out in these three foundational documents is outstanding in spirit and content, yet it has remained abstract in practice. Concrete exemplary models reflecting those standards have been lacking, and implementations of the recommendations could not be realized without a significant commitment of resources. Recognizing an opportunity to stimulate improvement in student learning, the National Science Foundation (NSF) made a strong commitment to bring life to the documents' messages by supporting several K–12 mathematics curriculum development projects (standards-based curriculum) and a number of other related dissemination and implementation projects.

Standards-based middle school curricula are designed to engage students in a variety of mathematical experiences, including thoughtfully planned explorations that provide and reinforce fundamental skills while illuminating the power and utility of mathematics in our world. These materials integrate central concepts in algebra, geometry, data analysis and probability, and mathematics of change, and they focus on important unifying ideas such as proportional reasoning.

The mathematical content of standards-based middle grade mathematics materials is challenging and relevant to our technological world. Its effective classroom implementation is dependent upon teachers having strong and appropriate mathematical preparation. *The Connecting Middle School and College Mathematics Project (CM)^2* is a three-year (2001–2004) NSF funded project addressing the need for improved teacher qualifications and viable recruitment plans for middle grade mathematics teachers through the development of four foundational mathematics courses with accompanying support materials and the creation and implementation of effective teacher recruitment models.

The $(CM)^2$ materials are built upon a framework laid out in the *CBMS Mathematical Education of Teachers Book* (MET) (2001). This report outlines recommendations for the mathematical preparation of middle grade teachers that differ significantly from those for the preparation of elementary teachers and provides guidance to those developing new programs. Our books are designed to provide middle grade mathematics teachers with a strong mathematical foundation and connect the mathematics they are learning with the mathematics they will be teaching. Their focus is on algebraic and geometric structures, data analysis and probability, and mathematics of change, and they employ standards-based middle grade mathematics curricular materials as a springboard to explore and learn mathematics in more depth. They have been extensively piloted in Summer Institutes, courses offered at school-based sites, a variety of professional development programs, and in courses offered at a number of universities throughout the nation.

CALCULUS CONNECTIONS

Although the majority of in-service or pre-service middle school mathematics teachers clearly realize the importance of developing a solid understanding of algebra, data analysis, probability, and geometry as a necessary preparation for the teaching of mathematics in middle school, the subject of calculus is often viewed in a very different light. For many of these teachers, calculus is seen as a lofty and abstract subject which is "higher" than the mathematics they will teach in the classroom and therefore of little practical value. Many fail to understand the importance of learning calculus as a preparation for the teaching of mathematics in the middle school even though the ideas and principles of calculus lie at the very heart and soul of middle school mathematics. The reason is quite clear: calculus courses do not address these issues!

One of the recommendations outlined in MET for the preparation of middle school mathematics teachers is the development and incorporation of a calculus course that focuses on concepts and applications. In this spirit, we have sought to create a calculus course specifically designed to better prepare middle school mathematics teachers. In this course, we aim to illuminate the connections that exist between the various topics of calculus and the mathematics curricula taught in the middle school.

Philosophy and Structure.   Our underlying approach to learning is one of discovery and inquiry. Instead of the usual formula-proof-example approach to calculus, here we initiate each new topic with a Classroom Discussion or a Classroom Connection. The Classroom Discussions are carefully structured sequences of questions designed to lead students to discover the underlying principles of calculus. Detailed answers have been provided for all of the Classroom Discussions at the end of the book which gives students the opportunity to independently revisit the ideas discussed in the classroom. The Classroom Connections are activities from the middle school curricula related to the material to be discussed, and which teachers might someday use in their classrooms. They provide excellent motivation for studying the calculus concepts that are subsequently introduced. The Classroom Discussions are followed by examples and Practice Problems. We also give projects and extensions which either focus on applications of the material or provide explorations into new material. The projects and extensions are important tools for further involving students with more in-depth reasoning by sometimes addressing topics that go beyond what is in the text. Each chapter closes with a summary of the topics covered in the chapter and a set of review exercises.

Content.   Our selection of topics for this book is based upon a careful study of the existing middle school mathematics curricula. In Chapter 1 we cover sequences and series. Although some calculus textbooks do not include these topics, it is our belief that they should be an integral part of a calculus course offered to future middle school teachers, a view supported by the large number of connections with the middle school curricula included in this chapter. In Chapter 2 we deal with functions—a concept well anchored in the middle school curricula—starting from concrete examples, then fully developing the notion of limit of a function, and closing with a section on continuous functions. Chapter 3 is devoted to differentiation. We start by analyzing average rates of change, move on to instantaneous rates of change and connections with slopes of tangent lines and motion along a straight line, and then prove the main formulas for differentiation. In Chapter 4 we cover optimization, graph sketching, and exponential change. The problems in this chapter underscore the power of calculus when dealing with real-life problems. In Chapter 5, we introduce the notion of integration and discuss basic techniques for evaluating integrals. Integration (or antidifferentiation) is the process that reverses differentiation, and it naturally has a wide variety of practical applications. In this chapter, we discuss how integration can be used to solve various problems related to the study of motion. Chapters 6 and 7 are devoted to exploring applications of integration to various topics from the middle school mathematics curricula: finding areas, lengths, surface areas, and volumes. In these chapters, we begin by computing the area, surface area, and volume of simple figures using Euclidean geometry. Then, building on the results obtained from geometry, we use calculus to produce formulas which hold for more general shapes. Our presentation stresses the interplay between geometry and calculus, and it shows that calculus is a powerful and essential tool for computing areas, lengths, surface areas, and volumes.

Throughout the book there is an emphasis on making a clear distinction between what constitutes a proof and what constitutes an informal explanation. The expectation is that, in the process, students will gradually grow accustomed to a certain standard of rigor that they will, in turn, take back to their classrooms.

Intended Audience.   This book is primarily designed for the mathematical preparation of pre-service and in-service middle school teachers. It can be used to prepare those students who will not take the traditional sequence of calculus courses, as well as those who have already taken the standard sequence of calculus courses but need to strengthen their understanding of calculus and get exposure to its connections to middle school curricula.

By design, the book can serve as the basis for either an undergraduate course or a graduate course, depending on the structure of the course and the manner in which the connections and projects are incorporated. In particular, the projects can be used to provide supplementary material when the course is offered for graduate credit. While piloting this material, the authors assigned some of the projects contained in the book as individual papers and as team projects. The individual papers were aimed at providing students with a deeper understanding of a variety of calculus concepts and/or exposure to topics not covered in traditional calculus courses. The team projects were aimed at connecting a college level calculus topic to middle/secondary school mathematics topics chosen from standards-based curricula.

To the Instructor.   While we envision the Classroom Discussions as classroom activities to be completed in groups or individually, due to time constraints instructors may choose to assign some of these to be done outside the classroom. Depending on the goals of the course and the mathematical background of the students, instructors might choose to use calculators more extensively, particularly to avoid spending time on computations involving differentiation and integration when the focus is on concepts. Throughout the book, exercises requiring a computing device are preceded by the symbol $\langle T \rangle$.

To the Student.   Although the answers to the questions raised in the Classroom Discussions and Practice Problems are presented in full detail at the end of the book, it would be a mistake to view them until after you have spent a sufficient amount of time thinking about the questions on your own. Keep in mind that the best way to learn is by trial and error, and perseverance. Ideally, the answers should be consulted only to double-check your own results. This is the main reason that answers have been included in the book.

Acknowledgments.   The authors would like to thank all of the instructors and faculty members who helped with piloting the materials or providing valuable suggestions, and the graduate students David Barker, Dustin Foster, Dustin Jones, Jessica Ostrom, Anna Skripka, and Chris Thornhill for their contributions. Dorina Mitrea would also like to thank Matthew Wright for his help. Finally, we thank the reviewers for their constructive comments, Brenda Frazier for providing technical assistance, Heather Meledin at Progressive Publishing Alternatives, and Petra Recter and Michael Bell at Prentice-Hall for their help in getting this book into print.

REFERENCES FOR THE PREFACE

Conference Board of Mathematical Sciences (2001). *The Mathematical Education of Teachers Book.* Washington, DC: Mathematical Association of America.

National Council of Teachers of Mathematics (1989). *Curriculum and Evaluation Standards for School Mathematics.* Reston, VA: National Council of Teachers of Mathematics.

National Council of Teachers of Mathematics (1991). *Professional Standards for School Mathematics.* Reston, VA: National Council of Teachers of Mathematics.

National Council of Teachers of Mathematics (1995). *Assessment Standards in School Mathematics.* Reston, VA: National Council of Teachers of Mathematics.

National Council of Teachers of Mathematics (2000). *Principles and Standards for School Mathematics.* Reston, VA: National Council of Teachers of Mathematics.

U.S. Department of Education (2000). *Before It's Too Late: A Report to the Nation from the National Commission on Mathematics and Science Teaching for the 21st Century.* John Glenn, Commission Chairman, Washington, DC: U.S. Department of Education.

List of Classroom Connections

List of Classroom Discussions

List of Projects

Calculus Connections

Sequences and Series

1.1 SEQUENCES
1.2 SERIES

Sequences and series have both fascinated and bewildered mankind for centuries. They are indispensable tools in the study of the mathematics of change, particularly in the context of differential and integral calculus. Both of these fundamental concepts involve the idea of carrying out a process "forever." Informally, a sequence is the output of a process involving infinitely many steps, while series formalize the act of summing up infinitely many numbers. In this section, we introduce key words such as *convergence*, *limit*, and *sum* through intuitive examples; we then define them rigorously.

1.1 SEQUENCES

Sequences • Arithmetic sequences • Geometric sequences • Limits of sequences • Convergent sequences

You have worked with many number patterns in your life. The most common example is the list of counting numbers, 1, 2, 3, ... This is an example of a sequence. In general, a sequence is an ordered list of objects (most typically numbers) called **terms**.

Classroom Connection 1.1.1: Number Strips and Dot Patterns

The following exploration is from student pages 1–4 in the eighth-grade textbook *Mathematics in Context, Patterns and Figures*. Working in groups, answer questions 1–8.

Student Page 1

A. PATTERNS

Patterns are at the heart of mathematics, and you can find patterns by looking at shapes, numbers, and many other things. In this unit you will discover and explore patterns and describe them with numbers and formulas.

Number Strips

Below, numbers starting with 0 are shown on a paper strip. The strip has alternating red and white colors.

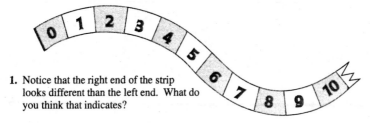

1. Notice that the right end of the strip looks different than the left end. What do you think that indicates?

2. **a.** What do the white numbers have in common?

   b. Think of a large number not shown on the strip. How can you tell the color for your number?

Student Page 2

Here is a different strip made with the repeating pattern red-white-blue-red-white-blue.

Red `0`

White `1`

Blue ■

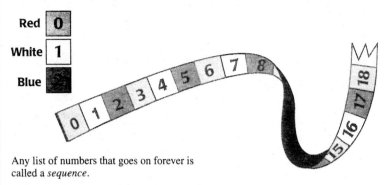

Any list of numbers that goes on forever is called a *sequence*.

3. How can you figure out the color for 253,679?

One way to "see" a pattern is to use dots to represent numbers. For example, the red numbers from the red and white strip on page 1 can be drawn like this:

Dot Pattern:	••	•• ••	•• •• ••	•• •• •• ••	•• •• •• •• •• etc.
Pattern Number:	0	1	2	3	4	5

Below each dot pattern is a pattern number. The pattern number tells you where you are in a sequence. (Notice that the pattern number starts with 0, and there are no dots for pattern number 0.) Pattern number 1 shows two dots, pattern number 2 shows four dots, and so on.

4. a. Look at the dot pattern for the red numbers. When the pattern number is 37, how many dots are there?

    b. Someone came up the formula $R = 2n$ for the red numbers. What do you think R and n stand for?

    c. Does the formula work? Explain your answer.

Student Page 3

The white numbers from the red-white strip on page 1 can be drawn like this:

Dot Pattern:	o	o oo	o oo oo	o oo oo oo	o oo oo oo oo	o oo oo oo oo oo etc.
Pattern Number:	0	1	2	3	4	5

5. a. Now look at the pattern for the white numbers. How many dots are in pattern number 50?

   b. Write a formula for the white numbers.

> MIRIAM TOLD ME THAT IF YOU ADD TWO ODD NUMBERS, YOU GET AN EVEN NUMBER.

6. a. Use dots to explain Miriam's rule.

   b. Make up some other rules like Miriam's, and use dots to explain them.

Student Page 4

The sequence of even numbers (0, 2, 4 , 6, 8, . . .) can be described
by the formula:

> START NUMBER = 0
> NEXT EVEN NUMBER = CURRENT EVEN NUMBER + 2

You have seen these NEXT-CURRENT formulas in previous *Mathematics in Context* units.
They are also called *step-by-step formulas* or *recursive formulas*.

7. **a.** Write a step-by-step formula for the sequence of odd numbers (1, 3, 5, 7, . . .).

  b. Compare the step-by-step formulas for even and odd numbers. What is the same
  and what is different?

> A formula such as $R = 2n$ for even numbers or $W = 2n + 1$ for odd
> numbers is called a *direct formula*.

8. Why do you think these are called direct formulas?

> Some graphing calculators let you create a sequence of numbers
> using either direct or recursive formulas. You can use **Student
> Activity Sheet 1** to investigate sequences with a calculator.

12 Section A Patterns   **Britannica Mathematics System**

◆

  A good way to keep track of the order of terms in a sequence is to use subscripts.
In the examples from Classroom Connection 1.1.1, we can denote the sequence of
even numbers by $x_0 = 0, x_1 = 2, x_2 = 4, x_3 = 6, \ldots$ and the sequence of odd numbers
by $y_0 = 1, y_1 = 3, y_2 = 5, y_3 = 7, \ldots$. We denote the **general terms** by x_n and y_n,

respectively. You can write the formulas you have discovered for the general terms of these sequences as $x_n = 2n$, and $y_n = 2n + 1$, where $n = 0, 1, 2, \ldots$. We refer to these formulas as **direct formulas** because they allow us to compute the nth term of a sequence explicitly. A condensed way of denoting a sequence is $\{x_n\}_{n \geq 0}$.

The set of counting numbers is denoted by $\mathbb{N}$, also called the **set of natural numbers**. We use the notation $n \in \mathbb{N}$ if n is an element of the set of natural numbers $\mathbb{N}$. As such, a shorthand notation for $n = 1, 2, 3, 4, \ldots$ is $n \in \mathbb{N}$.

It is important to point out that there are many ways to denote a given sequence. For example, the sequence of even whole numbers $0, 2, 4, 6, \ldots$ can also be denoted by $z_1 = 0$, $z_2 = 2$, $z_3 = 4$, $z_4 = 6$, $\ldots$, with the formula for the general term being $z_n = 2(n - 1)$, $n \in \mathbb{N}$. In this case, a condensed notation for the sequence is $\{z_n\}_{n \geq 1}$ or simply $\{z_n\}_n$.

We generally say that we have a **recursive formula** for a sequence whenever we can compute a given term from some or all of the previous terms. For example, the sequence of even numbers $x_n = 2n$, $n = 0, 1, 2, 3, \ldots$ from Classroom Connection 1.1.1 has the recursive formula "next even number = current even number + 2." If the next even number is x_{n+1}, then the recursive formula can be written as $x_{n+1} = x_n + 2$, $n \geq 0$. Observe that if we want to recover the sequence $x_n = 2n$, $n = 0, 1, 2, 3, \ldots$ from the recursive formula $x_{n+1} = x_n + 2$, $n \geq 0$, we also have to specify that $x_0 = 0$. Any other value for x_0 in combination with the same recursive formula yields another sequence. For example, $x_0 = 1$ and $x_{n+1} = x_n + 2$, $n \geq 0$ yields the sequence $1, 3, 5, 7, 9, 11, \ldots$, while $x_0 = 0.5$ and $x_{n+1} = x_n + 2$, $n \geq 0$ yields the sequence $0.5, 2.5, 4.5, 6.5, 8.5, 10.5, \ldots$. For the rest of the book, in order to ensure that the sequences described by recursive formulas are completely determined, we will specify the first few terms of a sequence (also called the **initial values** of the sequence); this is an integral part of the task of determining a recursive formula for each sequence.

EXAMPLES

1. Let $\{x_n\}_n$ be the sequence whose first few terms are $1, 8, 27, 64, 125, \ldots$. Determine both a direct formula for the general term and a recursive formula for a sequence whose first few terms coincide with the given ones.

2. Let $\{x_n\}_n$ be the sequence whose first few terms are $1, \frac{1}{2}, \frac{1}{3}, \frac{1}{4}, \ldots$. Determine both a direct formula for the general term and a recursive formula for a sequence whose first few terms coincide with the given ones.

3. Write the first five terms of the sequence whose recursive formula is $x_{n+2} = x_{n+1} - x_n$, $n \geq 1$, with initial values $x_1 = 1$ and $x_2 = 4$.

Solutions

1. A formula for the general term is $x_n = n^3, n \geq 1$. A recursive formula for this sequence is $x_{n+1} = (\sqrt[3]{x_n} + 1)^3$ for $n \geq 1$, with initial value $x_1 = 1$.

2. A formula for the general term is $x_n = \frac{1}{n}$ for $n \geq 1$. To determine a recursive formula for the sequence, we observe that $x_1 = 1$ and that

$$x_{n+1} = \frac{1}{n+1} = \frac{1}{\frac{1}{x_n} + 1} = \frac{x_n}{x_n + 1} \text{ for } n \geq 1.$$

3. We set $n = 1, n = 2$, and $n = 3$ in the recursive formula to obtain $x_3 = x_2 - x_1 = 3$, $x_4 = x_3 - x_2 = -1$, and $x_5 = x_4 - x_3 = -4$, respectively. Hence, the first five terms of the sequence are $1, 4, 3, -1$, and -4. ∎

Practice Problems

1. Let $\{x_n\}_n$ be the sequence whose first few terms are $2, \frac{3}{2}, \frac{4}{3}, \frac{5}{4}, \frac{6}{5}, \ldots$. Determine a direct formula for a sequence whose first few terms coincide with the given ones.

2. Let $\{y_n\}_n$ be the sequence whose first few terms are $1, 0, -2, -5, -9, -14, \ldots$. Determine a recursive formula for a sequence whose first few terms coincide with the given ones.

3. Write the first five terms of the sequence whose recursive formula is $x_{n+2} = 3x_{n+1} - 2x_n, n \geq 1$, with initial values $x_1 = 2$ and $x_2 = 3$.

Historical Note: Fractals.   In 1975, in an article published in *Scientific American* magazine, Benoit Mandelbrot used the term **fractal** to describe shapes that are "self-similar," meaning they look the same at different magnifications. For a fractal, the final level of detail is never fully observed by the human eye and is only gradually revealed by increasing the scale at which observations are made. To create a fractal, one starts with a simple shape and duplicates it successively according to a set of fixed rules. While the procedure has a simple formula, it can create very complex structures that sometimes resemble real-world objects such as trees, coastlines, and so forth.

Classroom Connection 1.1.2: Fractals and Sequences

The following exploration is from pages 239–240 and page 4-52 in the eighth-grade textbook *Math Thematics, Book 3*. Complete tasks 3–6 and task 8 in Exploration 1. For each sequence discussed, write a recursive formula and determine a direct formula for the general term.

Exploration

Fractals & Sequences

SET UP *You will need: • Labsheet 1A • ruler • 4 different-colored pencils*

GOAL

LEARN HOW TO...
- write rules for sequences

AS YOU...
- explore fractals

KEY TERMS
- fractal
- sequence
- term
- self-similar

Use Labsheet 1A for Questions 3–5.

3 Follow the steps below to draw a *Fractal Tree*. Fill in the table on the labsheet through Step 3.

Step 1

The first branch of the tree has been drawn for you on the labsheet.

Step 2

Add three branches to the first branch. Make each of them 1 in. long.

Step 3

Add three branches, each $\frac{1}{2}$ in. long, to each branch added in Step 2.

Section 1 Fractals, Sequences, and Triangles **239**

4 Compare the length of the branch in Step 1 and the length of a branch added in Step 2. How are they related? How is the length of each branch added in Step 3 related to the length of a branch added in Step 2?

5 Add the new branches for Steps 4 and 5 to your fractal tree. Then fill in the table on the labsheet for these steps.

▶ **Describing Sequences** You may have noticed some number patterns in the table on Labsheet 1A. These patterns are called *sequences*. A **sequence** is an ordered list of numbers or objects called **terms**. You can write rules to describe some sequences.

EXAMPLE

Rule: To find a term of the sequence below, add 4 to the previous term.

Term number	1	2	3	4	5	6
Term	3	7	11	15	19	23

$$7 + 4 = 11$$

6 Suppose the first term in a sequence is 3, and the rule for finding a term is to multiply the previous term by 4. Give the first 6 terms of this sequence.

7 **Use Labsheet 1A.** Write a rule for the sequence in each column on the labsheet. Then use your rules to complete Steps 6 and 7 on the labsheet.

...checks that you can write a rule for a sequence.

8 ✔ **CHECKPOINT** Write a rule for finding a term of each sequence. Then give the next three terms of the sequence.

a. 2, 4, 8, 16, ...   b. $3x$, $3x^2$, $3x^3$, $3x^4$, ...

c. 0.3, 0.03, 0.003, ...   d. 400, 299, 198, ...

▶ **A fractal is self-similar if it is made of smaller pieces that are similar to the whole figure. The self-similar figure you created on Labsheet 1A is not a complete fractal. The fractal is the figure you would get if the sequence was carried out forever. The human eye could never see all the parts of a fractal.**

(240) **Module 4**   Patterns and Discoveries

Name _____   Date _____

MODULE 4	LABSHEET 1A

Fractal Tree (Use with Questions 3–5 and Question 7 on pages 239–240.)

Directions Follow the directions in your book to create a fractal tree and complete the table below. Use a different color for the branches at each new step.

Step	Number of new branches added	Length of each new branch (inches)	Height of tree (inches)	Total length of new branches (inches)	Total length of all branches (inches)
1	1	2	2	2	2
2	3	1	3	3	5
3					
4					
5					
6					
7					

◆

Two types of sequences are particularly important: arithmetic sequences and geometric sequences.

Arithmetic Sequences.   A sequence is an **arithmetic sequence** if the next term in the sequence is found by adding a fixed number d to the current term. Hence, $\{x_n\}_{n \geq 1}$ is an arithmetic sequence provided $x_{n+1} = x_n + d$ for $n \geq 1$. Both x_1 (the first term in the sequence) and d are given. Why do you think d is called the **common difference**?

Classroom Discussion 1.1.1: Direct Formula for Arithmetic Sequences

1. Return to the sequences in Classroom Connection 1.1.2 and find a recursive relation for each sequence. Determine which ones are arithmetic. Explain.

2. John asked Michelle, "Is there a way to compute any term in an arithmetic sequence if you know the first term and the common difference?" Michelle answered "You can do that by adding to the first term a suitable multiple of the common difference." Do you agree with Michelle?

3. If the sequence $\{x_n\}_{n \geq 1}$ is an arithmetic sequence with common difference d, explain how to obtain the formula

$$x_n = x_1 + (n - 1) d \text{ for each } n \in \mathbb{N}.$$

4. What does this formula give for each arithmetic sequence from Classroom Connection 1.1.2? ◆

Geometric Sequences.   A sequence is a **geometric sequence** if the next term in the sequence is found by multiplying the current term by a fixed number r. Thus, $\{x_n\}_{n \geq 1}$ is a geometric sequence provided $x_{n+1} = x_n \cdot r$ for $n \geq 1$. Both x_1 and r are given. Of course, if $x_1 = 0$ or $r = 0$, then $x_n = 0$ for every n. Thus, it is typically assumed that $x_1 \neq 0$ and $r \neq 0$, in which case $x_n \neq 0$ for every n. Why do you think r is called the **common ratio**?

Classroom Discussion 1.1.2: Direct Formula for Geometric Sequences

1. Return to the sequences in Classroom Connection 1.1.2 and determine which ones are geometric. Explain.

2. Alice asked Bob, "Is there a way to compute any term in a geometric sequence if you know the starting term and the common ratio?" Bob answered "You can do that by multiplying the starting term by the common ratio raised to a suitable power." Do you agree with Bob?

3. If $\{y_n\}_{n \geq 1}$ is a geometric sequence with common ratio r, explain how to obtain the formula

$$y_n = y_1 \cdot r^{n-1} \text{ for each } n \in \mathbb{N}.$$

4. What does this formula give for each geometric sequence from Classroom Connection 1.1.2? ◆

Limits of Sequences.   When analyzing sequences, we are often interested in understanding what happens to the value of x_n as n gets larger and larger. As you can imagine, the possibilities are numerous. However, sequences whose general term x_n gets closer and closer to some fixed number as n gets larger and larger turn out to be particularly important. In order to better understand this idea, work through the tasks in Classroom Discussion 1.1.3.

Classroom Discussion 1.1.3: The Circumference of a Circle and Perimeters of Inscribed Polygons

1. Consider a circle of radius r. What is the standard formula in terms of π and r for its circumference L?

2. Take $r = 4$ inches and use the formula you determined from Problem 1 to find an approximate value for L rounded to four decimal places.

3. Now analyze the following construction:

    a. Draw a circle C of radius $r = 4$ and center O, and draw a square inscribed in C. Denote the perimeter of this square by p_4.

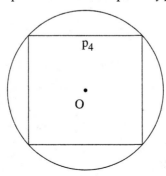

b. Inscribe an octagon in the circle C whose vertices contain the square's vertices. Denote the octagon's perimeter by p_8.

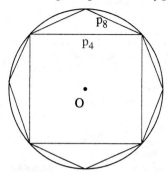

c. Continue this procedure of inscribing regular polygons in the circle C. At each step, the polygon you construct should have twice as many sides and should include all the vertices of the previous polygon. If a polygon has n vertices, its perimeter is denoted by p_n. The following figure shows the result of the first three steps of the procedure.

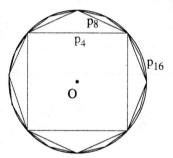

Describe what you expect to happen to the inscribed polygons as the number of sides increases.

4. Consider the general case of a regular n-gon inscribed in the circle C. Write a formula for its perimeter p_n. Your answer should depend on n. Hint: Look at the following figure.

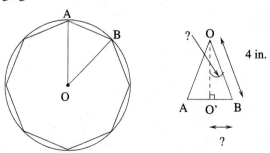

5. Use your calculator to generate a table with columns n and p_n for $n = 3, 4, 5, \ldots, 102$. Create a scatter plot with the first forty values in the table. On the same graph, draw the line $y = L$, where L is the circumference of the circle you computed in Problem 2.

6. As the number of sides n increases, use the table and the scatter plot to describe what happens to the values of the perimeters of the regular n-gons inscribed in the circle C. How do these values compare to the circle's circumference? Is your conclusion in line with the prediction you made in Problem 3?

7. What is the best approximation for the circumference of the circle you get from the table in Problem 5? What would you suggest doing to get an even better approximation? ◆

Convergent Sequences

Consider what would happen if you "forever" continued constructing inscribed polygons with more and more sides. This nonstop process of letting n become larger and larger is denoted by $n \to \infty$ (n goes to infinity). The values p_n of the perimeters of the regular n-gons inscribed in the circle C would approach the circumference L of the circle C. The larger n is, the better p_n approximates L, the circle's circumference. Think of how the scatter plot looked in Problem 6. You can make p_n as close to L as you want by choosing n sufficiently large. We say that *the limit of the sequence* $\{p_n\}_n$ *as n goes to infinity is L*, and we write $\lim_{n\to\infty} p_n = L$. Alternatively, we say that the sequence $\{p_n\}_n$ **converges** to L.

Important Observation

Return to the scatter plot from Problem 6, that is, the plot of points (n, p_n), $n = 3, 4, 5, \ldots, 40$. Imagine continuing to plot these points for all $n \geq 41$. Next, choose an open interval on the y-axis containing the value $y = L = 8\pi$. We want to know how many p_n's will remain outside this interval. To do so, imagine drawing the horizontal band containing all the points whose y coordinates are in the interval. Given our plot, at most a finite number of the points (n, p_n) will be outside the band. No matter how narrow the band you select, as long as it contains the line $y = 8\pi$, all the points (n, p_n) will be in the band starting from some value $n = N_0$. Thus, any open interval containing L will have all but a finite number of p_n's. This idea is formalized in the following definition.

> **Definition 1.1.1** A sequence $\{x_n\}_n$ is said to **converge** to a real number L provided any open interval centered at L contains all but finitely many terms of the sequence. In this case, L is called **the limit** of the sequence $\{x_n\}_n$, and we write
>
> $$\lim_{n\to\infty} x_n = L.$$

A sequence that converges to a real number L is called **convergent**; a sequence that does not converge to any real number is called **divergent**.

Remarks

1. A sequence $\{x_n\}_n$ with the property that $x_n < x_{n+1}$ for all n is called **increasing**. For example, the previous sequence $\{p_n\}_n$ is increasing. If a sequence $\{x_n\}_n$ satisfies the condition that $x_n \leq x_{n+1}$ for all n, it is called **nondecreasing**.

2. A sequence $\{x_n\}_n$ with the property that $x_n > x_{n+1}$ for all n is called **decreasing**. The sequence $\{\frac{1}{2^n}\}_n$ is an example of a decreasing sequence. If a sequence $\{x_n\}_n$ satisfies the condition that $x_n \geq x_{n+1}$ for all n, it is called **nonincreasing**.

Classroom Discussion 1.1.4: Limits of Sequences

1. Consider the sequence $x_n = \frac{1}{n}, n \geq 1$.

   a. What happens to the x_n's as $n \to \infty$?

   b. Consider the open intervals that are centered at 1 and that have lengths 4, 3, and 2, respectively. How many terms of the sequence $\{x_n\}_n$ are left outside each interval? Can you conclude that $\lim_{n \to \infty} x_n = 1$?

   c. Now look at the intervals $(\frac{1}{2}, \frac{3}{2})$, $(\frac{9}{10}, \frac{11}{10})$, $(\frac{99}{100}, \frac{101}{100})$ and decide how many x_n's are outside each interval. Do your findings agree with your previous conclusion regarding $\lim_{n \to \infty} x_n$?

   d. What candidate would you suggest for $\lim_{n \to \infty} \frac{1}{n}$? Use Definition 1.1.1 to prove that your guess is correct.

2. Consider the sequence $y_n = n, n \geq 1$.

   a. What happens to the y_n's as $n \to \infty$?

   b. Consider the open intervals $(-10, 10)$, $(-100, 100)$, and $(-1,000, 1,000)$. How many terms of the sequence $\{y_n\}_n$ are left outside each of these intervals? Can you conclude that $\lim_{n \to \infty} y_n = 0$?

   c. Is there a number L that satisfies $\lim_{n \to \infty} n = L$? To prove your conclusion, use Definition 1.1.1 and reason by contradiction.

   d. Show that the sequence $\{y_n\}_n$ has the property that for each fixed positive integer M, all but finitely many terms are larger than M. This type of behavior for the sequence $\{y_n\}_n$ is denoted by $\lim_{n \to \infty} y_n = \infty$.

   e. What can you say about $\lim_{n \to \infty} (2n + 100)$?

   f. Is there a number L that satisfies $\lim_{n \to \infty} (-n) = L$?

   g. Show that the sequence $z_n = -n, n \geq 1$ has the property that for each fixed negative integer M, all but finitely many terms are smaller than M. This type of behavior for the sequence $\{z_n\}_n$ is denoted by $\lim_{n \to \infty} z_n = -\infty$.

   h. What can you say about $\lim_{n \to \infty} (-n^2 + 1)$? ◆

EXAMPLE   Use Definition 1.1.1 to prove that $\lim_{n \to \infty} \frac{1}{n+1} = 0$.

Solution   Let a be an arbitrary positive real number. If $a \geq 1$, then all the terms of the sequence $x_n = \frac{1}{n+1}, n \geq 1$ are contained in the interval $(-a, a)$. If $0 < a < 1$, then $\frac{1}{a} - 1 > 0$, and for all natural numbers $n > \frac{1}{a} - 1$ we have $\frac{1}{n+1} \in (-a, a)$. Hence, only the x_n's with $n \leq \frac{1}{a} - 1$ are left outside the interval $(-a, a)$, and there are only finitely many such n's. Since $a > 0$ was arbitrary, it follows that any open interval centered at 0 contains all but finitely many x_n's. By Definition 1.1.1, we can conclude that $\lim\limits_{n \to \infty} \frac{1}{n+1} = 0$. ∎

Practice Problem

Use Definition 1.1.1 to prove that $\lim\limits_{n \to \infty} \frac{2}{n+1} = 0$.

Some Properties of Limits of Sequences.   Relying only on Definition 1.1.1 when studying the convergence of sequences can lead to involved computations. To simplify matters, whenever possible we will use the following properties. Except for P4 (Property 4), all these properties can be proved based on Definition 1.1.1. We omit their proofs.

P1 The property of being convergent is not affected by changing/omitting a finite number of terms in the sequence. In particular, the sequence $\{x_n\}_{n \in \mathbf{N}}$ converges to L if and only if the sequence $\{x_n\}_{n \geq k}$ converges to L, where k is any positive integer.

P2 A convergent sequence has a unique limit.

P3 If a sequence $\{x_n\}_n$ is convergent, then it is **bounded**, i.e., there exists $M > 0$ such that $|x_n| \leq M$ for all $n \in \mathbf{N}$. A sequence that is not bounded is called **unbounded**. An **unbounded** sequence is necessarily divergent.

Historical Note: Karl Theodor Wilhelm Weierstrass (1815–1897; from Ostenfelde, Westphalia [now Germany]).   Karl Weierstrass was one of the rigor in analysis leaders and eventually became known as the "father of modern analysis." In addition, he is considered one of the greatest mathematics teachers of all time. To satisfy his father's wishes, Karl entered the University of Bonn in 1834 to pursue a career as an accountant. However, he soon realized that this was not his true vocation and returned home after four years without a degree. His father was not pleased and had Karl enroll at the Theological and Philosophical Academy at Munster where he received a teacher's certificate in 1841. For the next fifteen years, he was employed as a secondary mathematics teacher, during which time he continued to work mathematics. In 1854, he published a paper on Abelian functions that startled the mathematical community. Almost immediately, he received an honorary doctorate from the University of Konigsberg and, in 1856, was offered an associate professorship at the University of Berlin. Weierstrass is famous in mathematics for numerous accomplishments. He was the first person to construct a continuous function that is not differentiable at any point. He developed a general theory of Abelian integrals and functions. He used power series to describe analytic functions and introduced the sequential definition of irrational numbers based on convergent series. However, he is most famous for his insistence on rigor in all his works, especially analysis, by demanding that mathematics be based on clear and correct proofs. In particular, he would not release his work until he was sure it was on a firm mathematical foundation.

P4 Weierstrass's theorem: A nondecreasing sequence is convergent if and only if it is bounded. A nonincreasing sequence is convergent if and only if it is bounded.

P5 Let $\{x_n\}_n$, $\{y_n\}_n$ be two convergent sequences, and let c be a real number. Then the following are true:

a. The sequence $\{x_n + y_n\}_n$ is convergent with

$$\lim_{n \to \infty} (x_n + y_n) = \lim_{n \to \infty} x_n + \lim_{n \to \infty} y_n.$$

b. The sequence $\{x_n - y_n\}_n$ is convergent with

$$\lim_{n \to \infty} (x_n - y_n) = \lim_{n \to \infty} x_n - \lim_{n \to \infty} y_n.$$

c. The sequence $\{c \cdot y_n\}_n$ is convergent with

$$\lim_{n \to \infty} (c \cdot y_n) = c \cdot \lim_{n \to \infty} y_n.$$

d. The sequence $\{x_n \cdot y_n\}_n$ is convergent with

$$\lim_{n \to \infty} (x_n \cdot y_n) = \lim_{n \to \infty} x_n \cdot \lim_{n \to \infty} y_n.$$

e. If, in addition, $y_n \neq 0$ for all n and $\lim_{n \to \infty} y_n \neq 0$, then the sequence $\{\frac{x_n}{y_n}\}_n$ is convergent with

$$\lim_{n \to \infty} \frac{x_n}{y_n} = \frac{\lim_{n \to \infty} x_n}{\lim_{n \to \infty} y_n}.$$

P6 The sandwich theorem for sequences: Suppose the sequences $\{x_n\}_n$, $\{y_n\}_n$, and $\{z_n\}_n$ are such that

a. $x_n \leq y_n \leq z_n$ for each n, and
b. $\lim_{n \to \infty} x_n = \lim_{n \to \infty} z_n = L$ for some $L \in \mathbb{R}$.

Then, the sequence $\{y_n\}_n$ is convergent and $\lim_{n \to \infty} y_n = L$.

P7 Assume two convergent sequences $\{x_n\}_n$, $\{y_n\}_n$ are such that $x_n \leq y_n$ for each n. Then, $\lim_{n \to \infty} x_n \leq \lim_{n \to \infty} y_n$.

Note that even if two convergent sequences $\{x_n\}_n$, $\{y_n\}_n$ are such that $x_n < y_n$ for each n, we can conclude only that $\lim_{n \to \infty} x_n \leq \lim_{n \to \infty} y_n$ and not that $\lim_{n \to \infty} x_n < \lim_{n \to \infty} y_n$. Think of $x_n = \frac{1}{n}$ and $y_n = \frac{1}{n+1}$, $n = 1, 2, 3 \ldots$.

EXAMPLES

1. Compute $\lim_{n \to \infty} (-1)^n \frac{1}{n}$.

2. In a–d, decide if the given sequence is convergent or divergent. If convergent, determine its limit.

    a. $a_n = \frac{1}{n^2}$ **b.** $b_n = \frac{2n}{n+7}$

    c. $c_n = 1 + (-1)^n$ **d.** $d_n = \frac{1+(-1)^n}{n}$

3. Assume that $\{x_n\}_n$ is a sequence with the property that $\lim_{n\to\infty} |x_n| = 0$. Prove that $\lim_{n\to\infty} x_n = 0$.

Solutions

1. Since $-\frac{1}{n} \le (-1)^n \frac{1}{n} \le \frac{1}{n}$ for each n, $\lim_{n\to\infty} \frac{1}{n} = 0$ (proved in Classroom Discussion 1.1.4), and $\lim_{n\to\infty} (-\frac{1}{n}) = -\lim_{n\to\infty} \frac{1}{n} = 0$ (we used P5), by the sandwich theorem for sequences, we can conclude that $\lim_{n\to\infty} (-1)^n \frac{1}{n} = 0$.

2.   **a.** We apply the sandwich theorem for sequences: $0 \le a_n \le \frac{1}{n}$, for all $n \ge 1$, $\lim_{n\to\infty} 0 = 0$, $\lim_{n\to\infty} \frac{1}{n} = 0$, thus $\lim_{n\to\infty} \frac{1}{n^2} = 0$.

    b. We apply P5 after factoring and simplifying out n from the numerator and denominator:

$$\lim_{n\to\infty} \frac{2n}{n+7} = \lim_{n\to\infty} \frac{2}{1 + \frac{7}{n}} = \frac{\lim_{n\to\infty} 2}{\lim_{n\to\infty} (1 + \frac{7}{n})}$$

$$= \frac{2}{\lim_{n\to\infty} 1 + 7 \lim_{n\to\infty} \frac{1}{n}} = \frac{2}{1 + 7 \cdot 0} = 2.$$

    c. We have

$$c_n = \begin{cases} 0 & \text{if } n \text{ is odd} \\ 2 & \text{if } n \text{ is even.} \end{cases}$$

The sequence $\{c_n\}_n$ is divergent. Here is a proof of this fact. Suppose there exists a real number L, such that $\lim_{n\to\infty} c_n = L$. If $L = 0$, consider the interval $(-1, 1)$ centered around 0. Then, infinitely many terms of the sequence will be outside this interval: $c_n \notin (-1, 1)$ for $n = 2, 4, 6, 8, \ldots$. According to Definition 1.1.1, the sequence $\{c_n\}_n$ does not converge to 0. Hence, $L \ne 0$. In this case, we choose either the interval $(0, 2L)$ if $L > 0$ or the interval $(2L, 0)$ if $L < 0$, and the terms c_n will be outside both intervals for $n = 1, 3, 5, 7, \ldots$. This contradicts the fact that L is the limit of the sequence $\{c_n\}_n$. Consequently, there is no real number L such that $\lim_{n\to\infty} c_n = L$, which means that $\{c_n\}_n$ is divergent.

    d. We have $0 \le d_n \le \frac{2}{n}$ for all n. By P5 it follows that $\lim_{n\to\infty} \frac{2}{n} = 2 \cdot \lim_{n\to\infty} \frac{1}{n} = 2 \cdot 0 = 0$. Next, we apply the sandwich theorem for sequences to conclude that $\lim_{n\to\infty} d_n = 0$.

3. Since $-|x_n| \le x_n \le |x_n|$ for each n, $\lim_{n \to \infty} |x_n| = 0$ (given), and $\lim_{n \to \infty} (-|x_n|) = \lim_{n \to \infty} [(-1) \cdot |x_n|] = (-1) \cdot \lim_{n \to \infty} |x_n| = -\lim_{n \to \infty} |x_n| = 0$ (we used P5), by the sandwich theorem for sequences, we conclude that $\lim_{n \to \infty} x_n = 0$. ∎

Practice Problems

In a–d, decide if the sequence is convergent or divergent. If convergent, determine its limit.

a. $a_n = \dfrac{1}{n+2}$

b. $b_n = \dfrac{-n^2}{3n^2 + n + 5}$

c. $c_n = (-1)^n 2$

d. $d_n = \dfrac{(-1)^n 2}{n^2}$

Classroom Discussion 1.1.5: Convergence Criteria for Arithmetic and Geometric Sequences

1. Give three examples of arithmetic sequences: one with common difference $d = 0$, one with $d > 0$, and one with $d < 0$.

2. Determine which sequences are convergent and, in each case, determine the limit. You may, using a calculator, generate a table with two columns: n and the nth term of the sequence, for $n = 1, 2, \ldots, 100$.

3. Fill in the blanks:

   a. An arithmetic sequence $\{x_n\}_n$ with positive common difference is _____ and $\lim_{n \to \infty} x_n = $ _____.

   b. An arithmetic sequence $\{x_n\}_n$ with negative common difference is _____ and $\lim_{n \to \infty} x_n = $ _____.

   c. An arithmetic sequence $\{x_n\}_n$ with common difference equal to zero is _____ and $\lim_{n \to \infty} x_n = $ _____.

   Explain your reasoning.

4. Give three examples of geometric sequences with common ratio r in each of the following cases a–d. For each example, you may use a calculator to generate a table with two columns: n and the nth term of the sequence, for $n = 1, 2, \ldots, 100$.

   a. $r = 1$

   b. $r = -1$

   c. $|r| > 1$ (consider both $r > 0$ and $r < 0$)

   d. $|r| < 1$ (consider both $r > 0$ and $r < 0$)

5. Determine which sequences are convergent.

6. Fill in the blank:

A geometric sequence converges if and only if the common ratio _____.
Explain your reasoning. ◆

EXAMPLES   In a–f, determine the limit.

a. $\lim\limits_{n\to\infty} [2 - 3(n - 1)]$

b. $\lim\limits_{n\to\infty} 4$

c. $\lim\limits_{n\to\infty} \left(\frac{1}{2} + \frac{n-1}{3} \right)$

d. $\lim\limits_{n\to\infty} (-3)^{n+1}$

e. $\lim\limits_{n\to\infty} \frac{(-3)^n}{5^{n+1}}$

f. $\lim\limits_{n\to\infty} \frac{7^{n-1}}{2^n}$

Solutions

a. The sequence $x_n = 2 - 3(n - 1)$, $n \geq 1$ is arithmetic with the first term $x_1 = 2$ and common difference $d = -3$. Since $d \neq 0$, the sequence is not convergent. In addition, $\lim\limits_{n\to\infty} x_n = -\infty$.

b. The sequence $x_n = 4$, $n \geq 1$ is arithmetic with the first term $x_1 = 4$ and common difference $d = 0$, thus it is convergent with $\lim\limits_{n\to\infty} x_n = 4$.

c. The sequence $x_n = \frac{1}{2} + \frac{n-1}{3}$, $n \geq 1$ is arithmetic with the first term $x_1 = \frac{1}{2}$ and common difference $d = \frac{1}{3}$. Since $d \neq 0$, the sequence is not convergent. In addition, $\lim\limits_{n\to\infty} x_n = \infty$.

d. The sequence $x_n = (-3)^{n+1}$, $n \geq 1$ is geometric with the first term $x_1 = 9$ and common ratio $r = -3$. Since $r < -1$, the sequence is divergent, and $\lim\limits_{n\to\infty} x_n$ does not exist.

e. The sequence $x_n = \frac{(-3)^n}{5^{n+1}}$, $n \geq 1$ is geometric with the first term $x_1 = -\frac{3}{25}$ and common ratio $r = -\frac{3}{5}$. Since $|r| < 1$, the sequence is convergent with $\lim\limits_{n\to\infty} x_n = 0$.

f. The sequence $x_n = \frac{7^{n-1}}{2^n}$, $n \geq 1$ is geometric with the first term $x_1 = \frac{1}{2}$ and common ratio $r = \frac{7}{2}$. Since $r > 1$, the sequence is not convergent. In addition, $\lim\limits_{n\to\infty} x_n = \infty$.   ∎

Practice Problems

In a–f, determine the limit.

a. $\lim\limits_{n\to\infty} (-2)$

b. $\lim\limits_{n\to\infty} \left(-\frac{3}{4} - \frac{2n-2}{5} \right)$

c. $\lim\limits_{n\to\infty} [1 + 2(n - 1)]$

d. $\lim\limits_{n\to\infty} \frac{(-5)^{n+1}}{6^{n-1}}$

e. $\lim\limits_{n\to\infty} \frac{(-7)^n}{4^{n+2}}$

f. $\lim\limits_{n\to\infty} \frac{-4^n}{3^n}$

Historical Note: Johann Carl Friedrich Gauss (1777–1855; from Brunswick, Duchy of Brunswick [now Germany]).   The mathematical potential of Carl Friedrich Gauss was noticed in elementary school. It was there that he instantly summed the integers from 1 to 100 by spotting that the sum was 50 pairs of numbers, each summing to 101. By the time Gauss received a degree from Brunswick in 1799, he had made one of his most important discoveries—the construction of a regular 17-gon by ruler and compass, the most major advance in this field since the time of Greek mathematics. This was published as Section VII of Gauss's monumental work *Disquisitiones Arithmeticae*. Shortly after that, Gauss submitted a doctoral dissertation to the University of Helmstedt, which was a collection of proofs of the fundamental theorem of algebra. Gauss's contributions are numerous and include results in number theory, differential equations, conic sections, series, hypergeometric functions, potential theory, and terrestrial magnetism. He is considered to be one of the greatest mathematicians of all times, alongside Archimedes (287–212 BC) and Isaac Newton (1643–1727).

Classroom Discussion 1.1.6: Adding Consecutive Terms of an Arithmetic Sequence

Is there a fast way to add a given number of consecutive terms in an arithmetic sequence? To answer this question, try the following investigation:

1. Start with the sum $1 + 2 + 3 + 4 + \cdots + 999 + 1,000$. Are the numbers we want to add consecutive terms of an arithmetic sequence?
2. Compute the sums $1 + 1,000, 2 + 999, 3 + 998, \ldots$. What do you observe?
3. Use the observation in Problem 2 to evaluate the original sum without using a calculator. Look at the following diagram:

$$
\begin{array}{ccccccccc}
1 & + & 2 & + & 3 & + \cdots + & 999 & + & 1,000 \\
1,000 & + & 999 & + & 998 & + \cdots + & 2 & + & 1 \\
\hline
1,001 & + & 1,001 & + & 1,001 & + \cdots + & 1,001 & + & 1,001
\end{array}
$$

4. In his preteen years, the mathematician C. F. Gauss (1777–1855) discovered the idea you used in Problem 3 to add consecutive terms in an arithmetic sequence. Use similar reasoning to compute the sum $1 + 3 + 5 + 7 + 9 + \cdots + 997 + 999$ without a calculator.
5. Let $x_1, \ldots, x_n$ be the first n terms in an arithmetic sequence with first term $x_1 = a$ and common difference d. Recall from Classroom Discussion 1.1.1 that $x_n = a + (n - 1)d$, for each $n \in \mathbb{N}$. Is it possible to express the sum $x_1 + x_2 + \cdots + x_n$ entirely in terms of a, d, and n? Look at the following diagram:

$$
\begin{array}{ccccc}
a & + & (a + d) & + \cdots + & [a + (n - 1)d] \\
[a + (n - 1)d] & + & [a + (n - 2)d] & + \cdots + & a \\
\hline
[2a + (n - 1)d] & + & [2a + (n - 1)d] & + \cdots + & [2a + (n - 1)d]
\end{array}
$$

6. Compute the following sums without using a calculator:

   $2 + 4 + 6 + 8 + 10 + \cdots + 1,998 + 2,000$

   $4 + 7 + 10 + 13 + 16 + \cdots + 3,001 + 3,004 + 3,007$

   $97 + 95 + 93 + 91 + \cdots + 9 + 7 + 5$ ◆

The Summation Notation. The mathematical symbol used for writing sums in a condensed way is the Greek letter sigma, Σ. If $\{x_n\}_{n\geq1}$ is an arbitrary sequence, the sum $x_4 + x_5 + \cdots + x_{20}$ can be written as $\sum_{i=4}^{20} x_i$. In general, if $n \leq m$ are whole numbers, then

$$x_n + \cdots + x_m = \sum_{i=n}^{m} x_i.$$

The index i can be replaced by any other letter, and the sum does not change. In this sense, $\sum_{i=4}^{20} x_i = \sum_{j=4}^{20} x_j = \sum_{k=4}^{20} x_k$, etc. Other examples include

$$x_1 + x_2 + \cdots + x_{15} = \sum_{i=1}^{15} x_i,$$

$$x_2 + x_4 + \cdots + x_{14} = \sum_{i=1}^{7} x_{2i},$$

$$x_3 + x_5 + x_7 + \cdots + x_{23} = \sum_{i=1}^{11} x_{2i+1}.$$

In some cases, sums are written explicitly with no reference to a particular sequence. Take $3 + 5 + 7 + \cdots + 99$. In order to use the summation notation, we must identify the numbers added with terms of a particular sequence for which a direct formula exists. In this case, $3 + 5 + 7 + \cdots + 99 = \sum_{j=1}^{49} (2j + 1)$. Here are a few more examples:

$$1 + 2 + 3 + 4 + \cdots + 1{,}000 = \sum_{i=1}^{1{,}000} i,$$

$$5 + 10 + 15 + 20 + \cdots + 135 = \sum_{i=1}^{27} 5i.$$

Classroom Discussion 1.1.7: Adding Consecutive Terms of a Geometric Sequence

Is there a fast way to add a given number of consecutive terms in a geometric sequence? To answer this question, try the following investigation.

1. Start with the sum $S = 3 + 3 \cdot 2 + 3 \cdot 2^2 + \cdots + 3 \cdot 2^{2{,}002} + 3 \cdot 2^{2{,}003}$. Are the numbers we want to add consecutive terms of a geometric sequence?
2. Multiply the identity defining S by 2 and then show that $2S + 3 = S + 3 \cdot 2^{2{,}004}$.
3. Solve for S in the equation in Problem 2 without computing $2^{2{,}004}$.
4. Use similar reasoning to write the sum $S = 2 + 1 + \frac{1}{2} + \cdots + \frac{1}{2^{100}} + \frac{1}{2^{101}}$ in a condensed form.

5. Let $y_1, y_2, \ldots, y_n$ be the first n terms of a geometric sequence with starting term $y_1 = a$ and common ratio r. Recall that in this case, $y_n = ar^{n-1}$ for each $n \in \mathbb{N}$ (see Classroom Discussion 1.1.2). The goal is to express the sum $\sum_{i=1}^{n} y_i$ in terms of a, r, and n. Start by showing that

$$r(y_1 + y_2 + \cdots + y_n) = y_2 + y_3 + \cdots + y_n + y_{n+1}.$$

6. Denote the sum $\sum_{i=1}^{n} y_i$ by S and prove that $rS + a = S + ar^n$.

7. Assuming $r \neq 1$, solve for S in the equation in Problem 6.

8. Express the sum $\sum_{i=1}^{n} y_i$ in terms of a and n in the case $r = 1$.

9. Evaluate the sum $\sum_{i=1}^{2,002} 2 \cdot 3^{i-1}$. ◆

Here is a summary of the results you proved in this section for arithmetic and geometric sequences.

Arithmetic Sequences

If $\{x_n\}_{n \geq 1}$ is an arithmetic sequence with common difference d, then

- $x_{n+1} = x_n + d$, $n \geq 1$, where x_1 given, is the recursive formula for this sequence;
- $x_n = x_1 + (n - 1)d$, $n \geq 1$ is the direct formula for this sequence;
- the sequence is convergent if and only if $d = 0$;
- if $x_1 = a$, then the sum of the first n consecutive terms in the sequence is

$$\sum_{i=1}^{n} x_i = \sum_{i=1}^{n} [a + (i - 1)d] = na + \frac{n(n-1)}{2} d.$$

Geometric Sequences

If $\{y_n\}_{n \geq 1}$ is a geometric sequence with common ratio r, then

- $y_{n+1} = y_n \cdot r$, $n \geq 1$, where y_1 given, is the recursive formula for this sequence;
- $y_n = y_1 \cdot r^{n-1}$, $n \geq 1$ is the direct formula for this sequence;
- the sequence is convergent if and only if $r = 1$, $|r| < 1$, or $y_1 = 0$;
- if $y_1 = a$, then the sum of the first n consecutive terms in the sequence is

$$\sum_{i=1}^{n} y_i = \sum_{i=1}^{n} a \cdot r^{i-1} = \begin{cases} a \dfrac{r^n - 1}{r - 1} & \text{if } r \neq 1 \\ na & \text{if } r = 1. \end{cases}$$

EXERCISES 1.1

In Exercises 1–6, determine a recursive formula for a sequence whose first few terms coincide with the given ones.

1. $5, 15, 45, 135, 405, \ldots$

2. 7, 14, 28, 56, 112, . . .

3. 4, 10, 16, 22, 28, . . .

4. 5, 2, −1, −4, −7, . . .

5. 3, 6, 9, 15, 24, 39, 63, . . .

6. 1, 2, 3, 5, 8, 13, 21, . . .

In Exercises 7–12, determine a recursive formula for the given sequence. Give the value of the sequence's first term.

7. An arithmetic sequence $\{x_n\}_{n\geq1}$ with $x_2 = 2$ and a common difference of -3

8. An arithmetic sequence $\{x_n\}_{n\geq1}$ with $x_3 = 3$ and a common difference of $\frac{4}{3}$

9. A geometric sequence $\{y_n\}_{n\geq1}$ with $y_3 = 1$ and a common ratio of $-\frac{1}{4}$

10. A geometric sequence $\{y_n\}_{n\geq1}$ with $y_2 = 1$ and a common ratio of $\frac{3}{4}$

11. A geometric sequence $\{y_n\}_{n\geq1}$ with $y_3 = 6$ and $y_5 = 24$

12. An arithmetic sequence $\{x_n\}_{n\geq1}$ with $x_8 = 40$ and $x_{20} = -20$

In Exercises 13–18, determine a direct formula for a sequence whose first few terms coincide with the given ones.

13. 100, 98, 96, 94, 92, . . .

14. $\frac{4}{3}, \frac{5}{3}, 2, \frac{7}{3}, \frac{8}{3}, \ldots$

15. 1,024, 512, 256, 128, 64, . . .

16. $2, -\frac{2}{3}, \frac{2}{9}, -\frac{2}{27}, \frac{2}{81}, \ldots$

17. 9, 16, 25, 36, 49, . . .

18. $-2, 4, -8, 16, -32, \ldots$

In Exercises 19–24, determine a direct formula for the given sequence.

19. An arithmetic sequence $\{x_n\}_{n\geq1}$ with $x_4 = 21$ and a common difference of -2

20. An arithmetic sequence $\{x_n\}_{n\geq1}$ with $x_3 = 10$ and a common difference of -5

21. A geometric sequence $\{y_n\}_{n\geq1}$ with $y_2 = 1$ and a common ratio of 3

22. A geometric sequence $\{y_n\}_{n\geq1}$ with $y_4 = 2$ and a common ratio of 5

23. A geometric sequence $\{y_n\}_{n\geq1}$ with $y_5 = 10$ and $y_8 = -10$

24. An arithmetic sequence $\{x_n\}_{n\geq1}$ with $x_3 = 7$ and $x_5 = 13$

25. Let S_n denote the sum of the first n terms of the geometric sequence $\{y_n\}_{n\geq1}$. Find S_9 if $S_3 = 80$ and $S_6 = 90$.

26. Let S_n denote the sum of the first n terms of the arithmetic sequence $\{y_n\}_{n\geq1}$. Find S_{15} if $S_5 = 2$ and $S_{10} = 0$.

In Exercises 27–42, determine if the given sequence is convergent or divergent; if convergent, determine its limit.

27. $x_n = 2 + (-1)^n$

28. $x_n = 1 + \left(-\frac{1}{2}\right)^n$

29. $x_n = (-0.3)^n$

30. $x_n = (0.7)^n$

31. $x_n = 5^n$

32. $x_n = 2^n$

33. $a_n = -\frac{1}{n}$

34. $a_n = -\frac{2}{n+2}$

35. $a_n = \frac{n}{n^2+n}$

36. $a_n = -\frac{3+n}{n^3+2}$

37. $a_n = \dfrac{5n-100}{2n+1}$ **38.** $a_n = \dfrac{99-3n}{n+2}$

39. $x_n = \dfrac{1}{n^2+n+2}$ **40.** $x_n = \dfrac{1}{n^2-n+1}$

41. $a_n = \dfrac{n!}{n^n}$ **42.** $a_n = \dfrac{1+2+\cdots+n}{n^2}$

43. How can you express the condition that three given numbers are consecutive terms in an arithmetic sequence? Use this characterization to show that if a^2, b^2, c^2 are consecutive terms in an arithmetic sequence, then so are $\dfrac{1}{b+c}, \dfrac{1}{a+c}$, and $\dfrac{1}{a+b}$, provided the denominators are nonzero.

44. Show that if $a, b, c > 0$ and if a^2, b^2, c^2 are consecutive terms in an arithmetic sequence, then so are $\dfrac{a}{b+c}, \dfrac{b}{a+c}$, and $\dfrac{c}{a+b}$, provided the denominators are nonzero.

In Exercises 45–52, use summation notation to rewrite the given sums.

45. $8 + 10 + 12 + 14 + \cdots + 2{,}000$

46. $9 + 12 + 15 + 18 + \cdots + 3{,}486{,}784{,}410$

47. $9 + 16 + 25 + 36 + \cdots + 10{,}000$

48. $1 + 8 + 27 + 64 + \cdots + 1{,}000{,}000$

49. $11 + 14 + 17 + 20 + \cdots + 1{,}010$

50. $1{,}000 + 998 + 996 + 994 + \cdots + 18$

51. $10 + 20 + 40 + 80 + \cdots + 2{,}560$

52. $5 + 10 + 20 + 40 + \cdots + 5{,}120$

53. Expand $\displaystyle\sum_{i=1}^{27} 5i$ and $\displaystyle\sum_{j=0}^{26}(5 + 5j)$. What do you observe? What conclusion can you draw?

54. Expand $\displaystyle\sum_{i=1}^{30}(3i - 1)$ and $\displaystyle\sum_{j=0}^{29}(2 + 3j)$. What do you observe? What conclusion can you draw?

In Exercises 55–58, rewrite the given sum without using the summation notation.

55. $\displaystyle\sum_{i=3}^{20}(3i - 1)$ **56.** $\displaystyle\sum_{i=5}^{9} 2 \cdot 3^{i-1}$ **57.** $\displaystyle\sum_{i=1}^{1} 3\dfrac{i}{i+1}$ **58.** $\displaystyle\sum_{i=4}^{10} 5(2i + 1)$

59. Compute the sum $S = 11 + 14 + 17 + 20 + \cdots + 1{,}010$ using the following steps:

   a. Find a direct formula for a sequence whose first few terms are 11, 14, 17, 20, $\ldots$, 1,010.

   b. Determine how many terms the sum S has. What is the twelfth term in the sum?

   c. Compute S without using a calculator.

60. Compute the sum $S = 100 + 96 + 92 + 88 + \cdots - 296$ using the following steps:

   a. Find a direct formula for a sequence whose first few terms are 100, 96, 92, 88, $\ldots$, -296.

b. Determine how many terms the sum S has. What is the thirtieth term in the sum?

c. Compute S without using a calculator.

61. Compute the sum $S = 9 + 3 + 1 + \frac{1}{3} + \cdots + \frac{1}{729}$ using the following steps:

   a. Find a direct formula for a sequence whose first few terms are $9, 3, 1, \frac{1}{3},$
   $\ldots, \frac{1}{729}.$

   b. Determine how many terms the sum S has. Find the fifth term in the sum.

   c. Compute S without using a calculator.

62. Compute the sum $S = 1 + \sqrt{2} + 2 + 2\sqrt{2} + \cdots + 1{,}024$ using the following steps:

   a. Find a direct formula for a sequence whose first few terms are $1, \sqrt{2}, 2, 2\sqrt{2},$
   $\ldots, 1{,}024.$

   b. Determine how many terms the sum S has. Find the eighth term in the sum.

   c. Compute S without using a calculator.

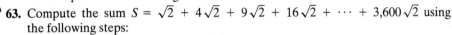 **63.** Compute the sum $S = \sqrt{2} + 4\sqrt{2} + 9\sqrt{2} + 16\sqrt{2} + \cdots + 3{,}600\sqrt{2}$ using the following steps:

   a. Find a direct formula for a sequence whose first few terms are $\sqrt{2}, 4\sqrt{2}, 9\sqrt{2},$
   $16\sqrt{2}, \ldots, 3{,}600\sqrt{2}.$

   b. Determine how many terms the sum S has.

   c. Use your calculator to compute S.

64. Compute the sum $S = 1 + 8 + 27 + \cdots + 1{,}728$ using the following steps:

   a. Find a direct formula for a sequence whose first few terms are $1, 8, 27,$
   $\ldots, 1{,}728.$

   b. Determine how many terms the sum S has.

   c. Use your calculator to compute S.

65. Use Definition 1.1.1 to prove that the sequence $\{\frac{1}{n+4}\}_n$ converges to 0.

66. Use Definition 1.1.1 to prove that the sequence $\{\frac{1}{2n-1}\}_n$ converges to 0.

67. Consider a sequence $\{x_n\}_n$ generated by randomly picking a number x_n in the interval $\left(\frac{1}{n+1}, \frac{1}{n}\right)$ for each integer $n \geq 1$. Is it true that any such sequence will converge to 0? Explain. Can you create a similar problem?

Historical Note: Leonardo Pisano Fibonacci (1175–1240; from Pisa, Italy).
Leonardo Pisano is better known by his nickname Fibonacci, or "son of Bonaccio."
His father was engaged in business in northern Africa, and Leonardo studied under
a Muslim teacher and traveled to Egypt, Syria, and Greece. Around 1200, Fibonacci
returned to Pisa. There he wrote a number of texts that played an important role in
reviving ancient mathematical skills, such as the Arabic algebraic methods, including
the Hindu-Arabic place-valued decimal system and the use of Arabic numerals. He
also made significant contributions of his own. Copies of his books *Liber Abaci*,
Practica Geometriae, *Flos*, and *Liber Quadratorum* survive to this day.

PROJECTS AND EXTENSIONS 1.1

I. Classroom Connection 1.1.3: The Fibonacci sequence

Sometimes deriving a formula for the general term of a sequence might not be easy, even when the sequence is described in a way that makes it possible to compute each term easily. The following exploration is a classic example and is taken from pages 256–257 in the eighth-grade textbook *Math Thematics, Book 3*. It describes the construction of the Fibonacci sequence. Denote the Fibonacci sequence by $\{f_n\}_n$. Determine the recursive relation for this sequence.

Section ② **Rotations and Rational Numbers**

IN THIS SECTION

EXPLORATION 1
◆ Rotational Symmetry

EXPLORATION 2
◆ Rational and Irrational Numbers

Nature's Sequences

····*Setting the Stage*

SET UP   *Work with a partner.*

His neighbors called him *Bigollone*, which means "the blockhead." His real name was Leonardo Fibonacci, and he was not a blockhead at all. He was a mathematician. He was born in Italy around 1170, but grew up in the city of Bougie on the Barbary Coast of North Africa.

▲ Roman and Arabic numerals

Fibonacci learned the Arabic number system in Africa. He realized that Arabic numerals were much easier to use than the Roman numerals then used in Europe. In 1202 Fibonacci published a book that introduced Arabic numerals to the European world. In his book he presented a mathematical problem that has fascinated people for centuries.

Suppose two newborn rabbits, male and female, are put in a cage. How many pairs of rabbits will there be at the end of one year if this pair of rabbits produces another pair every month, and every new pair of rabbits produces another new pair every month? All rabbits must be two months old before they can produce more rabbits.

Here is what happens in the first three months after the first pair of newborn rabbits are put in the cage:

Start   Start with **1st pair** of newborn rabbits.
Month 1   **1st pair** are growing.
Month 2   **1st pair** are adults. They produce **2nd pair** of rabbits.
Month 3   **1st pair** produce a 3rd pair of rabbits.
             2nd pair are growing.

Think About It ▸▸▸

1 Solve Fibonacci's rabbit problem. Start by creating a model or diagram that shows the total number of rabbit pairs over the first six months of the year. Find a way to show new-born, growing, and adult rabbits. Use your model to help complete a table like the one shown.

Month	Number of rabbit pairs			Total number of rabbit pairs
	Newborn	Growing	Adult	
1	1	0	0	1
2	0	1	?	?
3	?	?	?	?

2 The number pattern in the last column of the table you made in Question 1 is known as the *Fibonacci sequence*. The numbers in the sequence are sometimes called Fibonacci numbers.

Fibonacci sequence				
1	1	2	3	5...
↓	↓	↓		
1st term	2nd term	3rd term		

a. How are any two consecutive terms of the Fibonacci sequence used to find the next term in the sequence**?**

b. What is the answer to Fibonacci's rabbit problem**?** Explain how you got your answer.

▸ **Fibonacci's rabbit problem is not a realistic model of how rabbits reproduce. Nevertheless, in this section you'll see that the Fibonacci sequence has a strange way of showing up in nature.**

Section 2 Rotations and Rational Numbers   257

◆

II. Classroom Connection 1.1.4: Number Patterns

Look at other number patterns from the eight-grade textbook *Connected Mathematics, Frogs, Fleas, and Painted Cubes*. In Problems 2–3 (pages 45–46) and Problems 16, 18, and 19 (pages 49–50), identify the sequence described by the given pattern and provide the formula for the general term.

As you work on these ACE questions, use your calculator whenever you need it.

Applications

1. In a school math league, each team has six student members and two coaches.

   a. At the start of a match, the coaches and student members of one team exchange handshakes with the coaches and student members of the other team. How many handshakes will be exchanged?

   b. At the end of the match, the members and coaches of the winning team exchange high fives to celebrate their victory. How many high fives will be exchanged?

2. The dot patterns below represent the first four *square numbers*, 1, 4, 9, and 16.

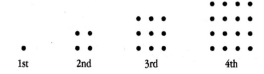

| 1st | 2nd | 3rd | 4th |

   a. What are the next two square numbers?

   b. Write an equation for calculating the *n*th square number.

   c. Make a table and a graph of the first ten square numbers. Describe the pattern of change from one square number to the next.

3. The dots below are arranged in rectangular patterns. The numbers of dots in the figures are called the *rectangular numbers*.

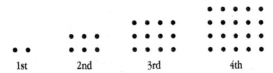

| 1st | 2nd | 3rd | 4th |

a. What are the first four rectangular numbers?

b. Find the next two rectangular numbers.

c. Describe the pattern of change from one rectangular number to the next.

d. Use the pattern of change to predict the 7th and 8th rectangular numbers.

e. Write an equation for calculating the nth rectangular number.

4. In what ways are triangular numbers, square numbers, and rectangular numbers similar to the cases given in Problem 3.1?

In 5–8, tell whether the number is a triangular number, a square number, a rectangular number, or none of these, and explain how you know you are correct.

5. 110   **6.** 66   **7.** 121   **8.** 60

9. Graphs i–iv on the next page represent situations you have looked at in this unit. Study the graphs and then answer the following questions.

a. Which graph might represent the number of high fives exchanged among a team with n players? Justify your choice.

b. Which graph might represent areas of rectangles with a fixed perimeter? Justify your choice.

c. Which graph might represent the area of a rectangle formed by increasing one dimension of a square by 2 centimeters and decreasing the other dimension by 3 centimeters? Justify your choice.

16. The dots below are arranged in patterns that represent the first three *star numbers*.

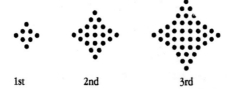

1st   2nd   3rd

a. What are the first three star numbers?

b. Use the dot patterns and your answers for part a to find the next three star numbers.

c. Write an equation you could use to calculate the nth star number.

17. a. Ten former classmates attend their class reunion. They all shake hands with each other. How many handshakes are exchanged? Explain your answer by drawing a picture or writing a convincing argument.

b. A little later, two more classmates arrive. If these two people shake hands with each other and the ten other classmates, how many new handshakes are exchanged? Explain your answer by drawing a picture or writing a convincing argument.

18. The dots below are arranged in hexagonal patterns. The numbers of dots in the patterns represent the *hexagonal numbers.*

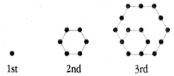

1st   2nd   3rd

a. What are the first three hexagonal numbers?

b. Use the dot patterns and your answers from part a to find the next two hexagonal numbers.

c. Which equation below could you use to calculate the *n*th hexagonal number?

$$b = \frac{n}{2}(n + 1) \qquad b = n(3n - 2) \qquad b = n(2n - 1)$$

19. If you look carefully, you can find 30 squares of various sizes in this 4-by-4 grid.

a. Sixteen of the 30 squares are the identical small squares that make up the grid. Find the other 14 squares. Draw pictures or give a description that would help someone else see all 30 squares.

b. How many squares can you find in an *n*-by-*n* grid? (Hint: Start with some simple cases and search for a pattern.)

◆

III. The Number Called π

This project's goal is to define the number π and, as a by-product, derive a formula for the circumference of a circle. Start by assuming that you have no prior knowledge of such a formula.

1. Construct two circles of centers O and O' with different radii r and r', respectively, and assume $r < r'$. Denote their circumferences by L and L'. In each circle, inscribe a regular hexagon. Denote the side lengths of these hexagons by l_6 and l'_6 as shown here.

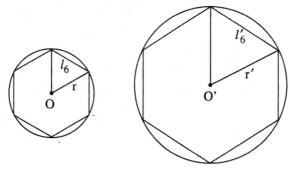

2. Use similarity of triangles to show that $\frac{l_6}{l'_6} = \frac{r}{r'}$. Deduce that $\frac{p_6}{2r} = \frac{p'_6}{2r'}$, where p_6 and p'_6 are the hexagons' perimeters.

3. Does the value of the quotient $\frac{p_6}{2r}$ depend on the particular circle you selected for the construction?

4. Now for each $n \geq 3$, consider regular n-gons inscribed in the circles of centers O and O' with perimeters p_n and p'_n, respectively. Use the idea employed in the case $n = 6$ to show that $\frac{p_n}{2r} = \frac{p'_n}{2r'}$ holds for each n.

5. For each fixed n, does the value of the fraction $\frac{p_n}{2r}$ depend on r? Explain.

6. Does the value of the fraction $\frac{p_n}{2r}$ depend on n? Explain.

7. Define $b_n = \frac{p_n}{2r}$ for each $n \geq 3$. Is it true that $b_3 < b_4 < b_5 < b_6 \ldots$? Is the sequence $\{b_n\}_n$ convergent? Explain.

8. What is the value of $\lim_{n \to \infty} p_n$?

9. Combine all of the previous problems to show that the ratio "circumference of circle/diameter of circle" is a constant that is independent of r.

The constant you obtained in Problem 9 is called π. The name comes from the first letter of the Greek word $\pi\varepsilon\rho\iota\phi\varepsilon\rho\iota o\nu$ ("periferion"), meaning circumference. In this way, you arrive at the following formula for the circumference of a circle

$$\text{Circumference} = \pi \cdot \text{diameter} = 2\pi \cdot \text{radius}.$$

Remark.   Informally, the conclusion that $\frac{L}{2r} = \frac{L'}{2r'}$ can be viewed as a result of the fact that any two circles are "similar." The proof, however, is much more involved, as you have previously seen.

Historical Note: Archimedes of Syracuse (287–212 BC; from Syracuse, Sicily). Archimedes is regarded as the greatest mathematician and scientist of antiquity and is one of the three greatest mathematicians of all time (together with Isaac Newton [1643–1727] and Carl Friederich Gauss [1777–1855]). Archimedes performed numerous geometric proofs using the geometric formalism outlined by Euclid, and he excelled at computing areas and volumes using the method of exhaustion. Two well-known mathematical accomplishments are the methods for approximating π and the proof that if a sphere is inscribed in a right circular cylinder, then the ratio of the cylinder and sphere's volumes are the same as the ratio of their areas. Some of Archimedes's works that have survived are *On Plane Equilibriums* (two books), *Quadrature of the Parabola, On the Sphere and Cylinder* (two books), *On Spirals, On Conoids and Spheroids, On Floating Bodies* (two books), *Measurement of a Circle*, and *The Sandreckoner*. Archimedes is also well-known for discovering his famous theory of buoyancy, the screw, the compound pulley, and many war machines used in the defense of Syracuse, which the Romans besieged in 214–212 BC. In the sack of the city, a Roman soldier slew Archimedes, despite orders from the Roman general Marcellus that his life be spared.

IV. Archimedes's Computation of π

Did you ever wonder how people came up with the exact values for the first few decimals of the number π? In fact, this was done thousands of years ago. The following outline is due to Archimedes. See for yourself how he computed the first few decimals of π.

1. Consider a circle of radius $r = \frac{1}{2}$. According to the formula deduced in Project and Extension Problem III, its circumference is π.

2. For each $n \geq 3$, denote by p_n and P_n the perimeters of the regular inscribed and circumscribed n-gons, respectively. Look at the following picture for the case $n = 6$.

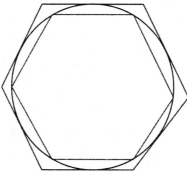

   Is the sequence $\{p_n\}_n$ increasing, decreasing, or neither? Is the sequence $\{P_n\}_n$ increasing, decreasing, or neither?

3. For each fixed n, how do p_n, P_n, and π compare?

4. What do you expect to happen to the values of p_n and P_n as n gets larger and larger? Are the sequences $\{p_n\}_n$ and $\{P_n\}_n$ convergent?

5. Use trigonometric functions to find explicit formulas for $\{p_n\}_n$ and $\{P_n\}_n$. Recall the computation of p_n from Classroom Discussion 1.1.3.

6. For a circle of radius $\frac{1}{2}$, here is a table containing some of the first few values of p_n and P_n,

n	p_n	P_n
3	2.59808	5.19615
6	3.00000	3.46410
12	3.10583	3.21539
24	3.13263	3.15966
48	3.13935	3.14609
96	3.14103	3.14271

Based on the values in the table, how many exact decimals do you obtain for π? Explain.

7. Use your calculator to expand the table in Problem 6 in order to compute the first three exact decimals of π.

V. Real Numbers as Limits of Sequences of Rational Numbers

The purpose of this exploration is to shed new light on the concept of *real number*, one made possible by the tools of calculus developed so far. When defining the set of real numbers, it is customary to start with the set of *natural* numbers $\mathbb{N} = \{1, 2, 3, 4, \ldots\}$. Next, the number 0 is defined; we need a symbol for what we obtain when counting "nothing"—that is, the number of elements of the empty set. Bringing into play the opposites of the natural numbers leads to the set of *integers* $\mathbb{Z} = \{\ldots, -3, -2, -1, 0, 1, 2, 3, \ldots\}$. Integers, in turn, are used to define fractions, thus leading to the set of *rational* numbers $\mathbb{Q} = \{\frac{p}{q} : p, q \in \mathbb{Z}, q \neq 0\}$ (read "$\mathbb{Q}$ is the set of all numbers $\frac{p}{q}$ with the property that p and q are integers and $q \neq 0$"). How we pass from rational numbers to real numbers is not as simple. While each rational number in decimal representation has a repeated pattern, there are certain numbers, called *irrational* numbers, whose decimal representation does not have a repeating pattern and which cannot be represented as quotients of integers. The set of rational numbers together with the set of irrational numbers yields the set of real numbers. The following property gives you an idea of how we pass from rational numbers to real numbers, namely by taking limits of convergent sequences of rational numbers.

Property

For every real number x, there exists a sequence of rational numbers convergent to x. In fact, for each real number x, we can select two sequences $\{y_n\}_n$ and $\{z_n\}_n$ satisfying the following:

a. y_n and z_n are rational numbers for each n;

b. $y_1 \leq y_2 \leq y_3 \leq \cdots$ (that is, y_n is nondecreasing) and $z_1 \geq z_2 \geq z_3 \geq \cdots$ (that is, z_n is nonincreasing);

c. $y_n \le x \le z_n$ for each n;

d. $\lim\limits_{n \to \infty} y_n = \lim\limits_{n \to \infty} z_n = x.$

In order to better understand why this property holds, carry out the following outline.

1. Let x be a rational number. Show that the constant sequences $y_n = z_n = x$ will satisfy the properties a–d.

2. Suppose now that x is the irrational number whose decimal representation is of the form $x = 0.1223334444555556666667777777\ldots$, i.e., the digits after the decimal point are one 1, two 2s, three 3s, four 4s, etc. In this case, define

$$y_1 = 0.1, \; y_2 = 0.12, \; y_3 = 0.122, \; y_4 = 0.1223, \; y_5 = 0.12233, \; y_6 = 0.122333, \; \ldots$$

$$z_1 = 0.2, \; z_2 = 0.13, \; z_3 = 0.123, \; z_4 = 0.1224, \; z_5 = 0.12234, \; z_6 = 0.122334, \; \ldots$$

- Determine the procedure that was used to construct these sequences.
- Do these sequences satisfy a–d? Explain.

3. Take $x = \pi = 3.14159265358979323846264338327950288419\ldots$. Use the method in Problem 2 to write the first 20 terms of two sequences $\{y_n\}_n$, $\{z_n\}_n$, satisfying a–d for π.

4. Take now the case when x is an arbitrary real number with decimal representation

$$x = b.a_1 a_2 a_3 a_4 a_5 \ldots$$

Here, $a_1, a_2, \ldots$ are the digits on the tenths, hundredths, etc., position, and b denotes the integer in the decimal representation of x. Follow the procedure used in Problem 2 to construct $\{y_n\}_n$ and $\{z_n\}_n$, satisfying a–d.

- Describe how you obtain y_n for each n.
- Is it true or false that $z_n = y_n + \dfrac{1}{10^n}$ for each n?
- How do you prove that d holds for the sequences you constructed?

VI. Operations with Real Numbers

Adding, subtracting, multiplying, and dividing real numbers are considered to be basic skills in today's world. By the end of middle school, the majority of the students will have a good understanding of these operations when performed on rational numbers, but the issue is much more subtle when dealing with irrational numbers. The notion of convergence plays an essential role in the process of passing from operations with rational numbers to operations with real numbers.

Take two rational numbers. The addition, subtraction, multiplication, and division (recall, however, that division by zero is not allowed) of these numbers

is defined naturally. For example, $\frac{2}{3} + \frac{5}{6} = \frac{3}{2}$, $\frac{2}{3} - \frac{5}{6} = -\frac{1}{6}$, $\frac{2}{3} \cdot \frac{5}{6} = \frac{5}{9}$, $\frac{\frac{2}{3}}{\frac{5}{6}} = \frac{4}{5}$.
What if the two numbers are irrational?

1. Consider the numbers

$$\pi \text{ and } x = 0.1223334444555556666667777777\ldots$$

   How does one compute $\pi + x$, $\pi - x$, πx, and $\frac{\pi}{x}$? To answer this question, recall the property from Project and Extensions Problem V and define

   $u_1 = 3.1$, $u_2 = 3.14$, $u_3 = 3.141$, $u_4 = 3.1415$, $u_5 = 3.14159$, $u_6 = 3.141592,\ldots$

   $v_1 = 0.1$, $v_2 = 0.12$, $v_3 = 0.122$, $v_4 = 0.1223$, $v_5 = 0.12233$, $v_6 = 0.122333,\ldots$

2. What are the values of $\lim_{n\to\infty} u_n$ and $\lim_{n\to\infty} v_n$? Explain.
3. Enter the values of u_n and v_n in a table for $n = 1,\ldots,14$. Generate the corresponding values for $u_n + v_n$, $u_n - v_n$, $u_n \cdot v_n$, and $\frac{u_n}{v_n}$ for $n = 1,\ldots,14$. What do you observe?
4. Based on P5 of limits of sequences, what can you say about $\lim_{n\to\infty}(u_n + v_n)$, $\lim_{n\to\infty}(u_n - v_n)$, $\lim_{n\to\infty}(u_n \cdot v_n)$, and $\lim_{n\to\infty} \frac{u_n}{v_n}$?
5. Take now

   $u_1 = 3.2$, $u_2 = 3.15$, $u_3 = 3.142$, $u_4 = 3.1416$, $u_5 = 3.1416$, $u_6 = 3.141593,\ldots$

   $v_1 = 0.2$, $v_2 = 0.13$, $v_3 = 0.123$, $v_4 = 0.1224$, $v_5 = 0.12234$, $v_6 = 0.122334,\ldots$

   Complete the tasks in Problem 3 for these new values of the sequences $\{u_n\}_n$ and $\{v_n\}_n$. What do you observe?
6. Will the values of the limits in Problem 4 change for these choices of the sequences $\{u_n\}_n$ and $\{v_n\}_n$? Explain.
7. Alain makes the following statement:

If $\{u_n\}_n$ and $\{v_n\}_n$ are arbitrary sequences of rational numbers satisfying $\lim_{n\to\infty} u_n = \pi$ and $\lim_{n\to\infty} v_n = x$, then we can define the addition, subtraction, product, and quotient of π and x in the following way,

$$\pi + x = \lim_{n\to\infty}(u_n + v_n), \quad \pi - x = \lim_{n\to\infty}(u_n - v_n), \quad \pi \cdot x = \lim_{n\to\infty}(u_n v_n),$$

$$\frac{\pi}{x} = \lim_{n\to\infty} \frac{u_n}{v_n}.$$

Do you agree with Alain? Explain.

Observation.   Your calculator performs operations with real numbers based on this approach. For example, the approximation returned by the calculator for π^2 will be the value of x_n^2 for some n (with n depending on the calculator's accuracy), where $\{x_n\}_n$ is a sequence of rational numbers converging to π, whose first few terms are stored in the calculator's memory.

VII. Computing $\sqrt{x}$, for x a Real Positive Number

By definition, if x is a real positive number, then $\sqrt{x}$ is the real positive number whose square is x. For this definition to make sense, it is important to know that for a given $x > 0$, $\sqrt{x}$ exists (existence statement), and that there are no two different positive numbers whose squares equal x (uniqueness statement). Complete the following outline to address these two issues.

1. Start with the case $x = 2$. The goal is to construct a convergent sequence of rational numbers $\{x_n\}_n$ for which $\lim_{n \to \infty} x_n^2 = 2$. There are many ways in which this can be done. Here is one: Set x_1 to be the largest rational number with at most one nonzero decimal located to the right of the decimal point and whose square is at most 2. Thus, $x_1 = 1.4$, since $(1.1)^2 < 2, (1.2)^2 < 2, (1.3)^2 < 2$, $(1.4)^2 < 2$, while $(1.5)^2 > 2$. Observe that you can rewrite the conditions for x_1 as

$$x_1^2 \leq 2, \quad (x_1 + 0.1)^2 > 2.$$

2. Continue the procedure in Problem 1 for constructing x_2, x_3, x_4, and x_5. As the subscript of the term of the sequence increases, so does the number of decimals after the decimal point that are allowed to be nonzero. Fill in the blanks:

$$x_2^2 \leq 2 \qquad (x_2 + 0.01)^2 > 2$$

$$x_3^2 \leq 2 \qquad (x_3 + 0.001)^2 > 2$$

$$x_4^2 \leq \underline{\quad} \qquad (x_4 + \underline{\quad})^2 > \underline{\quad}$$

$$x_5^2 \leq \underline{\quad} \qquad (x_5 + \underline{\quad})^2 > \underline{\quad}$$

What are the values of x_2, x_3, x_4, and x_5?

3. There is no reason to stop this construction at $n = 5$. Continuing indefinitely, you obtain a nondecreasing sequence $\{x_n\}_n$ with the general term x_n, satisfying the following:

- x_n is a rational number with at most n nonzero decimals,
- $x_n^2 \leq 2$,
- $(x_n + 10^{-n})^2 > 2$.

Use P1–P7 to show that the sequence $\{x_n\}_n$ is convergent.

4. Let $L = \lim\limits_{n\to\infty} x_n$. Prove that $L > 0$ and $L^2 = 2$. Hint: Since $x_n^2 \leq 2$ and $(x_n + 10^{-n})^2 > 2$ for all $n \geq 1$, what do these imply about L^2?

5. The previous outline for the construction of $\sqrt{2}$ can be applied to any real positive number x. Rewrite the properties from Problem 3 of the sequence $\{x_n\}_n$ when x replaces 2. Show that the corresponding sequence $\{x_n\}_n$ is convergent and that if $L = \lim\limits_{n\to\infty} x_n$, then $L > 0$ and $L^2 = x$. This gives a proof for the fact that $\sqrt{x}$ exists for any real positive number x.

6. If $y_1^2 = x = y_2^2$, for some $y_1 > 0$, $y_2 > 0$, show that necessarily $y_1 = y_2$. This addresses the uniqueness issue.

1.2 SERIES

Series • Convergent series • Geometric series

What is a series? To answer this question, begin with Classroom Discussion 1.2.1.

Classroom Discussion 1.2.1: The Intriguing Number $0.\overline{9}$

Take $x = 0.\overline{9} = 0.9999\ldots$. The computation

$$
\begin{aligned}
10x &= 9.\overline{9} \\
-x &= -0.\overline{9} \\
\hline
9x &= 9
\end{aligned}
\tag{1}
$$

shows that $x = 1$. To better understand the identity $0.\overline{9} = 1$, analyze the number $0.\overline{9}$ from another perspective.

1. Starting with $0.9 = \frac{9}{10}$, $0.09 = \frac{9}{100}$, $0.009 = \frac{9}{1,000}$, it is natural to write

$$
0.\overline{9} = \frac{9}{10} + \frac{9}{100} + \frac{9}{1,000} + \frac{9}{10,000} + \cdots
$$

For each whole number n, define $a_n = \frac{9}{10^n}$. What type of sequence is $\{a_n\}_n$?

2. Denote the sum of the first n terms in this sequence by s_n, i.e., $s_n = \sum\limits_{i=1}^{n} \frac{9}{10^i}$. It is clear that $s_1 = 0.9$, $s_2 = 0.99$, ... In general, each s_n has n copies of 9 after the decimal point. What can you say about the sequence $\{s_n\}_n$?

3. Using the formulas developed in Section 1.1, compute s_n for arbitrary n.

4. Is the sequence $\{s_n\}_n$ convergent? If yes, compute its limit.

5. How does this conclusion compare with the computation in (1)? ◆

Classroom Connection 1.2.1: Ballots and Geometric Series

The following exploration is from pages 45 and 58 in the eighth-grade textbook *Connected Mathematics, Growing, Growing, Growing*. Work Problem 9. Make the connection with Classroom Discussion 1.1.6.

Exponential Decay

The exponential patterns you have studied so far in this unit have all involved variables with increasing values. In this investigation, you will explore variables with values that decrease, or *decay*, exponentially as time passes.

 ### Making Smaller Ballots

In Problem 1.1, you read about the ballots that Alejandro was making for an election. Recall that Alejandro cut a sheet of paper in half, stacked the two pieces and cut them in half, and then stacked the resulting four pieces and cut them in half.

In Problem 1.1, you investigated the pattern in the number of ballots created by each cut. In this problem, you will look at the pattern in the areas of the ballots.

9. Study this pattern:

row 1: $\frac{1}{2} =$

row 2: $\frac{1}{2} + \left(\frac{1}{2}\right)^2 =$

row 3: $\frac{1}{2} + \left(\frac{1}{2}\right)^2 + \left(\frac{1}{2}\right)^3 =$

row 4: $\frac{1}{2} + \left(\frac{1}{2}\right)^2 + \left(\frac{1}{2}\right)^3 + \left(\frac{1}{2}\right)^4 =$

a. Find the sum for each row.

b. Suppose the pattern continued. Write the expression that would be in row 5, and find its sum.

c. What would be the sum of the expression in row 10? In row 20?

d. Describe the pattern of sums in words and with an equation.

e. For which row does the sum first exceed 0.9?

f. As the row number increases, the sum gets closer and closer to what number?

g. Celeste claims that the pattern is related to the pattern of the areas of the ballots cut in Problem 4.1. She drew this picture to explain her thinking.

$$\text{row } 6 = \frac{1}{2} + \left(\frac{1}{2}\right)^2 + \left(\frac{1}{2}\right)^3 + \left(\frac{1}{2}\right)^4 + \left(\frac{1}{2}\right)^5 + \left(\frac{1}{2}\right)^6$$

What relationship do you think Celeste has observed?

◆

In the examples discussed so far, you wanted to define a sum of the type $x_1 + x_2 + x_3 + \cdots$ for various choices of the terms $x_1, x_2, x_3, \ldots$. In each case, you have defined a new sequence $\{s_n\}_n$ by computing partial sums of the infinite sum. This is the same as looking at the sum of "more and more" x_i's. Since, in the end, you wanted to add all the x_i's, $i \geq 1$, you should look at the behavior of the partial sums s_n as $n \to \infty$. Thus, the following definition is natural.

Definition 1.2.1 An **infinite series** is an expression of the form

$$x_1 + x_2 + x_3 + \cdots, \qquad (2)$$

where the numbers $x_1, x_2, \ldots,$ are the **terms** of the series. The sequence $\{s_n\}_n$ of **partial sums** is

$$s_1 = x_1$$
$$s_2 = x_1 + x_2$$
$$s_3 = x_1 + x_2 + x_3$$
$$\vdots$$
$$s_n = x_1 + x_2 + x_3 + \cdots + x_n = \sum_{i=1}^{n} x_i \quad \text{for each } n \geq 1.$$

If the sequence $\{s_n\}_n$ converges to a real number L, that is, $\lim_{n \to \infty} s_n = L$, then the series (2) is called **convergent**, and L is called the **sum** of the series. In this case, we write $x_1 + x_2 + x_3 + \cdots = L$ or $\sum_{i=1}^{\infty} x_i = L$. Otherwise, we say that the series **diverges**.

Classroom Discussion 1.2.2:

The Mysterious Series 1 − 1 + 1 − 1 + 1 − 1 + · · ·

Dan and John are trying to compute the infinite sum

$$1 - 1 + 1 - 1 + 1 - 1 + \cdots$$

Dan writes

$$\underbrace{1 - 1}_{=0} + \underbrace{1 - 1}_{=0} + \underbrace{1 - 1}_{=0} + \cdots$$

and concludes that the sum of the series is zero. John writes

$$1 + \underbrace{(-1 + 1)}_{=0} + \underbrace{(-1 + 1)}_{=0} + \underbrace{(-1 + 1)}_{=0} + \cdots$$

and concludes that the sum must be equal to 1. Who do you think is right? Explain. ◆

Geometric Series.   In the following discussion, we restrict our analysis to a particular type of series whose terms form a geometric sequence of ratio r. Such a series is called **geometric series of ratio** r. It can be written as $a + ar + ar^2 + \cdots$, where a is the first term of the series. The issue of the convergence of geometric series is well understood. Use the following outline to derive the main results for geometric series.

Classroom Discussion 1.2.3: Convergence of Geometric Series

The strategy dictated by Definition 1.2.1 is to first compute the general term of the sequence of partial sums, $s_n = a + ar + ar^2 + \cdots + ar^{n-1}$. Recall that $s_n = \frac{a - ar^n}{1-r}$ if $r \neq 1$ (see Classroom Discussion 1.1.7). Clearly, $s_n = 0$ if $a = 0$. When answering the following questions 1–5, assume that $a \neq 0$. In each case, decide if the series $\sum\limits_{i=1}^{\infty} a r^{i-1}$ converges or diverges.

1. Compute $\lim\limits_{n \to \infty} s_n$ if $|r| < 1$.
2. What can you say about $\lim\limits_{n \to \infty} s_n$ if $r > 1$?
3. What can you say about $\lim\limits_{n \to \infty} s_n$ if $r < -1$?
4. What is the value of s_n if $r = 1$? What can you say about $\lim\limits_{n \to \infty} s_n$ for $r = 1$?
5. What is the value of s_n if $r = -1$? What can you say about $\lim\limits_{n \to \infty} s_n$ for $r = -1$? ◆

You have proved the following result.

Theorem 1.2.1 (Geometric Series Convergence Theorem). For any nonzero real number a, the geometric series

$$a + ar + ar^2 + \cdots$$

converges to the sum $\frac{a}{1-r}$ if $|r| < 1$ and diverges if $|r| \geq 1$.

EXAMPLES

1. Return to the geometric series $\frac{9}{10} + \frac{9}{100} + \frac{9}{1,000} + \frac{9}{10,000} + \cdots$. Use Theorem 1.2.1 to determine if the series converges or diverges. If the series converges, determine its sum.
2. Return to Classroom Discussion 1.2.2 and explain what happens from the perspective of Theorem 1.2.1.

Solutions

1. The series $\frac{9}{10} + \frac{9}{100} + \frac{9}{1,000} + \frac{9}{10,000} + \cdots$ is geometric with the first term $a = \frac{9}{10}$ and the ratio $r = \frac{1}{10}$. Since $|r| < 1$, Theorem 1.2.1 guarantees its convergence to the sum $\frac{\frac{9}{10}}{1 - \frac{1}{10}} = 1$. In this light, you see that real numbers with infinitely many nonzero decimals are understood as sums of series (not always geometric!). The expression $0.\overline{9}$ denotes the sum of the series $\frac{9}{10} + \frac{9}{100} + \frac{9}{1,000} + \frac{9}{10,000} + \cdots$. According to what you have discovered so far, the sum of this series is indeed 1. This is why you can write $0.\overline{9} = 1$.
2. The series $1 - 1 + 1 - 1 + 1 - 1 + \cdots$ is geometric with the ratio $r = -1$. According to Theorem 1.2.1, it diverges. ∎

Practice Problems

1. Let $x = 0.\overline{41}$. Show that $x = \frac{41}{99}$, first by using Theorem 1.2.1 and then by a computation similar to computation (1) at the beginning of Section 1.2.

2. Consider the series $2 + 4 + 8 + 16 + \cdots$. Is this a convergent or a divergent series? Explain.

Classroom Connection 1.2.2: Pattern

The following exploration is from page 59 in the eight-grade textbook *Connected Mathematics, Growing, Growing, Growing*. Discuss the questions in Problem 10 from the series perspective.

---

10. Study this pattern:

row 1: $\frac{1}{3} =$

row 2: $\frac{1}{3} + (\frac{1}{3})^2 =$

row 3: $\frac{1}{3} + (\frac{1}{3})^2 + (\frac{1}{3})^3 =$

row 4: $\frac{1}{3} + (\frac{1}{3})^2 + (\frac{1}{3})^3 + (\frac{1}{3})^4 =$

a. Find the sum for each row.

b. Suppose the pattern continued. Write the expression that would be in row 5, and find its sum.

c. What would be the sum of the expression in row 10? In row 20?

d. Describe the pattern of sums in words and with an equation.

e. For which row does the sum first exceed 0.9?

f. As the row number increases, the sum gets closer and closer to what number?

Investigation 4: Exponential Decay   59

Return to $0.\overline{9} = 1$

Here are two more proofs of the fact that $0.\overline{9} = 1$.

a. Most middle-school students will agree with the fact that $\frac{1}{3} = 0.\overline{3}$. They can perform the division $1 \div 3$ and see that the decimal 3 keeps on repeating. Now they just need to multiply this last equality by 3 to conclude that

$$1 = 3 \cdot \frac{1}{3} = 3 \cdot 0.\overline{3} = 0.\overline{9}.$$

b. This proof involves reasoning by contradiction. If $0.\overline{9} < 1$, then the difference $c = 1 - 0.\overline{9}$ is a number contained in the interval $(0, 1)$. This means that c is of the form $c = 0.0000 * \ldots$, where $*$ denotes the first nonzero digit in the decimal representation of x. Thus, $* \geq 1$. However, even if $* = 1$, when you add c back to $0.\overline{9}$, you obtain a number that is larger than 1. Indeed, if we take $c = 10^{-n}$ for some natural number n, then

$$c + 0.\overline{9} = 1.\underbrace{0\cdots0}_{n \text{ zeroes}}99\ldots > 1.$$

This contradicts the fact that $c + 0.\overline{9} = 1$; thus our assumption that $0.\overline{9} < 1$ is false. Since clearly $0.\overline{9} \not> 1$, we must have $0.\overline{9} = 1$.

EXERCISES 1.2

In Exercises 1–6, write the given numbers as fractions by first employing Theorem 1.2.1 and then by using a computation similar to computation (1) at the beginning of Section 1.2. All patterns are assumed to repeat indefinitely.

1. $0.535353\ldots$ **2.** $0.212121\ldots$ **3.** $0.025025025\ldots$

4. $0.45045045\ldots$ **5.** $1.123434343\ldots$ **6.** $2.32818181\ldots$

In Exercises 7–18, decide whether the geometric series converges or diverges. Justify your answer. If the series converges, compute its sum.

7. $\frac{2}{3} + \frac{2}{9} + \frac{2}{27} + \frac{2}{81} + \cdots$

8. $\frac{5}{4} + \frac{5}{8} + \frac{5}{16} + \frac{5}{32} + \cdots$

9. $\frac{5}{4} + \frac{25}{16} + \cdots + \left(\frac{5}{4}\right)^n + \cdots$

10. $1 - \frac{3}{2} + \frac{9}{4} - \frac{27}{8} + \cdots$
$\qquad + \left(-\frac{3}{2}\right)^{n-1} + \cdots$

11. $1 - \frac{1}{2} + \frac{1}{4} - \frac{1}{8} + \cdots$
$\qquad + \left(-\frac{1}{2}\right)^{n-1} + \cdots$

12. $\frac{2}{3} + \frac{4}{9} + \cdots + \left(\frac{2}{3}\right)^n + \cdots$

13. $1 - \frac{\pi}{3} + \frac{\pi^2}{9} - \frac{\pi^3}{27} + \frac{\pi^4}{81} - \cdots$

14. $1 - \frac{\pi}{4} + \frac{\pi^2}{16} - \frac{\pi^3}{64} + \frac{\pi^4}{256} - \cdots$

15. $\frac{3}{4} - \frac{9}{16} + \frac{27}{64} - \frac{81}{256} + \cdots$

16. $1 - \frac{\sqrt{2}}{3} + \frac{2}{9} - \frac{2\sqrt{2}}{27}$
$\qquad + \frac{4}{81} - \cdots$

17. $2 - 2 + 2 - 2 + 2 - 2 + \cdots$

18. $1 - \sqrt{3} + 3 - 3\sqrt{3} + 9 - 9\sqrt{3} + \cdots$

In Exercises 19–20, decide whether the series converges or diverges.

19. $\sum\limits_{i=1}^{\infty} (3i - 2)$

20. $\sum\limits_{i=1}^{\infty} (4 - 3i)$

21. A ball is dropped from a height of 10 feet. After it hits the floor, the ball bounces back and reaches a height equal to $\frac{3}{4}$ of the previous height. The ball bounces again and again. For each bounce, it reaches a height equal to $\frac{3}{4}$ of the previous bouncing height. The following is the graph of the ball's height (in feet) after t seconds. Only the first few bounces are shown in the graph, but assume that the ball continues to bounce forever. (Notice that the graph shows the ball's height over time and not its trajectory.)

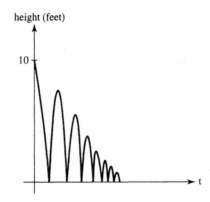

What is the total vertical distance traveled by the ball?

22. Compute the sum of the geometric series
$$x - x^3 + x^5 - x^7 + x^9 - x^{11} + \cdots \text{ for } |x| < 1.$$

23. Compute the sum of the series $S = \sum\limits_{i=1}^{\infty} ix^{i-1}$ for $|x| < 1$ using the following outline.

  a. Write out the series $S - xS$ and cancel the like terms.
  b. Do you have a formula for the sum of the series obtained?
  c. How does this last series compare to $(1 - x)S$?

PROJECTS AND EXTENSIONS 1.2

I. Classroom Connection 1.2.3: Fractals and Series

The following exploration is from page 241 in the eighth-grade textbook *Math Thematics, Book 3*. Problem 9 refers to the construction in Classroom Connection 1.1.2. Identify the series that arise in Problem 9; using the results proved in this section, answer the questions raised.

9 Discussion Suppose you could continue adding branches to your fractal tree forever.

a. How tall would the final fractal tree be? How wide would it be? Would the tree ever be too large to fit on your paper? Explain your thinking.

b. Can you find the total length of all the branches in the final fractal tree? Explain.

◆

II. Classroom Connection 1.2.4: Series and the Sierpinski Triangle

Exploration 3 is from pages 244–246 and page 4-53 in the eighth-grade textbook *Math Thematics, Book 3*. In the construction of the Sierpinski triangle, denote by A the area of the original triangle. For each $i = 1, 2, 3, \ldots$, we denote by a_i and A_i the area *removed* and the area *left*, respectively, at step i.

1. Create a table with the values of a_i and A_i for $i = 1, 2, 3, 4, 5$. Use this table to determine formulas for a_i and A_i, respectively, that depend on A and i only.

2. How do the series relate to the Sierpinski triangle?

3. Use the results proved in this section to answer the questions raised in Problems 25b and c.

GOAL

LEARN HOW TO...
+ use a compass and ruler to construct a perpendicular bisector

AS YOU...
+ make a Sierpinski triangle

KEY TERM
+ midpoint

Exploration

The Sierpinski TRIANGLE

SET UP   You will need: • Labsheet 1B • compass • protractor • ruler • plain white paper • black paper • marker

▶ You'll use a compass and the straight edge of a ruler to explore patterns in a Sierpinski triangle.

20 Follow these steps to begin creating a Sierpinski triangle.

Step 1 In the middle of a piece of plain white paper draw a segment that is $2\frac{1}{2}$ in. long. Label the endpoints A and B.

Step 2 Adjust a compass to a radius that is the same length as the segment. Place the compass point on A and draw a circle. Draw a second circle with the same radius centered at B.

Step 3 Draw a line through the two points where the circles intersect. Use a ruler and a protractor to check that this line is the perpendicular bisector of $\overline{AB}$. Label the point C where the circles intersect above $\overline{AB}$.

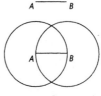

▶ In Question 20 you used a perpendicular bisector to find the *midpoint* of the segment. The **midpoint** of a segment divides it into two congruent segments.

21 Follow the steps below to continue creating your Sierpinski triangle.

Step 4 Draw segments from *A* to *C* and from *B* to *C*. Use a marker to trace the triangle.

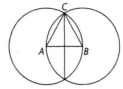

Step 5 Place the compass point on *C* and draw a third circle the same size as the first two. Locate the midpoints of $\overline{AC}$ and $\overline{BC}$ as you did in Step 3 of Question 20. Mark the midpoints but do not draw a line.

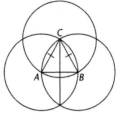

Step 6 Connect the midpoints by drawing segments inside the triangle. You now have one large triangle divided into four smaller triangles. Shade the middle triangle.

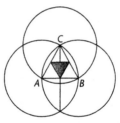

22 What type of triangles have you constructed? How is the area of each small triangle related to the area of the large triangle?

23 **a.** Measure with a ruler to find the midpoints of the sides of each of the three unshaded triangles. Connect the midpoints to form new triangles. Shade the middle triangle in each group.

b. How is the area of each of these smaller shaded triangles related to the area of the largest triangle?

Name _____   Date _____

Sierpinski Triangle Table (Use with Question 25 on page 246.)

Steps	Number of unshaded triangles	Area of each unshaded triangle (in square units)	Total area of unshaded triangles (in square units)
	1	1	1
	3	$\frac{1}{4} = 0.25$	$\frac{3}{4} = 0.75$

Math Thematics, Book 3 **4-53**

▶ You could continue the process of shading smaller and smaller triangles, but your original triangle is too small to show what a Sierpinski triangle would look like.

24 Work with two other students. Cut out the triangle you constructed in Questions 21 and 23. Place it on black paper with two other triangles, as shown.

25 **Use Labsheet 1B.** Each time you shade a part of your Sierpinski triangle, it is as if you are putting a "hole" in the figure.

    a. To see what would happen if you kept putting smaller and smaller holes in the figure, complete the *Sierpinski Triangle Table.*

    b. **Discussion** How does the total area of the shaded triangles change as the step numbers increase?

    c. **Discussion** Remember that a fractal is the result of carrying out steps forever. Will there be any unshaded areas left in the Sierpinski triangle if you could continue the process of making smaller and smaller holes forever? Explain.

◆

III. Zeno's Paradoxes

Zeno of Elea (490–425 BC) was an ancient philosopher. Little is known about his life, but he has made mathematical history with his four paradoxes: *Dichotomy, Achilles, Arrow*, and *Stadium.* Zeno originally produced a book that contained forty paradoxes, but the book has not survived to modern times, only these four of his paradoxes are known today, the best known of which is the paradox of Achilles and the Tortoise. The following is a description of the paradox, taken from PRIME articles at http://www.mathacademy.com/pr/prime/articles/zeno_tort/index.asp.

    The Tortoise challenged Achilles to a race, claiming that he would win as long as Achilles gave him a small head start. Achilles laughed at this, for of course he was a mighty warrior and swift of foot, whereas the Tortoise was heavy and slow. "How big a head start do you need?" he asked the Tortoise with a smile. "Ten meters," the latter replied. Achilles laughed louder than ever. "You will surely lose, my friend, in that case," he told the Tortoise, "but let us race, if you wish it." "On the contrary," said the Tortoise, "I will win, and I can prove it to you by a simple argument." "Go on then," Achilles replied, with less confidence than he felt before. He knew he was the superior athlete, but he also knew the Tortoise had the sharper wits, and he had lost many a bewildering argument with him before this. "Suppose," began the Tortoise, "that you give me a 10-meter head start. Would you say that you could cover that 10 meters between us very quickly?" "Very quickly," Achilles affirmed. "And in that time, how far should I have gone, do you think?" "Perhaps a meter—no more," said Achilles after a moment's thought. "Very well," replied the Tortoise, "so now there is a meter between us. And you would catch up that distance very quickly?" "Very quickly indeed!" "And yet, in that time I shall have gone a little way farther, so that now you must catch that distance up, yes?" "Ye-es," said Achilles slowly. "And while you are doing so, I shall have gone a little way farther, so that you must then catch up the new distance," the Tortoise continued smoothly. Achilles said nothing. "And so you see, in each moment you must be catching up the distance between us, and yet I—at the same time—will be adding a new distance, however small, for you to catch up again." "Indeed, it must be so," said Achilles wearily. "And so you can never catch up," the Tortoise concluded sympathetically. "You are right, as always," said Achilles sadly, and conceded the race.

Is the Tortoise's reasoning correct? To answer this question, use the following outline:

1. Denote by d_1 the initial distance between Achilles and the Tortoise, and denote by v_A and v_T the running speed of Achilles and the Tortoise, respectively. What does $v_A - v_T$ represent? How about $\dfrac{d_1}{v_A - v_T}$?

   Next, analyze this paradox from the perspective of sequences and series.

2. Let t_1 be the time Achilles needs to cover the distance d_1. Determine the expression of t_1 in terms of d_1 and v_A.

3. What is the distance covered by the Tortoise during the time t_1? Denote by d_2 this distance. Observe that this is precisely the distance between Achilles and the Tortoise after the time t_1 has passed. Determine the expression of d_2 in terms of t_1 and v_T.

4. Let t_2 be the time Achilles needs to cover the distance d_2. Determine the expression of t_2 in terms of t_1, v_T, and v_A. What is the ratio between t_2 and t_1?

5. Let d_3 be the distance covered by the Tortoise during the time t_2. Determine the expression of d_3 in terms of t_2 and v_T.

6. Let t_3 be the time Achilles needs to cover the distance d_3. Determine the expression of t_3 in terms of t_2, v_T, and v_A. What is the ratio between t_3 and t_2?

7. If we continue this line of reasoning, we construct two sequences $\{d_n\}_{n \geq 1}$ and $\{t_n\}_{n \geq 1}$: d_1 is the initial distance between Achilles and the Tortoise; for each natural number $n \geq 2$, t_{n-1} represents the time Achilles needs to cover the distance d_{n-1}, while d_n represents the distance traveled by the Tortoise during the time t_{n-1}. Fix $n \geq 1$, and express d_{n+1} in terms of t_n and v_T, and t_{n+1} in terms of t_n, v_T, and v_A. What is the ratio between t_{n+1} and t_n?

8. What type of sequence is $\{t_n\}_{n \geq 1}$?

9. What does the series $\sum\limits_{n=1}^{\infty} t_n$ represent? Is this series convergent? If yes, compute its sum.

10. Compare the results in Problems 1 and 9. What do you observe? Explain the flaw in the Tortoise's reasoning.

IV. Snowflakes

Consider one more example of series related to fractals. Start with an equilateral triangle with a side of length b. Divide each side into three congruent parts. On each middle segment, construct an equilateral triangle (toward the exterior of the original one) and then erase the old side of each new equilateral triangle. Next, again divide each exterior segment into three equal parts; on each middle segment, construct an equilateral triangle (toward the exterior); then erase the old side of each new equilateral triangle, and so on. This construction gives rise to polygons whose number of sides keeps increasing. The following figure is the result of the first four steps of such a construction.

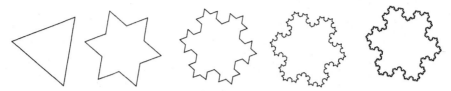

Imagine now that this construction is continued indefinitely.

1. For each $i = 1, 2, 3, \ldots$, denote by p_i the polygon's perimeter at step i. Create a table with the values of p_i for $i = 1, 2, 3, 4, 5$. Use this table to determine a formula for p_i that depends on b and i only.
2. What is the perimeter of the resulting fractal?
3. For each $i = 1, 2, 3, \ldots$, denote by A_i the polygon's area at step i. Show that the original triangle's area is $A = \dfrac{b^2 \sqrt{3}}{4}$. Create a table with the values of A_i for $i = 1, 2, 3, 4, 5$. Use this table to determine a formula for A_i that depends on A and i only.
4. What is the area of the resulting fractal?

Historical Note: Leonhard Euler (1707–1783; from Basel, Switzerland).   Euler's work in mathematics is extremely vast and he was one of the most prolific writers of mathematics of all time, publishing more than 500 books and papers during his lifetime. Euler completed his studies at the University of Basel in 1726. He held positions at the St. Petersburg Academy of Sciences (1927–1741, 1766–1783) and the Berlin Academy of Science (1741–1766). Even after he became blind in 1771, Euler was able to continue with his work. Amazingly, after his return to St. Petersburg, he produced almost half his total works in spite of being totally blind. After his death in 1783 the St. Petersburg Academy continued to publish Euler's manuscripts for nearly 50 more years, bringing the bibliographical list of Euler's works to nearly 900 entries. In addition to making significant contributions to virtually every branch of pure and applied mathematics, Euler was also the most successful notation builder of all times. We owe to Euler the notation $f(x)$ for a function, e for the sum of the series discussed below, i for the square root of -1, π, Σ for summation, and many others.

V. Euler's Number e

For each natural number n we denote by $n!$, and read "n factorial," the product $1 \cdot 2 \cdot 3 \cdots n$. By convention, $0! = 1$. The goal is to investigate the series

$$1 + \frac{1}{1!} + \frac{1}{2!} + \frac{1}{3!} + \frac{1}{4!} + \frac{1}{5!} + \cdots$$

a. Compute the first eleven partial sums of this series using a calculator. What do you observe?

b. Let $a_n = \frac{1}{n!}$ for $n = 0, 1, 2, \ldots$. Show that

$$a_2 \leq \frac{1}{2}, \ a_3 \leq \frac{1}{2^2}, \ a_4 \leq \frac{1}{2^3}, \ a_5 \leq \frac{1}{2^4}.$$

More generally, show that $a_n \leq \frac{1}{2^{n-1}}$ for $n = 2, 3, 4, \ldots$.

c. Let $s_n = \sum\limits_{i=0}^{n} a_i$, $n \geq 0$ be the sequence of partial sums for this series. Is it true that this sequence is increasing?

d. Show that $s_0 < 3$, $s_1 < 3$, $s_2 < 3$, $s_3 < 3$. Next, use b to show that $s_n < 3$ for every $n \geq 0$.

Hint: Sum up the series with the ratio $\frac{1}{2}$.

e. How can you explain the fact that $\lim\limits_{n \to \infty} s_n$ exists? The value of this limit is denoted by e and is called Euler's number. This symbol has been chosen in homage to the mathematician Leonard Euler, who discovered this number in 1748.

f. Observe that from c and d it follows that $2 < e < 3$. However, we want to find approximations for the number e. It is immediate from the definition that any partial sum of the series $1 + \frac{1}{1!} + \frac{1}{2!} + \frac{1}{3!} + \cdots$ is an approximation for e. However, you do not know the magnitude of the error for each such approximation. What you know is that

$$e - s_1 = \frac{1}{2!} + \frac{1}{3!} + \frac{1}{4!} + \cdots$$

$$e - s_2 = \frac{1}{3!} + \frac{1}{4!} + \frac{1}{5!} + \cdots$$

$$e - s_3 = \frac{1}{4!} + \frac{1}{5!} + \frac{1}{6!} + \cdots$$

$$\vdots$$

$$e - s_n = \frac{1}{(n+1)!} + \frac{1}{(n+2)!} + \cdots$$

$$\vdots$$

are the errors made when approximating e by $s_1, s_2, s_3, \ldots, s_n, \ldots$, respectively, so, if you can estimate the series in the right-hand sides of the identities above, then you will have estimates for the corresponding errors. To illustrate this, consider the case $n = 3$.

$$\frac{1}{4!} + \frac{1}{5!} + \frac{1}{6!} + \cdots < \frac{1}{4!}\left(1 + \frac{1}{4} + \frac{1}{4^2} + \cdots\right)$$

$$= \frac{1}{4!}\frac{1}{1 - \frac{1}{4}} = \frac{1}{3!}\frac{1}{3} < 0.06.$$

Explain each step in this estimation and then repeat the reasoning to see how many exact decimals you will get for e when using the approximation $e \approx s_{10}$.

g. Let's address the nature of the number e. Our claim is that e is an irrational number. Use the following outline to prove this claim. The starting point is a reasoning by contradiction.

- Since $2 < e < 3$, it follows that e is not an integer. Assume that e is rational. Then, there exist natural numbers p and q, such that $e = \frac{p}{q}$. You can assume that p and q are relatively prime; that is, 1 is the only common divisor for p and q. Also, since e is not an integer, $q \geq 2$.
- Next multiply by $q!$ the equality

$$\frac{p}{q} = 1 + \frac{1}{1!} + \frac{1}{2!} + \frac{1}{3!} + \frac{1}{4!} + \frac{1}{5!} + \cdots$$

In the left-hand side (LHS) of the resulting identity, you should obtain

$$\text{LHS} = p \cdot 2 \cdot 3 \cdots (q - 1).$$

Is this an integer?

- In the right-hand side (RHS) of the previous identity, you should obtain

$$\text{RHS} = q! + q! + 3 \cdot 4 \cdots q + 4 \cdot 5 \cdots q + \cdots + (q - 1)q + q + 1$$

$$+ \frac{1}{q + 1} + \frac{1}{(q + 1)(q + 2)} + \frac{1}{(q + 1)(q + 2)(q + 3)} + \cdots,$$

which is the sum of an integer with the series

$$\frac{1}{q + 1} + \frac{1}{(q + 1)(q + 2)} + \frac{1}{(q + 1)(q + 2)(q + 3)} + \cdots.$$

- Since $q \geq 2$, you can estimate the last series in the following way:

$$\frac{1}{q + 1} + \frac{1}{(q + 1)(q + 2)} + \frac{1}{(q + 1)(q + 2)(q + 3)} + \cdots$$

$$\leq \frac{1}{3} + \frac{1}{3^2} + \frac{1}{3^3} + \cdots = \frac{\frac{1}{3}}{1 - \frac{1}{3}} = \frac{1}{2}.$$

- Based on this last estimation, can you decide if the RHS sum is an integer? Explain.
- Combine all these to arrive at a contradiction, which then implies that the assumption that e is rational is false.

Remark.   The number π can also be expressed as the sum of a series. For instance, we have

$$\frac{\pi}{4} = 1 - \frac{1}{3} + \frac{1}{5} - \frac{1}{7} + \cdots$$

Here the general term of the series is $(-1)^n \frac{1}{2n+1}$, for each integer $n \geq 0$.

CHAPTER 1 REVIEW

A sequence $\{x_n\}_n$ of real numbers can be described using

- a direct formula when the general term x_n can be computed explicitly in terms of n;
- a recursive formula when the term x_{n+1} can be computed from some or all of the previous terms.

A sequence given by a recursive formula of the form $x_{n+1} = x_n + d, n \geq 1$ is called an *arithmetic sequence* with first term x_1 and common difference d. The direct formula for such a sequence is $x_n = x_1 + (n - 1)d, n \geq 1$.

A sequence given by a recursive formula of the form $y_{n+1} = y_n \cdot r, n \geq 1$ is called a *geometric sequence* with the first term y_1 and a common ratio r. The direct formula for such a sequence is $y_n = y_1 \cdot r^{n-1}, n \geq 1$.

The sum of the first n consecutive terms in an arithmetic sequence with first term a and common difference d is

$$\sum_{i=1}^{n} \left[a + (i - 1)d \right]$$

$$= na + \frac{n(n - 1)}{2} d \left(= \frac{\text{first term} + \text{last term}}{2} \cdot \text{number of terms} \right).$$

The sum of the first n consecutive terms in a geometric sequence with first term a and common ratio r is

$$\sum_{i=1}^{n} a \cdot r^{i-1} = \begin{cases} a \dfrac{r^{n-1}}{r-1} & \text{if } r \neq 1 \\ na & \text{if } r = 1. \end{cases}$$

A sequence $\{x_n\}_n$ converges to a real number L provided any open interval centered at L contains all but finitely many terms of the sequence. In this case we write $\lim_{n \to \infty} x_n = L$.

A sequence that does not converge to any real number is called divergent. Using this definition, you proved that $\lim_{n \to \infty} \frac{1}{n} = 0$.

A sequence $\{x_n\}_n$ is said to be bounded if there exists $M > 0$ such that $|x_n| \leq M$ for all n. A sequence $\{x_n\}_n$ is said to be unbounded if it is not bounded.

A sequence $\{x_n\}_n$ is called increasing if $x_n < x_{n+1}$ for all n and decreasing if $x_n > x_{n+1}$ for all n. A sequence $\{x_n\}_n$ is called nondecreasing if $x_n \leq x_{n+1}$ for all n and is called nonincreasing if $x_n \geq x_{n+1}$ for all n.

The definition of convergent sequences is used to prove various properties of limits of sequences. The following properties **P1–P7** have proved to be very useful tools in the investigation of the convergence of sequences and the computation of the corresponding limits whenever they exist.

P1 A sequence $\{x_n\}_{n \geq 1}$ converges to L if and only if the sequence $\{x_n\}_{n \geq k}$ converges to L for any fixed $k \in \mathbf{N}$.

P2 A convergent sequence has a unique limit.

P3 If a sequence $\{x_n\}_n$ is convergent, then it is bounded. Consequently, unbounded sequences are divergent.

P4 Weierstrass's theorem: A nondecreasing sequence is convergent if and only if it is bounded. A nonincreasing sequence is convergent if and only if it is bounded.

P5 Let $\{x_n\}_n$, $\{y_n\}_n$ be two convergent sequences and let c be a real number. Then the following are true:

   a. $\lim\limits_{n \to \infty} (x_n + y_n) = \lim\limits_{n \to \infty} x_n + \lim\limits_{n \to \infty} y_n$

   b. $\lim\limits_{n \to \infty} (x_n - y_n) = \lim\limits_{n \to \infty} x_n - \lim\limits_{n \to \infty} y_n$

   c. $\lim\limits_{n \to \infty} (c \cdot y_n) = c \cdot \lim\limits_{n \to \infty} y_n$

   d. $\lim\limits_{n \to \infty} (x_n \cdot y_n) = \lim\limits_{n \to \infty} x_n \cdot \lim\limits_{n \to \infty} y_n$

   e. If, in addition, $y_n \neq 0$ for all n and $\lim\limits_{n \to \infty} y_n \neq 0$, then

$$\lim_{n \to \infty} \frac{x_n}{y_n} = \frac{\lim\limits_{n \to \infty} x_n}{\lim\limits_{n \to \infty} y_n}.$$

P6 The sandwich theorem for sequences: If the sequences $\{x_n\}_n$, $\{y_n\}_n$, and $\{z_n\}_n$ are such that

   a. $x_n \leq y_n \leq z_n$ for each n, and

   b. $\lim\limits_{n \to \infty} x_n = \lim\limits_{n \to \infty} z_n = L$ for some $L \in \mathbf{R}$,

   then the sequence $\{y_n\}_n$ is convergent, and $\lim\limits_{n \to \infty} y_n = L$.

P7 If $\{x_n\}_n$, $\{y_n\}_n$ are convergent sequences and $x_n \leq y_n$ for each n, then

$$\lim_{n \to \infty} x_n \leq \lim_{n \to \infty} y_n.$$

An arithmetic sequence is convergent if and only if the common difference equals zero. A geometric sequence is convergent if and only if the common ratio r is equal to 1, $|r| < 1$, or if the first term of the sequence equals zero.

An infinite series $x_1 + x_2 + x_3 + \cdots$ is said to converge to a real number L, provided the sequence of partial sums $s_n = \sum_{i=1}^{n} x_i, n \geq 1$ converges to L. In this case, we write $\sum_{i=1}^{\infty} x_i = L$. If such an L does not exist, the series is said to be divergent.

If the terms $x_n, n \geq 1$ of a series form a geometric sequence, the series is called *geometric*.

If a is a nonzero real number, the series $a + ar + ar^2 + \cdots$ converges whenever $|r| < 1$, in which case $\sum_{i=1}^{\infty} a \cdot r^{i-1} = \frac{a}{1-r}$ and diverges whenever $|r| \geq 1$ (see Theorem 1.2.1).

CHAPTER 1 REVIEW EXERCISES

In Exercises 1–4, find a direct formula for a sequence whose first few terms coincide with the given ones.

1. $\frac{1}{2}, \frac{2}{3}, \frac{3}{4}, \frac{4}{5}, \frac{5}{6}, \ldots$

2. $3, -3, 3, -3, 3, -3, \ldots$

3. $1, \frac{1}{4}, \frac{1}{9}, \frac{1}{16}, \frac{1}{25}, \ldots$

4. $9, 5, 1, -3, -7, \ldots$

5. Determine a direct formula for the sequence $\{a_n\}_n$ having the recursive formula

$$(n + 2)(n + 1)a_{n+2} - n^2 a_n = 0, \ n \geq 1, \ a_1 = 0, \ a_2 = \frac{1}{2}.$$

In Exercises 6–9, find a recursive formula for the given sequence.

6. An arithmetic sequence $\{x_n\}_{n \geq 1}$ with $x_{44} = 13.5$ and a common difference of 0.5

7. A geometric sequence $\{y_n\}_{n \geq 1}$ with $y_3 = 0.5$ and a common ratio of $\sqrt{3}$

8. A geometric sequence $\{y_n\}_{n \geq 1}$ with $y_3 = 225$ and $y_5 = 81$ with $r < 0$

9. An arithmetic sequence $\{x_n\}_{n \geq 1}$ with $x_{47} = 74$ and $x_{24} = 47$

In Exercises 10–11, show that if the numbers a, b, c are consecutive terms in an arithmetic sequence, then so are the ones listed.

10. $a^2 - bc, b^2 - ca, c^2 - ab$

11. $b^2 + bc + c^2, c^2 + ca + a^2, a^2 + ab + b^2$

12. Use Definition 1.1.1 to prove that the sequence $\{\frac{n}{n+1}\}_n$ converges to 1.

In Exercises 13–20, determine if the sequence is convergent or divergent. If convergent, determine its limit.

13. $x_n = 2 + (-1)^n$

14. $x_n = \frac{(-1)^n}{2}$

15. $x_n = \frac{(-1)^n}{n}$

16. $x_n = \frac{3 + (-1)^n}{n}$

17. $a_n = \frac{2n^2 + 1}{n^2}$

18. $a_n = \frac{n^2 - 5n + 2}{3n^2 + 3n + 1}$

19. $a_n = \frac{n+2}{n^2 - 1}$

20. $a_n = \frac{5^n + 1}{5^n + 3}$

In Exercises 21–24, compute the sum without using a calculator.

21. $S = 2 + 8 + 14 + \cdots + 638$

22. $S = 198 + 194 + 190 + \cdots + 66$

23. $S = 10\sqrt{5} + 50 + 50\sqrt{5} + \cdots + 3{,}906{,}250$

24. $S = 6{,}561 - 2{,}187 + 729 + \cdots - 3$

25. If a is a nonzero real number, and $n \geq 2$, compute the sum

$$\left(a + \frac{1}{a}\right)^2 + \left(a^2 + \frac{1}{a^2}\right)^2 + \cdots + \left(a^n + \frac{1}{a^n}\right)^2.$$

26. If a is a nonzero real number, and $n \geq 2$, compute the sum

$$\left(a - \frac{1}{a}\right)^2 + \left(a^2 - \frac{1}{a^2}\right)^2 + \cdots + \left(a^n - \frac{1}{a^n}\right)^2.$$

27. If $\{y_n\}_n$ is a geometric sequence with common ratio r, compute the sum

$$S = y_1 y_2 y_3 + y_2 y_3 y_4 + \cdots + y_n y_{n+1} y_{n+2}$$

in terms of y_1 and r.

28. If $\{y_n\}_n$ is a geometric sequence with $y_1 \neq 0$ and with a common ratio $r \neq 0$, compute the sum

$$S = \frac{1}{y_1 y_2 y_3} + \frac{1}{y_2 y_3 y_4} + \cdots + \frac{1}{y_n y_{n+1} y_{n+2}}$$

in terms of y_1 and r.

29. If $\{x_n\}_n$ is an arithmetic sequence, compute the sum

$$\frac{1}{\sqrt{x_1} + \sqrt{x_2}} + \frac{1}{\sqrt{x_2} + \sqrt{x_3}} + \cdots + \frac{1}{\sqrt{x_{n-1}} + \sqrt{x_n}}$$

in terms of x_1, the common difference d, and n. All the terms in the sum are well defined.

30. If $\{y_n\}_n$ is a geometric sequence, compute the sum

$$\frac{\sqrt{y_2}}{\sqrt{y_2} - \sqrt{y_1}} + \frac{\sqrt{y_3}}{\sqrt{y_3} - \sqrt{y_2}} + \cdots + \frac{\sqrt{y_{n+1}}}{\sqrt{y_{n+1}} - \sqrt{y_n}}$$

in terms of n and the common ratio r. All the terms in the sum are well defined.

In Exercises 31–38, decide if the series converges or diverges. If the series converges, compute its sum.

31. $\frac{5}{6} + \frac{25}{36} + \frac{125}{216} + \cdots$

32. $-\frac{1}{4} + \frac{1}{16} - \frac{1}{64} + \frac{1}{256} + \cdots$

33. $2 + 4 + 8 + 16 + 32 + \cdots$

34. $3 + 9 + 27 + 81 + 243 + \cdots$

35. $2 + \frac{5}{2} + 3 + \frac{7}{2} + 4 + \frac{9}{2} + \cdots$

36. $\frac{1}{2} - \frac{1}{2} + \frac{1}{2} - \frac{1}{2} + \cdots$

37. $5 + \frac{5}{2} + \frac{5}{4} + \frac{5}{8} + \cdots$

38. $4 - \frac{4}{3} + \frac{4}{9} - \frac{4}{27} + \cdots$

39. Compute the sum

$$1 + 2 \cdot 2 + 3 \cdot 2^2 + 4 \cdot 2^3 + 5 \cdot 2^4 + \cdots + 100 \cdot 2^{99}.$$

40. Compute the sum

$$3 + 33 + 333 + 3{,}333 + \cdots + \underbrace{333\cdots3}_{n \text{ copies of } 3}.$$

41. Compute the limit

$$\lim_{n \to \infty} \frac{1 + \frac{1}{2} + \cdots + \frac{1}{2^n}}{1 + \frac{1}{3} + \cdots + \frac{1}{3^n}}.$$

Functions, Limits, and Continuity

2.1   FUNCTIONS

Functions • Function notation • Image • Composite functions

In your daily life, you continually encounter quantities that vary. In mathematics, we call such quantities *variables*, and we are often interested in establishing relationships between these variables. Consider the following examples.

Stocks.   The fluctuation in the average stock prices for the Dow and Nasdaq stock exchanges, at 10:00, 10:30, 11:00, 11:30, 12:00, 12:30, and 13:00 on Monday, February 25, 2002, are shown in the following tables.

Time	Fluctuation for Dow
10:00	84
10:30	90
11:00	98
11:30	87
12:00	117
12:30	90
13:00	90

Time	Fluctuation for Nasdaq
10:00	17
10:30	12
11:00	14
11:30	20
12:00	27
12:30	21
13:00	24

USPS. The cost of mailing a first-class letter via the United States Postal Service is $0.37 for the first ounce and $0.23 for each additional ounce, up to a total of 13 ounces. The following graph illustrates the relationship between the letter's weight and the cost. Some of the bullets at the ends of the segments in the graph are filled while others are empty. Why?

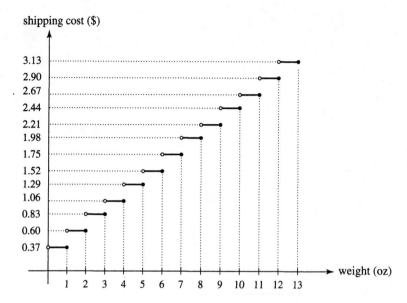

Volume of a Cube. We often use algebraic formulas to model real-life problems. For example, the relationship between the volume V of a cube and its side length l is given by the formula $V = l^3$.

In each of these examples, you can identify three entities: the collection of inputs, the collection of outputs, and the law that governs the relationship between inputs and outputs. In mathematical language, the inputs are called **independent variables**, and the outputs are called **dependent variables** (depending on the independent variables). For example, the volume V of a cube depends on the side length l of the cube. Here, l is the independent variable, V is the dependent variable, and the relationship between them is $V = l^3$.

> **Definition 2.1.1** A function is determined by two sets A, B, and a law, f (correspondence, assignment, rule) that associates to each element x in A a unique element y in B. The set A is called the **domain** of the function, while B is called the **range** of the function. Two functions are equal if they have the same domain, the same range, and identical laws.

Classroom Connection 2.1.1: Which Is a Function?

Use the definition of a function to answer the questions raised in Problems 14–18 from pages 473–474 in the eighth-grade textbook *Math Thematics, Book 3*.

14 For five days, Mei recorded the data shown at the right. She concluded that the amount of rainfall for any given day is a function of the high temperature for that day.

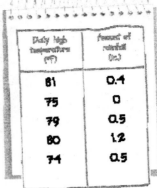

Daily high temperature (°F)	Amount of rainfall (in.)
81	0.4
75	0
79	0.5
80	1.2
74	0.5

a. How do you think Mei came to this conclusion?

b. Suppose Mei recorded these data for a year. Do you think she would still say that daily rainfall is a function of daily high temperature? Why or why not?

c. In general, do you think that daily rainfall is a function of daily high temperature? Explain.

15 **Discussion** Tell whether y is a function of x. Explain your thinking.

a. x = the amount of time that the sky is cloudy
y = the amount of rain that falls

b. x = the number of minutes someone drives at 55 mi/h
y = the distance that person drives

c. x = the time of day today
y = the number of people at your school at that time

d. x = a person's height
y = the foot length of someone with that height

▶ You can also think of a function as a *rule* that pairs each input value with exactly one output value. You can use an equation to write a rule for some functions.

16 Suppose it starts raining steadily at noon. The rain falls for the rest of the afternoon at a rate of 0.2 inches per hour.

a. Let y = the amount of rain that has fallen since noon. Let x = the number of hours since noon. Explain why the value of y is a function of the value of x.

b. Write an equation that models this function.

▶ You can use a table of values or a graph to tell whether an equation models a function.

EXAMPLE

Given an equation, you can tell whether y is a function of x by comparing input and output values in a table or a graph.

a. $y = x^2$ For every value of x, there is exactly one value of y. The equation models a function.

Input (x)	Output (y)
–2	4
–1	1
0	0
1	1
2	4

b. $x = y^2$ For some values of x, there are two different values of y. The equation does not model a function.

Input (x)	Output (y)
0	0
1	–1 and 1
4	–2 and 2

17 **Try This as a Class** Refer to the Example.

  a. How are the two graphs alike? How are they different?

  b. Use the tables to find the value(s) of y when $x = 1$ in both equations. Then use the graphs.

  c. How can you use a table of values to tell whether an equation models a function? How can you use a graph?

✔ QUESTION 18

...checks that you can identify a function.

18 **✔ CHECKPOINT** For each equation or graph, tell whether y is a function of x. Explain your thinking.

  a. $y = 7x$

  b. $2 + x = y^2$

  c.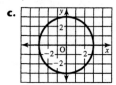

  d.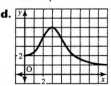

HOMEWORK EXERCISES ▶ See Exs. 10-26 on pp. 478-479.

◆

Notation and Further Terminology. For f, A, and B as in Definition 2.1.1, we write $f : A \to B$ and read "(the function) f maps A into B." If the dependent variable y corresponds to the independent variable x according to the relationship f, then we write $y = f(x)$ and read "y equals f of x," or "y equals the value of f at x." The domain of the function f will sometimes be denoted by Domain(f). The collection of all outputs of a given function f is called the **image** of f and is denoted by Image(f). That is,

$$\text{if } f : A \to B, \qquad \text{then} \qquad \text{Image}(f) = \{f(x) : x \in A\}.$$

Warning: The reader is advised that a slightly different terminology may be used in other textbooks. The set B is occasionally referred to as *codomain* rather than *range*, whereas the set Image(f) defined here is sometimes referred to as *range* rather than *image*.

A function whose range is a subset of $\mathbb{R}$ is called a **real valued** function.

Classroom Discussion 2.1.1: More on Functions

1. The relationship between the temperature x in degrees Fahrenheit and the temperature y in degrees Celsius is given by the formula $y = \frac{5(x-32)}{9}$. Denote by g the function modeling this relationship. In other words, $g(x) = \frac{5(x-32)}{9}$.

   a. What is the set of all possible inputs for g? We take this set to be the domain of g.

   b. Determine Image(g).

   c. Find the values for $g(32)$, $g(65)$, and $g(99)$. How do you interpret the values you obtained?

   d. Let x be a given real number. Find the values for $g(x + 1)$, $g(2x)$, and $g(3x - 5)$ in terms of x. How do you interpret the values you obtained?

   e. Write the algebraic identity corresponding to the following statement: "If the temperature of a body is increased by 1 degree Fahrenheit, its temperature measured in degrees Celsius also increases by 1 degree." Investigate its validity.

2. Return to the USPS example from the beginning of the section. Let f be the price function for mailing a first-class letter via the USPS.

   a. What is the domain of f? What is the image of f?

   b. Denote by x the letter's weight. Determine a formula for $f(x)$.

   Hint: You can make use of the ceiling function. By definition, the ceiling function of x, denoted by $\lceil x \rceil$, is the smallest integer that is greater than or equal to x, that is, $\lceil x \rceil = k$, where k is the integer value for which $k - 1 < x \le k$. ◆

People study functions because they are very useful in modeling physical phenomena. However, the definition of a function is not dependent upon a particular model of a "real-world" situation. The advantage of studying mathematics of change for general functions as opposed to particular examples is that once the theory is in place, one can apply it directly to specific examples. For this reason, we often work with abstract functions given by algebraic expressions. This is why we need to make the following conventions:

Unless otherwise specified, all functions considered in the remainder of the book are assumed to have subsets of the real line as their domains.

Whenever an algebraic expression $f(x)$ is referred to as being a function without further specifications regarding its domain or range, it is implicitly understood that this is the function consisting of

- domain = the set of all real numbers x for which the expression $f(x)$ is meaningful;
- range = the whole real line;
- law = x goes into $f(x)$.

Examples of Functions

1. If a function is of the form

$$f(x) = a_n x^n + a_{n-1} x^{n-1} + \cdots + a_1 x + a_0,$$

where n is a nonnegative integer and $a_n, a_{n-1}, \ldots, a_1, a_0$ are real numbers with $a_n \neq 0$, then f is called a **polynomial function of degree** n. Here are a few examples of polynomial functions: $f_1(x) = 13$, $f_2(x) = \frac{1}{2}x + 3$, $f_3(x) = -x^3$, and $f_4(x) = 5x^4 - 3x^2 - 4x - 19$. It is immediately apparent that polynomial functions are meaningful for any real value x. Thus, $(-\infty, \infty)$ is their domain.

2. Since the equation of a line is of the form $y = mx + b$, we call polynomial functions of degree 1 **linear functions**. Thus, a function f is linear if and only if there exist real numbers m and b such that $f(x) = mx + b$. A function that is not linear is called **nonlinear**.

3. The quotient of two polynomial functions defines a new function called a **rational function**. The following functions are examples of rational functions:

$$f_1(x) = \frac{1}{x}, \qquad f_2(x) = \frac{x + 2}{3 - 2x}, \qquad f_3(x) = \frac{x}{x^2 - 1},$$

$$f_4(x) = \frac{x^5 - 2x^3 + 4x + 5}{x^4 - 4x^2 + 4}.$$

Observe that if $p(x)$ and $q(x)$ are polynomial functions, then the domain of the rational function $f(x) = \frac{p(x)}{q(x)}$ is the set of real numbers for which $q(x) \neq 0$. In other words, the only real numbers x for which $f(x)$ is not meaningful are the ones for which the denominator equals zero. Let us determine the domains of the previous functions.

- The domain of $f_1(x)$ is the set of real numbers for which $x \neq 0$. Thus, the domain of $f_1(x)$ is the set of real numbers except 0, which is denoted by $\text{Domain}(f_1) = (-\infty, 0) \cup (0, \infty)$ or by $\text{Domain}(f_1) = \mathbb{R} \setminus \{0\}$.
- The domain of $f_2(x)$ is the set of real numbers for which $3 - 2x \neq 0$. Thus, the domain of $f_2(x)$ is $(-\infty, \frac{3}{2}) \cup (\frac{3}{2}, \infty)$. We can also write $\text{Domain}(f_2) = \mathbb{R} \setminus \{\frac{3}{2}\}$.

- The domain of $f_3(x)$ is the set of real numbers for which $x^2 - 1 \neq 0$, that is, $x \neq \pm 1$. Hence, the domain of $f_3(x)$ is $(-\infty, -1) \cup (-1, 1) \cup (1, \infty)$. We can also write Domain$(f_3) = \mathbb{R} \setminus \{-1, 1\}$.
- The domain of $f_4(x)$ is the set of real numbers for which $x^4 - 4x^2 + 4 \neq 0$. Since $x^4 - 4x^2 + 4 = (x^2 - 2)^2$, we see that $f_4(x)$ is not defined for $x = \pm\sqrt{2}$. As such, the domain of $f_4(x)$ is $(-\infty, -\sqrt{2}) \cup (-\sqrt{2}, \sqrt{2}) \cup (\sqrt{2}, \infty)$. We can also write Domain$(f_4) = \mathbb{R} \setminus \{-\sqrt{2}, \sqrt{2}\}$.

4. Recall that the absolute value of a real number x, denoted by $|x|$, is equal to x if $x \geq 0$, and equal to $-x$ if $x < 0$. The function $f(x) = |x|$ is called the **absolute value function**. Since it is meaningful to compute the absolute value of any real number, Domain$(f) = (-\infty, \infty)$, or Domain$(f) = \mathbb{R}$.

5. The square-root function $g(x) = \sqrt{x}$ is not defined for negative values of x, so the domain of $g(x)$ is $[0, \infty)$. We can also write Domain$(g) = \mathbb{R}_+$.

6. Let r be a rational number $r = \frac{m}{n}$, where m is an integer, n is a natural number, and m, n are relatively prime. Then, the domain of the rth root function $f(x) = \sqrt[r]{x}$ depends on the sign of r and on whether n is even or odd. To see why this is the case, recall that $\sqrt[r]{x} = \sqrt[n]{x^m}$. Thus, if $r > 0$ and n is odd, then Domain$(f) = \mathbb{R}$; if $r > 0$ and n is even, then Domain$(f) = [0, \infty)$; if $r < 0$ and n is odd, then Domain$(f) = \mathbb{R} \setminus \{0\}$; and if $r < 0$ and n is even, then Domain$(f) = (0, \infty)$.

Practice Problems

Determine the domain of each of the given functions.

a. $f(x) = \sqrt{x + 1}$ **b.** $g(x) = \frac{2x}{x+3}$ **c.** $h(x) = \frac{1}{\sqrt{2-x}}$.

Classroom Discussion 2.1.2: The Vertical Line Test

By definition, the **graph** of a function f is the collection of all points in the plane with coordinates $(x, f(x))$ for all values of x in the domain of f. This discussion's goal is to determine a necessary and sufficient condition where a given curve in the plane can be the graph of a function.

1. Suppose c is a fixed real number and the line $x = c$ intersects a given curve at two points. Can the curve be the graph of a function? Explain.
2. Do you agree with the following statement?

   A curve in the plane will represent the graph of a function if and only if any vertical line intersects the curve only once.

We will refer to this criterion as the **vertical line test**.   ◆

Classroom Connection 2.1.2: Graphs

The following exploration is from pages 16–17 and page 120 in the seventh-grade textbook *Mathematics in Context, Tracking Graphs*. Work Problems 17 and 18.

In Problem 17, after completing the graphs for the three possibilities a, b, and c, determine the domain and image of each function. Then, discuss Problem 19. Do you see any connections with the vertical line test?

Tracking Graphs

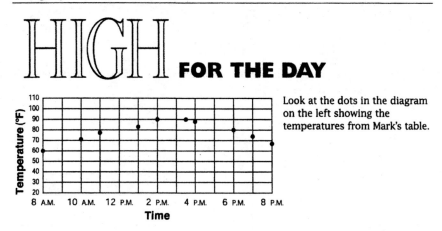

HIGH FOR THE DAY

Look at the dots in the diagram on the left showing the temperatures from Mark's table.

Below are three possibilities for what might have happened to the temperature between 2:00 P.M. and 3:30 P.M.

17. For each possibility, fill in the line graph between 2:00 P.M. and 3:30 P.M. and record the high. Use the three graphs on **Student Activity Sheet 8.**

Possibility a
Between 2:00 P.M. and 3:30 P.M., the temperature stayed at 90°F.

Possibility b
Between 2:00 P.M. and 3:30 P.M., it got even hotter than 90°F.

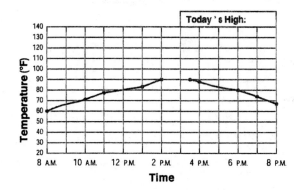

Possibility c
After 2:00 P.M., there was a severe thunderstorm, which caused the temperature to drop. When the storm was over, the temperature rose again. At 3:30 P.M., it was again 90°F.

Student Activity Sheet 8

Name _____

Use with *Tracking Graphs*, page 16.

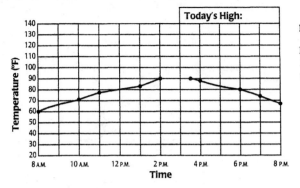

Possibility a:

Between 2:00 P.M. and 3:30 P.M., the temperature stayed at 90°F.

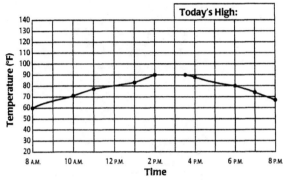

Possibility b:

Between 2:00 P.M. and 3:30 P.M., it got even hotter than 90°F.

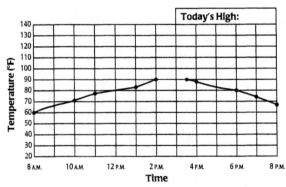

Possibility c:

After 2:00 P.M. there was a severe thunderstorm, which caused the temperature to drop. When the storm was over, the temperature rose again. At 3:30 P.M., it was again 90°F.

You have looked at three possibilities for the outside temperatures between 2:00 P.M. and 3:30 P.M. and the corresponding graphs.

18. Are there any other possibilities? If so, how many?

Each graph below shows just the part of the graph between noon and 5:00 P.M. Every graph shows a different situation between 2:00 P.M. and 3:30 P.M.

19. Explain why these situations are not likely to have occurred.

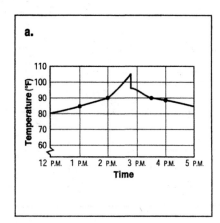

a.

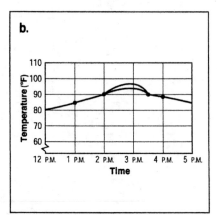

b.

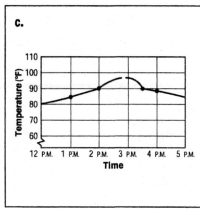

c.

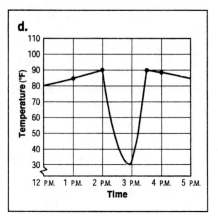

d.

3. In a–e, determine which equations in x and y describe y as a function of x. To do so, in each case, first solve for y in terms of x and then explain if you expect the vertical line test to hold.

a. $x^2 + y^2 = 1$ **b.** $xy = 3$ **c.** $y = 1$ **d.** $y^2 = 2x$ **e.** $2y - x = 5$. ◆

OPERATIONS WITH FUNCTIONS

1. Algebraic Operations with Functions

Algebraic operations with functions are performed in a very natural fashion. Specifically, given two real-valued functions f, g and a fixed real number c, we can create the sum $f + g$, the difference $f - g$, the constant multiple $c \cdot f$, the product $f \cdot g$, and the quotient $\frac{f}{g}$ by setting

$$(f + g)(x) = f(x) + g(x),$$
$$(f - g)(x) = f(x) - g(x),$$
$$(c \cdot f)(x) = c \cdot f(x),$$
$$(f \cdot g)(x) = f(x) \cdot g(x),$$
$$\left(\frac{f}{g}\right)(x) = \frac{f(x)}{g(x)}.$$

2. Compositions of Functions

Sometimes a function is the result of a "chain" of applications. Here is a real-life example of a chain of two applications:

Composite sales. A store in Columbia, MO, is planning to have a 45% sale on all sporting goods. The same day, Anna, the manager, learns that a new shipment is supposed to arrive. In order to increase sales and free some storage space, Anna realizes that the price reduction for the store's items should be larger than originally planned. She decides to reduce prices by an additional 20%. Instead of marking down the prices in two stages, first 45% and then 20%, Anna observes that it would be faster to apply just one reduction equivalent to the two consecutive ones. Verify that applying the two reductions in a row is equivalent to applying one reduction of 56%.

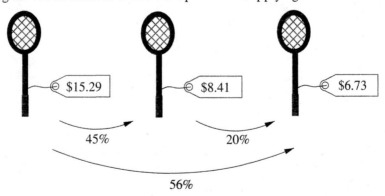

This idea of creating a new function that is the result of applying two functions back-to-back is at the core of **composite functions**. Consider the function $h(x) = (x^2 + 2x - 3)^3$. This function can be obtained by first applying the function g defined by $g(x) = x^2 + 2x - 3$, and then applying the function f defined by $f(y) = y^3$ to the output $y = g(x)$. In this way, the functions f and g act back-to-back.

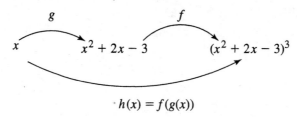

$$h(x) = f(g(x))$$

The notation for the resulting function h is $h(x) = f(g(x))$, which is read "f of g of x"; or $h(x) = (f \circ g)(x)$, which is read "f composed with g of x."

Definition 2.1.2 Suppose f and g are real-valued functions. Then the **composition** of f with g, denoted by $f \circ g$, is the function $(f \circ g)(x) = f(g(x))$, which is defined for all real values x in the domain of g for which $g(x)$ is in the domain of f.

Classroom Discussion 2.1.3: Examples of Composite Functions

1. Consider the functions $f(y) = \sqrt{y}$ and $g(x) = 2x + 5$. Recall that, by definition, for each $a > 0$, $\sqrt{a}$ is the positive number whose square is a.

   a. Determine the algebraic expression of $f(g(x))$.
   b. Is the number $g(-3)$ an element of the domain of f?
   c. What condition do you need to impose on x in order for $g(x)$ to be in the domain of f?
   d. Can you find a subset A of the real numbers so that the composition $(f \circ g)(x)$ makes sense for all $x \in A$? What is the largest such set?

2. Consider the following functions with domains and ranges as specified:
   $f : (0,\infty) \to (0,\infty)$ defined as $f(x) = x + \frac{1}{x}$, and $g : (0,\infty) \to (0,\infty)$ with $g(x) = x^2$.

   a. Does the composition $f \circ g$ make sense for any $x \in (0,\infty)$? Explain.
   b. Does the composition $g \circ f$ make sense for any $x \in (0,\infty)$? Explain.
   c. Show that there exists a real number c with the property that $(f \circ g)(x) - (g \circ f)(x) = c$ for all $x \in (0,\infty)$. What is the value of c?
   d. Is the order in which the composition is performed important?
   e. Why do you think we write $f \circ g$ and not $g \circ f$ for the composition in which g acts first and f acts on the output generated by g? ◆

EXAMPLES In a–c, determine the algebraic expressions and the domains of definition of the functions $f \circ g$ and $g \circ f$.

   a. $f(x) = 3x - 1$ and $g(x) = x^2$.
   b. $f(x) = \sqrt{x}$ and $g(x) = x^2$.
   c. $f(x) = x + 1$ and $g(x) = \frac{1}{x-2}$.

Solutions

   a. The functions f and g are both defined on the whole real line. This is why $f \circ g$ and $g \circ f$ are also defined on $\mathbb{R}$. In addition, for all $x \in \mathbb{R}$, we have $f(g(x)) = f(x^2) = 3x^2 - 1$ and $g(f(x)) = g(3x - 1) = (3x - 1)^2 = 9x^2 - 6x + 1$.

b. Since $g(x) \geq 0$ for all $x \in \mathbb{R}$, and the domain of f is $[0,\infty)$, we have that $(f \circ g)(x)$ is well defined for all $x \in \mathbb{R}$. In addition, for all $x \in \mathbb{R}$, we have

$$f(g(x)) = f(x^2) = \sqrt{x^2} = |x|.$$

Since the domain of g is $\mathbb{R}$, it follows that $(g \circ f)(x)$ is well defined for all x in the domain of f, which is $[0,\infty)$. In addition, for all $x \in [0,\infty)$, we have
$$g(f(x)) = g(\sqrt{x}) = (\sqrt{x})^2 = x.$$

c. Since the domain of f is $\mathbb{R}$, it follows that $(f \circ g)(x)$ is well defined for all x in the domain of g, which is $\mathbb{R} \setminus \{2\}$. In addition, for all $x \neq 2$, we have

$$f(g(x)) = f\left(\frac{1}{x-2}\right) = \frac{1}{x-2} + 1 = \frac{x-1}{x-2}.$$

Since the domain of g is $\mathbb{R} \setminus \{2\}$, $(g \circ f)(x)$ is well defined for all x in the domain of f for which $f(x) \neq 2$; that is, for $x \in \mathbb{R} \setminus \{1\}$. In addition, for all $x \in \mathbb{R} \setminus \{1\}$, we have
$$g(f(x)) = g(x + 1) = \frac{1}{x + 1 - 2} = \frac{1}{x - 1}. \qquad \blacksquare$$

Practice Problem

Determine the algebraic expressions and the domains of definition of the functions $f \circ g$ and $g \circ f$ if $f(x) = \sqrt{x + 2}$ and $g(x) = \frac{1}{x^2+1}$.

EXERCISES 2.1

In Exercises 1–4, determine which is a function. For each function, determine its image.

1. Domain = $\mathbb{Z}$, Range = the set of nonnegative integers, Correspondence = to each number, associate its square.

2. Domain = {Amy, Dan, Ellen, Ron} (students in sixth grade), Range = $\{A, B, C, D, F\}$, Correspondence = to each student, assign his or her letter grade in mathematics.

3. Domain = $[-2,2]$, Range = $[-2,2]$, Correspondence = to each x, associate y satisfying $x^2 + y^2 = 4$.

4. Domain = $\mathbb{N}$, Range = the set of all the elements in a given sequence $\{x_n\}_n$, Correspondence = to each whole number n, associate the element x_n.

5. The height in feet of an object falling from a cliff is given by the expression $h(t) = -16t^2 + 1,000$, where t is the time measured in seconds from the moment the object starts falling.
   a. What is the appropriate domain of h that models the object's fall?
   b. Compute $h(0), h(1), h(3), h(t + 1)$.
   c. How do you interpret the difference $h(t + 1) - h(t)$?
   d. Write the algebraic identity equivalent to the following statement: "The height of the falling object decreases each second by $h(1)$." Investigate its validity.

6. Classroom Connection 2.1.3: More on Graphs

The following exploration is from pages 39–42 in the seventh-grade textbook *Mathematics in Context, Tracking Graphs*. Work all the problems listed on these pages. In each case, identify the functions that arise and determine their domains and images.

TRY THIS!

Section A. Graphing a Race

The following graph represents a 50-kilometer bicycle race with three competitors.

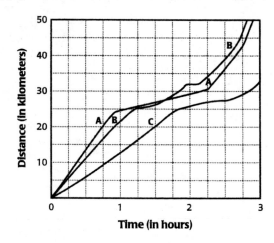

1. Study the above graph to answer the following questions. Be sure to explain all your answers.

   a. Who won the race and in how much time?

   b. Which competitor had the fastest start? How do you know?

   c. What might explain the graph for cyclist B at two hours into the race?

   d. There is a very steep hill in this course. How many kilometers into the race do the riders hit the hill? How do you know?

   e. What do you think competitor C's final time was?

2. Write a short article about the race. Be sure to include the important information shown on the graph.

Tracking Graphs

Section B. Filling in Temperatures

Due to unseasonably heavy rains, Nelson's town experienced severe flooding last spring. Nelson kept a journal in which he drew the following graphs showing the height of the river that runs through town.

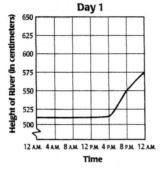

Day 1

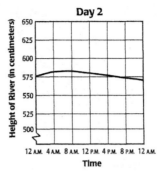

Day 2

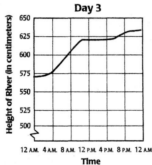

Day 3

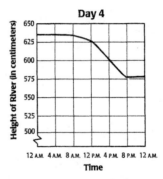

Day 4

1. Write journal entries for each day, describing what happened according to the graphs above.

On the fifth day of flooding, Nelson kept track of the height of the river using the chart on the right.

2. a. Draw a line graph using the data from this table.

   b. Draw another possible line graph based on the data from the table, with the additional information that it rained from 12:00 P.M. to 2:00 P.M. and then the sun came out for the rest of the day.

Day 5

Time	Height of River
4:00 A.M.	600 cm
8:00 A.M.	580 cm
12:00 P.M.	575 cm
4:00 P.M.	575 cm
8:00 P.M.	555 cm
12:00 A.M.	565 cm

Section C. Graphs over a Longer Time Period

Tameeka was born in January of 1987. Her doctor kept track of her weight during her first year and drew the following chart.

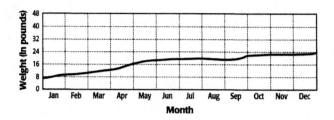

1. a. How much did Tameeka weigh in the middle of June?

   b. During which month did she gain the most weight?

As Tameeka grew older, her doctor changed her original weight chart to a five-year chart.

2. Copy the chart shown below and draw the line for 1987.

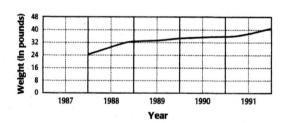

3. a. Why do you think Tameeka's doctor changed her original chart to a five-year chart?

   b. Can you still tell during which month she gained the most weight?

   c. Can you tell during which year she gained the most weight?

Tracking Graphs

Section D. Graphs Depicting Speed

Draw a speed graph for each of the following scenarios:

1. A subway train starts its route at 7:00 A.M. and makes a stop every 3 minutes.
The train stops at each station for 30 seconds. Between stops, the train travels at
35 mph. At 7:21 A.M., the train pulls into a station and has to remain there for
5 minutes because of a problem with the doors.

2. A plane takes off from an airport at 1:00 P.M., accelerates to a speed of 500 mph, and
maintains this speed until it lands 4 hours later.

3. At 5:00 P.M., a car leaves a parking lot and accelerates to the 35 mph speed limit.
The car stops at two red lights and then enters the highway at 5:10 P.M. The speed
limit on the highway is 55 mph. At 5:25 P.M., the car exits the highway and pulls
into a driveway.

Section E. Graphs of Tides

Below you see the tidal graph for a harbor on a certain day.

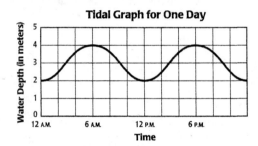

Tidal Graph for One Day

This harbor has a big pier off of which Pete likes to go fishing. The deeper the water,
the better the fishing.

1. What are the best times for Pete to fish? What are the worst times?

2. Pete finds that in order to catch squid, the water must be at least 4 meters deep.
When can Pete fish for squid from this pier?

3. Pete finds it more difficult to catch fish when the tide is going out (getting lower).
When is this happening?

4. On another day, the high tides are at 5:30 A.M. and 5:30 P.M. Approximately when is
low tide?

In Exercises 7–14, determine if the given equation describes y as a function of x.

7. $y + 2 = 4x$

8. $y - x^2 = 2$

9. $y^2 = x^2 + 1$

10. $y^2 = 4x - 1$

11. $3y^5 - 2x = 100$

12. $y^3 - 2 = x$

13. $y = -3$

14. $4y^2 + 3x^2 - 1 = 0$

15. Let $f(x)$ be equal to $x - 1$ if $x \leq 2$, and to $2x + 1$ if $x > 2$. One way to write this function in a condensed form is

$$f(x) = \begin{cases} x - 1 & \text{if } x \leq 2 \\ 2x + 1 & \text{if } x > 2. \end{cases}$$

Sketch the graph of f and determine its image.

16. Let $f(x)$ be equal to $1 - 2x$ if $x \leq -1$, and to $x + 2$ if $x > -1$. Write this function in a condensed form, sketch its graph, and determine its image.

In Exercises 17–32, determine the domain for the given function and its corresponding image.

17. $g(x) = x^2$

18. $g(x) = x^4 + 1$

19. $f(x) = x^{-\frac{1}{3}}$

20. $f(x) = (x - 1)^{-\frac{1}{5}}$

21. $h(x) = \frac{3}{x^2+1}$

22. $h(x) = \frac{4}{x^4+1}$

23. $f(x) = \frac{1}{x+1}$

24. $f(x) = \frac{x}{x-2}$

25. $g(x) = \sqrt{x^2}$

26. $g(x) = \sqrt{x^2 + 1}$

27. $h(x) = \sqrt{x - 1}$

28. $h(x) = \sqrt{x^2 + 2x}$

29. $g(x) = \sqrt[5]{x - 1}$

30. $g(x) = \sqrt[5]{x^3 - 1}$

31. $f(x) = \frac{x+2}{\sqrt{x+2}}$

32. $f(x) = \frac{2x-8}{\sqrt{x-4}}$

In Exercises 33–40, find the law and the domain of the specified composite function.

33. $f \circ g$ if $f(x) = 3 + 2x$ and $g(x) = -2x + 1$

34. $f \circ g$ if $f(x) = x^2 + 2x$ and $g(x) = x - 4$

35. $f \circ g$ if $f(x) = |x|$ and $g(x) = -x$

36. $f \circ g$ if $f(x) = |x + 1|$ and $g(x) = x^3 + 2$

37. $g \circ f$ if $f(x) = x + 3$ and $g(x) = \frac{1}{x}$

38. $g \circ f$ if $f(x) = \frac{x}{x^2+1}$ and $g(x) = \sqrt{x}$

39. $g \circ f$ if $f(x) = x^4 + 1$ and $g(x) = x^{\frac{1}{3}}$

40. $g \circ f$ if $f(x) = 2x + 5$ and $g(x) = x^{\frac{1}{4}}$

In Exercises 41–46, evaluate the composite function at the specified point, given that $f(1) = 2, f(3) = 0, f(-1) = 3, g(2) = 5, g(0) = -1,$ and $g(-1) = 1$.

41. $(f \circ g)(-1)$ **42.** $(f \circ g)(0)$

43. $(g \circ f)(1)$ **44.** $(g \circ f)(3)$

45. $(f \circ f)(-1)$ **46.** $(g \circ g)(0)$

In Exercises 47–52, find the conditions on f under which the composition is meaningful.

47. $g \circ f$ if $g(x) = \sqrt{x}$ **48.** $g \circ f$ if $g(x) = \sqrt{x - 2}$

49. $f \circ g$ if $g(x) = \sqrt{x}$ **50.** $f \circ g$ if $g(x) = \sqrt{x} - 2$

51. $g \circ f$ if $g(x) = \frac{1}{\sqrt{x+1}}$ **52.** $g \circ f$ if $g(x) = \frac{1}{\sqrt{1-x}}$

53. Given that $f : \mathbb{R} \rightarrow \mathbb{R}, f(x - 1) = x^2,$ determine $f(x + 1)$.

54. Given that $f : \mathbb{R} \rightarrow \mathbb{R}, f(2x + 1) = x^3 - 1,$ determine $f(x + 2)$.

55. Prove that there do not exist functions $f : \mathbb{R} \rightarrow \mathbb{R}$ and $g : \mathbb{R} \rightarrow \mathbb{R}$ with the property that $f(x) + g(y) = xy$ for all $x, y \in \mathbb{R}$.

56. Do there exist functions $f : \mathbb{R} \rightarrow \mathbb{R}$ and $g : \mathbb{R} \rightarrow \mathbb{R}$ with the property that $f(x) \cdot g(y) = x + y$ for all $x, y \in \mathbb{R}$?

57. Let $f : \mathbb{R} \rightarrow \mathbb{R}, \quad h : \mathbb{R} \rightarrow \mathbb{R}$ be defined by $f(x) = 2x - 3, h(x) = x + 6$ for all $x \in \mathbb{R}$. Determine two functions $g_1 : \mathbb{R} \rightarrow \mathbb{R}, g_2 : \mathbb{R} \rightarrow \mathbb{R}$ that verify the conditions $f \circ g_1 = h$ and $g_2 \circ f = h$.

PROJECTS AND EXTENSIONS 2.1

I. Using Graphs in Daily Life

Since graphs are central in the study of mathematics of change, it is important for you to be able to interpret graphs and to draw them when the relationship between variables is given by other means.

1. Classroom Connection 2.1.4: Graphing a Race

The following exploration is from pages 7–8 in the seventh-grade textbook *Mathematics in Context, Tracking Graphs*. Work Problems 19, 20, and 21. For Problem 19, use graph paper instead of the student activity sheet specified therein.

18. Write an article for the newspaper describing the race. Write it so the reader will refer to the graph while reading the article.

Last Year's *RACE*

Here is a description of last year's race:

The runners were not able to break away from each other during the first 3 minutes. In that time, they covered 650 meters. Then the runner from King stumbled, and the Jefferson runner took the lead. Six minutes into the race, Jefferson's runner was at 1,300 meters, and her lead was at its maximum (about 150 meters). The runner from King managed to regain her speed but could never catch up.

Results:

> First Place—Jefferson
> (9 minutes, 30 seconds)
>
> Second Place—King
> (10 minutes)

19. Make a graph for last year's race on another copy of **Student Activity Sheet 3.** Work in pairs or groups.

Suppose you were going to judge the graph of another group.

20. a. Why should you check to see that both lines start at the bottom left corner?

    b. Write three other things a graph of the race should have. Be sure your own graph fits your criteria.

This is the graph of a race held in a nearby town.

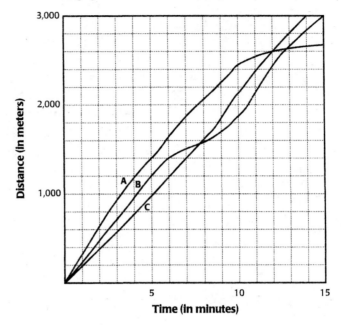

Study the graph shown above. There are some differences between it and the graph of the King vs. Jefferson race that you made for problem **19**.

21. **a.** What differences can you find?

  b. What dramatic thing happened in this race?

  c. From start to finish, one of the lines in the graph is almost straight. What does that tell you about the race of this runner?

  d. Write a short article about the race. Be sure to include all of the important facts.

◆

2. Classroom Connection 2.1.5: Speed Graphs

The following exploration is from pages 29–31 in the seventh-grade textbook *Mathematics in Context, Tracking Graphs*. Work Problems 1–8.

D. GRAPHS DEPICTING SPEED

Making SPEED Graphs

On the dashboard of every car, there is a speedometer. Most of them have a needle that moves along a row of numbers that tells how fast the car is moving. For example, if the needle is on the number 50, the car is moving at a speed of 50 miles per hour (mph).

1. What is the speed of a car, in miles per hour, if its speedometer looks like the one on the right?

2. Point to where the needle of the speedometer would be for each of the following situations:

   a. The car is standing still.

   b. The car is moving at a speed of 30 miles per hour.

3. a. What happens to the needle if the car speeds up?

   b. What happens to the needle if the car slows down?

   c. What is the car doing if the needle stays in the same position for a while?

   d. What happens to the needle if the car stops suddenly?

mph
80
70
60
50
40
30
20
10
0

Tracking Graphs

Activity

Pair up with another student to make a graph of several car trips.
For each line on the graph, one student will draw, while the other
student pulls the paper to the left. The student who draws should
move the pencil up when the car goes faster and down when the
car goes slower.

Record the following trips on the same graph. Use a different color
for each trip.

4. **a.** From a stop, gradually speed up to 55 miles per hour and
   stay at that speed until you reach the end of the paper.

   b. From a stop, quickly speed up to 40 miles per hour, then
   gradually slow down to a stop.

   c. Starting at 55 miles per hour, slow down and then speed
   up again. Do this repeatedly.

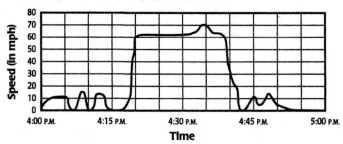

Below you see the speed graph for a car ride.

5. a. What might have been happening during the first 15 minutes of the ride?

   b. How often did the car stop during this period?

6. What type of road could the car have been on between 4:20 P.M. and 4:30 P.M.?

7. The line graph is almost horizontal between 4:20 P.M. and 4:30 P.M. What can you say about the speed during this period?

8. Suppose the speed limit on that road is 65 miles per hour. Did the car exceed the speed limit during the ride?

◆

II. Even and Odd Functions

1. A function f is called **even** if for any x in the domain of f, $-x$ also belongs to the domain of f, and $f(-x) = f(x)$. Give three examples of even functions.

2. A function f is called **odd** if for any x in the domain of f, $-x$ also belongs to the domain of f, and $f(-x) = -f(x)$. Give three examples of odd functions.

3. Are there functions that are neither even nor odd?

4. Can the interval $(-1, 2)$ be the domain of an even or odd function? Explain.

5. Is the sum of two even functions even, odd, or does it depend on the particular functions? Explain.

6. Is the sum of two odd functions even, odd, or does it depend on the particular functions? Explain.

7. What can you say about the difference, product, and quotient of two even functions?

8. What can you say about the difference, product, and quotient of two odd functions?

9. What type of symmetry does the graph of an even function exhibit?

10. What type of symmetry does the graph of an odd function exhibit?

11. Prove that if a function f is both odd and even, then f is identically zero.

12. As it turns out, if the domain D of a real valued function f is symmetric with respect to the point 0, then the function f can be represented as the sum of two functions: one even and one odd. Such a representation is unique; that is, there exists a unique even function f_1 and a unique odd function f_2 such that $f(x) = f_1(x) + f_2(x)$ for all $x \in D$. Prove the existence and uniqueness of such a representation.

Uniqueness of f_1 and f_2: Suppose that there exist two even functions f_1 and g_1 and two odd functions f_2 and g_2, such that $f(x) = f_1(x) + f_2(x)$ and $f(x) = g_1(x) + g_2(x)$ for all x in D.

    a. Is the function $f_1 - g_1$ even, odd, or neither?
    b. Is the function $g_2 - f_2$ even, odd, or neither?
    c. Compare $f_1 - g_1$ and $g_2 - f_2$.
    d. Why can you conclude that $f_1 = g_1$ and $f_2 = g_2$?

Existence of f_1 and f_2:

    a. Assume first that f_1 and f_2 with the given properties exist. Show that, for all x in D, $f(-x) = f_1(x) - f_2(x)$.
    b. Solve for f_1 and f_2 in the system

$$f(x) = f_1(x) + f_2(x)$$
$$f(-x) = f_1(x) - f_2(x)$$

    c. Do the expressions of f_1 and f_2 determined in b make sense without the assumption made in a?
    d. Show that the functions f_1 and f_2 obtained in b provide the required representation. In other words, show that f_1 is even, f_2 is odd, and $f(x) = f_1(x) + f_2(x)$ for all x in D.

Remark.   From the way f_1 and f_2 have been deduced, you can see that they necessarily have certain expressions and are hence unique.

13. Let $f(x) = 3x^5 - 2x^4 + x - 3$. Determine f_1 and f_2 as specified in Problem **12**.

III. Composing Odd and Even Functions

See Projects and Extensions, II for the definitions of even and odd functions.

1. Give two examples of even functions $f_1, f_2 : \mathbb{R} \to \mathbb{R}$ and two examples of odd functions $g_1, g_2 : \mathbb{R} \to \mathbb{R}$.

   a. Compute $f_1 \circ f_2$ and $f_2 \circ f_1$. Are the resulting functions even, odd, or neither?

   b. Compute $g_1 \circ g_2$ and $g_2 \circ g_1$. Are the resulting functions even, odd, or neither?

   c. Compute $f_1 \circ g_1$, $g_1 \circ f_1$. Are the resulting functions even, odd, or neither?

2. Fill in the blanks:

   a. Whenever meaningful, the composition of two even functions is always
   _____.

   b. Whenever meaningful, the composition of two odd functions is always
   _____.

   c. The composition of an even function with an odd function is always
   _____, regardless of the order of composition.

3. Prove the conjectures in Problem 2.

IV. Composing Reflections

Let L_1 and L_2 be two lines in the plane that intersect at the point A and form an angle $\alpha \in (0, \frac{\pi}{2})$ as seen in the following picture.

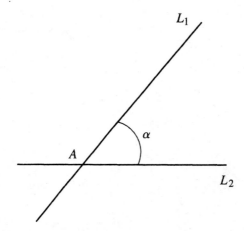

Denote by C the set of all points in the plane. Let $f_1 : C \to C$ be the function that reflects a point about the line L_1, and let $f_2 : C \to C$ be the function that reflects a point about the line L_2.

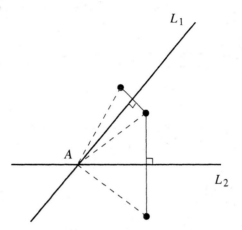

1. Describe the resulting function $f_1 \circ f_2$ as a single geometric transformation by using the following outline:

   a. Let P be a point in the plane, let $Q = f_2(P)$, and let P' be the point of intersection between the lines PQ and L_2. One particular case is shown in the next picture. Treat the other cases similarly. How do the measures of the angles $\widehat{PAP'}$ and $\widehat{P'AQ}$ compare? How do the lengths of the line segments PA and QA compare? Explain.

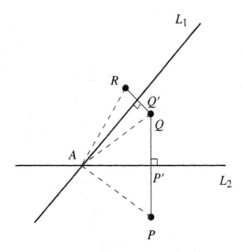

   b. Let $R = f_1(Q)$ and Q' be the point of intersection between the lines QR and L_1. In this context, ask and answer similar questions as in a.

    c. Express the measure of the angle $\widehat{PAR}$ in terms of α, the angle between L_1 and L_2. Then, find the point obtained from P by the counterclockwise rotation centered at A with angle 2α.

    d. What can you conclude about $f_1 \circ f_2$?

2. Describe the resulting function $f_2 \circ f_1$ as a single geometric transformation.

3. What can you say about the geometric transformation $(f_1 \circ f_2) \circ (f_2 \circ f_1)$? How about the transformation $(f_2 \circ f_1) \circ (f_1 \circ f_2)$?

V. Inverse Functions

What does it mean to "undo" a given function f? When can this be done? You answer these questions in this investigation. We expect that "undoing" a given function $f : A \to B$ amounts to finding another function $g : B \to A$, which sends any given output of f back to the original input. Unfortunately, this is not always possible! This point is further explored here.

1. Is it possible to "undo" a function $f : A \to B$ if $B \neq \text{Image}(f)$? In other words, if there is a number y in the $\text{Range}(f)$ that is not the result of applying f to some number in the $\text{Domain}(f)$, can we send y back to the $\text{Domain}(f)$ in a manner in which we would "undo" f? Explain.

2. A function $f : A \to B$ is said to be **onto** provided $B = \text{Image}(f)$. Determine which of the following functions are onto. In each case, sketch the graph of the function.

    a. $f : \mathbb{R} \to \mathbb{R}, f(x) = -2x + 6$

    b. $f : \mathbb{R} \to \mathbb{R}, f(x) = x^2 - x$

    c. $f : (-\infty, 0] \to \mathbb{R}, f(x) = x^2 - x$

    d. $f : (-\infty, 0] \to [0, \infty), f(x) = x^2 - x$

3. Is it possible to "undo" a function f for which there exist two different values x_1 and x_2 in the domain of f satisfying $f(x_1) = f(x_2)$? Explain your reasoning.

4. A function f is said to be **one-to-one** provided each element in its image is associated to only one element in its domain; or equivalently, if $x_1 \neq x_2$ are two distinct elements in the domain of f, then $f(x_1) \neq f(x_2)$. This is also equivalent to the condition that $f(x_1) = f(x_2)$ always implies $x_1 = x_2$. Fill in the blanks:

A function is one-to-one if and only if every _____ line intersects its graph _____. This is called the _____ line test for functions.

5. Use the graphs you sketched in Problem 2 to determine whether the corresponding functions are one-to-one.

In mathematical terms, a function that can be "undone" is called *invertible*. Here is the definition of an invertible function.

Definition 2.1.3 A function $f : A \to B$ is said to be **invertible** provided there exists a function $g : B \to A$ satisfying

(i) $g(f(x)) = x$ for each x in A,

(ii) $f(g(y)) = y$ for each y in B.

Such a function g is called the **inverse** of f and is denoted by f^{-1}.

6. Explain why if g is the inverse of f, then g is also invertible and f is its inverse. In particular, $(f^{-1})^{-1} = f$ whenever f is invertible.

7. Does an invertible function have a unique inverse? In other words, is it possible for an invertible function to have two inverses? To answer this question, start by assuming that an invertible function $f : A \to B$ has two inverses, say $g_1, g_2 : B \to A$, with $g_1(y) \neq g_2(y)$ for some $y \in B$. See how this leads to a contradiction.

8. At an informal level, you have seen that for a function f to be invertible, it is necessary that f be one-to-one and onto. These two conditions are also sufficient to guarantee invertibility. Therefore, a function is invertible if and only if it is one-to-one and onto.

   Complete the following steps to prove this statement.

   a. Suppose $f : A \to B$ is invertible with $g : B \to A$ its inverse.

   (i) To show that f is one-to-one, start by assuming that there exist x_1, x_2 in A such that $f(x_1) = f(x_2)$. Then use g to conclude that $x_1 = x_2$.

   (ii) To show that f is onto, check that Image$(f) = B$. The inclusion Image$(f) \subseteq B$ is always true. To prove also that $B \subseteq$ Image(f), take an arbitrary y in B. Find an element in A that is assigned to y by f. To which set does $g(y)$ belong? What do you obtain after applying f to $g(y)$?

   b. Suppose $f : A \to B$ is one-to-one and onto. Define $g : B \to A$ such that for each $y \in B$, $g(y)$ equals the unique $x \in A$, verifying $f(x) = y$. Show that the function g is the inverse of f.

9. The proof in Problem 8b sheds some light on how one might proceed in computing the inverse of an invertible function. If you know that $f : A \to B$ is invertible, finding an expression for the inverse $g : B \to A$ comes down to solving for x in terms of y in the equation $f(x) = y$, where y is in B. Use this idea to find an expression for the inverse of the invertible function $f : \mathbb{R} \to \mathbb{R}$, $f(x) = 2x + 3$ for $x \in \mathbb{R}$.

10. Use Problems 8 and 9 to decide which functions in Problems 2a–d are invertible and to compute the inverse whenever it exists.

11. In Problem 8, you saw that if $f : A \rightarrow B$ is invertible with $g : B \rightarrow A$ as its inverse, then for any $x \in A$ and any $y \in B$, $f(x) = y$ if and only if $g(y) = x$. Thus, the point (x, y) is on the graph of the function f if and only if the point (y, x) is on the graph of its inverse function g. From a geometric perspective, we can rephrase this statement as:

> The functions f and g are inverse to each other if and only if their graphs are symmetric with respect to the line $y = x$.

The graphs of three invertible functions are sketched here. Sketch the graphs of their inverses.

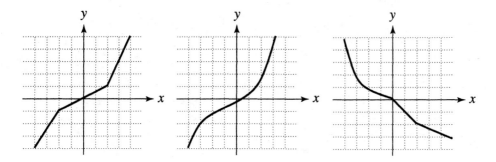

2.2 LIMITS OF FUNCTIONS

Limits of functions • Properties of limits of functions

From the calculus perspective, a basic aspect in the study of functions is the analysis of how changes in inputs affect changes in their corresponding outputs. Suppose we choose the inputs closer and closer to a given number. Will their corresponding outputs also get closer and closer to some number? In this section, we focus on answering this question.

Classroom Discussion 2.2.1: Graphical Approach for Finding Limits

1. It has been determined experimentally that, on average, the relationship between a female person's femur length x and her height $h(x)$ is $h(x) = 2.32x + 24$, where both x and $h(x)$ are measured in inches. Although the expression of $h(x)$ makes sense for every real number x, the length x of a femur is always positive. Thus, in order to have a more accurate model, we assume that x is positive, and consider the domain of h to be $(0, \infty)$. Here is the graph of h.

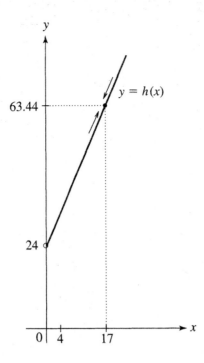

a. What can you say about the height values as the femur's length increases from 16.5 inches to 17 inches, while staying smaller than 17?

b. Does there exist a value L for which the following statement is true?

As the femur length x increases to 17, while staying smaller than 17, the values of the height $h(x)$ get closer and closer to L.

c. What can you say about the height values as the femur's length decreases from 17.5 inches to 17 inches, staying larger than 17?

d. Does there exist a value L' for which the following statement is true?

As the femur length x decreases to 17, while staying larger than 17, the values of the height $h(x)$ get closer and closer to L'.

e. How do L and L' compare?

f. What can you say about the values of $h(x)$ as $x \neq 17$ gets closer and closer to 17 from either side of 17?

Notation. When the values of x are getting closer and closer to a real number a but stay smaller than a, we write $x \to a^-$ and read "x approaches a from the left." Similarly, if the values of x are getting closer and closer to a real number a but stay larger than a, we write $x \to a^+$ and read "x approaches a from the right." If the values of x get closer and closer to a with $x \neq a$, we write $x \to a$ and read "x approaches a." In the previous discussion, you had $x \to 17^-$, then $x \to 17^+$, and then $x \to 17$.

The graph of h suggests that as $x < 17$ gets closer and closer to 17, the values of $h(x)$ get closer and closer to 63.44. In this case, we say that 63.44 is the **limit** of $h(x)$ as x approaches 17 **from the left**, and we write

$$\lim_{x \to 17^-} h(x) = 63.44.$$

Similarly, as $x > 17$ gets closer and closer to 17, the values of $h(x)$ get closer and closer to 63.44; we say that 63.44 is the **limit** of $h(x)$ as x approaches 17 **from the right**, and we write

$$\lim_{x \to 17^+} h(x) = 63.44.$$

In addition, since $\lim_{x \to 17^-} h(x) = \lim_{x \to 17^+} h(x)$, we say that their common value 63.44 is **the limit** of $h(x)$ as x approaches 17, and we write

$$\lim_{x \to 17} h(x) = 63.44.$$

Observe from the graph of h that $\lim_{x \to 17} h(x)$ is in fact equal to $h(17)$, the value of h at $x = 17$. Is this something we should expect for all functions? That is, is it true that $\lim_{x \to a} f(x) = f(a)$ for any function f? As it turns out, only a special class of functions has this type of behavior: the class of continuous functions, which we present in Section 2.3. The value of the limit of a function f at a point a is not determined by the value of f at a. What is important is the behavior of $f(x)$ as x approaches a, with $x \neq a$. In fact, there exist functions f that are not defined at a and for which $\lim_{x \to a} f(x)$ is well defined. Here is an example.

2. **The Function** $f(x) = \frac{x^2 - 4}{x - 2}$

    a. What is the domain of $f(x) = \frac{x^2 - 4}{x - 2}$?

    b. For what values of x is the equality $f(x) = x + 2$ true?

    c. Sketch the graph of f.

    d. Use the graph of f to fill in the blanks: $\lim_{x \to 2^-} f(x) = $ ____, $\lim_{x \to 2^+} f(x) = $ ____.

    e. What can you say about $\lim_{x \to 2} f(x)$?

    f. Did the fact that f is not defined at 2 play a role in the computation of $\lim_{x \to 2} f(x)$?

In the examples discussed so far, the limit you computed always existed. Is this something you should expect for arbitrary functions f and points a when considering $\lim\limits_{x\to a^-} f(x)$, $\lim\limits_{x\to a^+} f(x)$, or $\lim\limits_{x\to a^-} f(x)$? As it turns out, if a real number L does not exist, such that the values of $f(x)$ get closer and closer to L as $x \to a^-$, we say that $\lim\limits_{x\to a^-} f(x)$ does not exist. Similarly, if a real number L does not exist, such that the values of $f(x)$ get closer and closer to L as $x \to a^+$, we say that $\lim\limits_{x\to a^+} f(x)$ does not exist. Moreover, we say that $\lim\limits_{x\to a^-} f(x)$ does not exist if $\lim\limits_{x\to a^-} f(x)$ does not exist or $\lim\limits_{x\to a^+} f(x)$ does not exist, or $\lim\limits_{x\to a^-} f(x) \neq \lim\limits_{x\to a^+} f(x)$.

3. **USPS Prices**

   Return to the USPS example from Section 2.1. The function f defined there is $f(x) = 0.37 + 0.23\lceil x - 1 \rceil$ where $0 < x \leq 13$. Recall that $\lceil x \rceil$ denotes the ceiling function of x, and by definition, $\lceil x \rceil = k$, with k being the integer value for which $k - 1 < x \leq k$. Here is the graph of f:

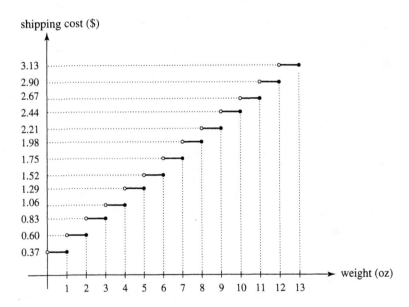

   a. Use the graph of f to fill in the blanks: $\lim\limits_{x\to 4^-} f(x) =$ _____, $\lim\limits_{x\to 4^+} f(x) =$ _____.

   b. Compare $\lim\limits_{x\to 4^-} f(x)$ and $\lim\limits_{x\to 4^+} f(x)$.

   c. Why do we say that $\lim\limits_{x\to 4} f(x)$ does not exist?

    d. Does it make sense to discuss $\lim\limits_{x \to 0^-} f(x)$ and $\lim\limits_{x \to 0^+} f(x)$? Why?

    e. Does it make sense to discuss $\lim\limits_{x \to 0} f(x)$? Why?

Definition 2.2.1 Let f be a real valued function whose domain is a subset of the real line. If $f(x)$ can be made arbitrarily close to a real number L by taking $x < a$ in the domain of f sufficiently close to a, then we say that L is the limit of f as x approaches a from the left. We write

$$\lim_{x \to a^-} f(x) = L.$$

    Similarly, if $f(x)$ can be made arbitrarily close to a real number L by taking $x > a$ in the domain of f sufficiently close to a, then we say that L is the limit of f as x approaches a from the right. We write

$$\lim_{x \to a^+} f(x) = L.$$

Moreover, if

$$\lim_{x \to a^-} f(x) = \lim_{x \to a^+} f(x),$$

we say that the limit of f as x approaches a exists, and we write

$$\lim_{x \to a} f(x) = L,$$

where L is the common value of $\lim\limits_{x \to a^-} f(x)$ and $\lim\limits_{x \to a^+} f(x)$.

If $\lim\limits_{x \to a^-} f(x)$ or $\lim\limits_{x \to a^+} f(x)$ does not exist or if they both exist but are different, we say that $\lim\limits_{x \to a} f(x)$ does not exist.

    From this definition we infer that $\lim\limits_{x \to a} f(x) = L$ if $f(x)$ can be made arbitrarily close to L by taking $x \neq a$ sufficiently close to a.

Remark.   In order to discuss the existence of $\lim\limits_{x \to a^-} f(x)$, we need a to be **approachable from the left within the domain of** f; that is, there exists a real number b, $b < a$, such that the interval (b, a) is contained in the domain of f. Similarly, in order to discuss the existence of $\lim\limits_{x \to a^+} f(x)$, we need a to be **approachable from the right within the domain of** f; that is, there exists a real number c, $c > a$, such that the interval (a, c) is contained in the domain of f. Take the two functions $g(x) = \sqrt{1 - x}$ and $h(x) = \sqrt{x}$. Then, Domain$(g) = (-\infty, 1]$ and Domain$(h) = [0, \infty)$. Since 1 is approachable from the left within the domain of g, it makes sense to discuss $\lim\limits_{x \to 1^-} g(x)$. On the other hand, 1 is not approachable from the right within the domain of g; hence, it does not makes sense to discuss $\lim\limits_{x \to 1^+} g(x)$. Similarly, within the domain of h, 0 is approachable from the right but not from the left; thus it makes sense to discuss $\lim\limits_{x \to 0^+} h(x)$ but not $\lim\limits_{x \to 0^-} h(x)$.

In order to study $\lim_{x \to a} f(x)$, we need a to be **approachable from both sides within the domain of** f; that is, there exist real numbers b and c, $b < a < c$, such that the intervals (b, a) and (a, c) are both contained in the domain of f. For the two examples previously considered, it does not make sense to study $\lim_{x \to 1} g(x)$ and $\lim_{x \to 0} h(x)$. On the other hand, it is meaningful to discuss $\lim_{x \to a} g(x)$ for any $a < 1$ and to discuss $\lim_{x \to a} h(x)$ for any $a > 0$.

These approachability conditions are natural, since, when studying the limit of a function f as x approaches a from either the left, right, or both sides, we need to see what happens to the values of $f(x)$ as x gets *arbitrarily close* to a, but not equal to a, from either the left, right, or both sides, respectively. The actual value of the function f at a, or whether the function f is defined at a, is irrelevant when studying the limit of f as x approaches a. What is important is that the point a be approachable from the left, right, or both sides within the domain of f, depending on which limit we consider. Thus, whenever we write $\lim_{x \to a^-} f(x) = L$ or $\lim_{x \to a^+} f(x) = L$ or $\lim_{x \to a} f(x) = L$, we will always assume that a is approachable within the domain of f from the left, right, or both sides, respectively.

4. Use the following graph to determine $\lim_{x \to a^-} f(x)$ and $\lim_{x \to a^+} f(x)$ for the specified values of a; then decide if $\lim_{x \to a} f(x)$ exists. If so, find its value.

   a. $a = -9$ **b.** $a = 3$ **c.** $a = 13$ **d.** $a = 15$

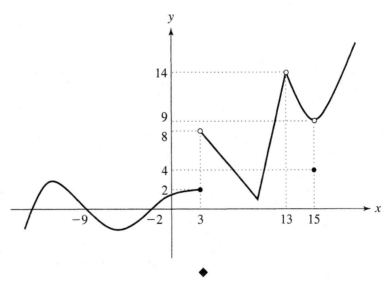

◆

Classroom Discussion 2.2.2: Numerical Approach for Finding Limits

1. Consider the function $f(x) = \frac{x+1}{x+3}$. The domain of f is $\mathbb{R} \setminus \{-3\}$. We want to see what happens to the values of f as $x \to 2$.

   a. Generate a table of values for $f(x)$ as $x \neq 2$ increases from 1.999 to 2, with increments of 10^{-4}. What do you observe?

b. Generate a table of values for $f(x)$ as $x \neq 2$ decreases from 2.001 to 2, with increments of 10^{-4}. What do you observe?

c. Use the tables from Problems a and b to fill in the blanks:

$$\lim_{x \to 2^-} f(x) = \underline{\hspace{1cm}} \qquad \lim_{x \to 2^+} f(x) = \underline{\hspace{1cm}} \qquad \lim_{x \to 2} f(x) = \underline{\hspace{1cm}}$$

2. Let $g(x) = \frac{1}{x}$. The domain of g is $\mathbb{R} \setminus \{0\}$. We are interested in the behavior of g as $x \to 0$.

a. As $x > 0$ gets closer and closer to 0, what can you say about the values of $\frac{1}{x}$? Look at the values of $g(x)$ at $x = 10^{-1}, x = 10^{-2}, x = 10^{-3}, x = 10^{-4}$, $x = 10^{-5}, x = 10^{-6}$, and $x = 10^{-7}$.

b. Why do you think we use the notation $\lim\limits_{x \to 0^+} \frac{1}{x} = \infty$ to describe the behavior of $g(x)$ as $x \to 0^+$?

c. As $x < 0$ gets closer and closer to 0, what can you say about the values of $\frac{1}{x}$? Look at the values of $g(x)$ at $x = -10^{-1}, x = -10^{-2}, x = -10^{-3}$, $x = -10^{-4}, x = -10^{-5}, x = -10^{-6}$, and $x = -10^{-7}$.

d. Why do you think we use the notation $\lim\limits_{x \to 0^-} \frac{1}{x} = -\infty$ to describe the behavior of $g(x)$ as $x \to 0^-$?

e. Does $\lim\limits_{x \to 0} \frac{1}{x}$ exist? Explain.

f. The graph of $\frac{1}{x}$ is sketched here. Give the geometric interpretation for

$$\lim_{x \to 0^+} \frac{1}{x} = \infty \text{ and } \lim_{x \to 0^-} \frac{1}{x} = -\infty.$$

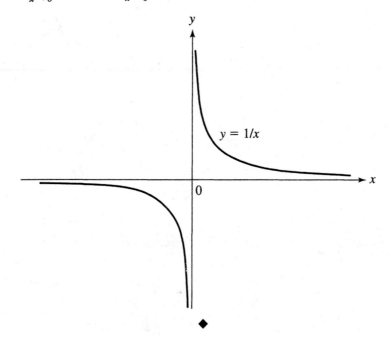

$y = 1/x$

◆

Definition 2.2.2 A function f defined on a set D of real numbers is called **bounded**, provided there exists $M > 0$ such that $|f(x)| \le M$ for all $x \in D$. Functions that are not bounded are called **unbounded**.

EXAMPLE Show that the function $g(x) = \frac{1}{x}$ is unbounded on $(0, \infty)$. More precisely, show that there does not exist a real number $M > 0$ with the property that $|g(x)| \le M$ for all $x > 0$.

Solution Suppose there exists $M > 0$ such that $|g(x)| \le M$ for all $x > 0$. The condition $|g(x)| \le M$ for all $x > 0$ is equivalent to $\frac{1}{x} \le M$ for all $x > 0$. If we select $0 < x < \frac{1}{M}$, then $\frac{1}{x} > M$. Hence, the assumption that there exists $M > 0$ such that $g(x) \le M$ for all $x > 0$ is false, and we can conclude that M, with the required property, does not exist. ∎

Practice Problem

Show that the function $g(x) = \frac{1}{x}$ is unbounded on $(-\infty, 0)$.

From now on, for an arbitrary function f and a real number a, we use the following notations:

$\lim_{x \to a^-} f(x) = \infty$ whenever $f(x)$ can be made arbitrarily large by choosing $x < a$ in the domain of f sufficiently close to a.

$\lim_{x \to a^-} f(x) = -\infty$ whenever $f(x)$ takes negative values that can be made arbitrarily large in absolute value by choosing $x < a$ in the domain of f sufficiently close to a.

$\lim_{x \to a^+} f(x) = \infty$ whenever $f(x)$ can be made arbitrarily large by choosing $x > a$ in the domain of f sufficiently close to a.

$\lim_{x \to a^+} f(x) = -\infty$ whenever $f(x)$ takes negative values that can be made arbitrarily large in absolute value by choosing $x > a$ in the domain of f sufficiently close to a.

Some Properties of Limits of Functions.   The following properties are useful for computing limits of a large variety of functions without having to rely on Definition 2.2.1 each time. To state them, fix arbitrary values $n \in \mathbb{N}$, $a, c \in \mathbb{R}$. Suppose f, g are two real valued functions. If $\lim_{x \to a} f(x)$ and $\lim_{x \to a} g(x)$ exist, then $\lim_{x \to a}[f(x) \pm g(x)]$, $\lim_{x \to a}[c \cdot f(x)]$, $\lim_{x \to a}[f(x) \cdot g(x)]$, $\lim_{x \to a} \frac{f(x)}{g(x)}$ (when $\lim_{x \to a} g(x) \ne 0$), and $\lim_{x \to a} \sqrt[n]{f(x)}$ (with the additional assumption that $\lim_{x \to a} f(x) > 0$ in the case where n is even) all exist. Moreover, we have

P1 $\lim_{x \to a}[f(x) \pm g(x)] = \lim_{x \to a} f(x) \pm \lim_{x \to a} g(x)$.

P2 $\lim_{x \to a}[c \cdot f(x)] = c \cdot \lim_{x \to a} f(x)$.

P3 $\lim_{x \to a}[f(x) \cdot g(x)] = \lim_{x \to a} f(x) \cdot \lim_{x \to a} g(x)$.

P4 $\lim\limits_{x \to a} \dfrac{f(x)}{g(x)} = \dfrac{\lim\limits_{x \to a} f(x)}{\lim\limits_{x \to a} g(x)}$ whenever $\lim\limits_{x \to a} g(x) \neq 0$.

P5 $\lim\limits_{x \to a} \sqrt[n]{f(x)} = \sqrt[n]{\lim\limits_{x \to a} f(x)}$ with the additional assumption that $\lim\limits_{x \to a} f(x) \geq 0$ in the case where n is even.

P6 If $f(x) \leq g(x)$ for all x in an open interval containing a, except possibly for $x = a$, then $\lim\limits_{x \to a} f(x) \leq \lim\limits_{x \to a} g(x)$.

Note that even if two functions f and g (for which $\lim\limits_{x \to a} f(x)$ and $\lim\limits_{x \to a} g(x)$ exist) verify $f(x) < g(x)$ for all x in an open interval containing a, we cannot conclude that $\lim\limits_{x \to a} f(x) < \lim\limits_{x \to a} g(x)$. To see this, take $f(x) = x^2$ and $g(x) = 2x^2$ and $a = 0$.

P1–P6 are intuitive and can be proved using Definition 2.2.1, however, we omit their proofs. The preceding statements remain true if $\lim\limits_{x \to a}$ is replaced by $\lim\limits_{x \to a^-}$ or by $\lim\limits_{x \to a^+}$. The following application underscores the usefulness of P1–P6.

Classroom Discussion 2.2.3: Limit of Rational Functions

Let c and a be arbitrary fixed real numbers.

1. Determine $\lim\limits_{x \to a} x$ and $\lim\limits_{x \to a} c$.

2. Use the results in Problem 1 and use P1 and P3 to compute $\lim\limits_{x \to a} c \cdot x^2$, $\lim\limits_{x \to a} c \cdot x^3$, and $\lim\limits_{x \to a} c \cdot x^4$.

3. Let n be a natural number. What is $\lim\limits_{x \to a} c \cdot x^n$?

4. Use the result in Problem 3 and use P1 to compute $\lim\limits_{x \to a} (2x^3 - \frac{4}{3}x^2 + x + \frac{1}{2})$.

5. Now let $f(x)$ be a *polynomial function* of degree n, i.e., $f(x) = b_n x^n + \cdots + b_1 x + b_0$ for all x, where $b_n, \ldots, b_1, b_0$ are fixed real numbers. Using Problem 3 and P1, compute $\lim\limits_{x \to a} f(x)$ and compare with $f(a)$. Fill in the blank:

   If $f(x)$ is a polynomial function, then for any real number a,
   $$\lim_{x \to a} f(x) = \underline{\hspace{1cm}}.$$

6. Let $p(x)$ and $q(x)$ be two polynomial functions and consider the rational function $\dfrac{p(x)}{q(x)}$. Suppose that $q(a) \neq 0$. Use Problem 5 and P4 to compute $\lim\limits_{x \to a} \dfrac{p(x)}{q(x)}$. Fill in the blank:

   If $p(x)$ and $q(x)$ are polynomial functions, then $\lim\limits_{x \to a} \dfrac{p(x)}{q(x)} = \underline{\hspace{1cm}}$ for any real number a satisfying $q(a) \neq 0$. ◆

EXAMPLES

Evaluate the limits using P1–P6 and the results from Classroom Discussion 2.2.3.

1. $\lim_{x \to -2} (4x^2 - x - 3)$

2. $\lim_{x \to 1} \sqrt{3x - 1}$

3. $\lim_{x \to 1} \frac{x+3}{2x-3}$

Solutions

1. The function $f(x) = 4x^2 - x - 3$ is a polynomial function. Applying the results from Classroom Discussion 2.2.3, we obtain

$$\lim_{x \to -2} (4x^2 - x - 3) = f(-2) = 15.$$

2. Using P5, we have that $\lim_{x \to 1} \sqrt{3x - 1} = \sqrt{\lim_{x \to 1} (3x - 1)}$. In addition, since $3x - 1$ is a polynomial function, we also have $\lim_{x \to 1} (3x - 1) = 3 \cdot 1 - 1 = 2$. Thus, $\lim_{x \to 1} \sqrt{3x - 1} = \sqrt{2}$.

3. The functions $p(x) = x + 3$ and $q(x) = 2x - 3$ are polynomial functions, and since $q(1) = -1 \neq 0$, the results from Classroom Discussion 2.2.3 imply that

$$\lim_{x \to 1} \frac{x + 3}{2x - 3} = \frac{p(1)}{q(1)} = -4. \qquad \blacksquare$$

Practice Problems

Evaluate the limits using P1–P6 and the results from Classroom Discussion 2.2.3.

1. $\lim_{x \to -1} (5x^4 + x^3 - \frac{1}{2}x + 5)$

2. $\lim_{x \to 0} \sqrt{1 + 6x - 5x^2}$

3. $\lim_{x \to 3} \frac{x^2-4}{x^2+2}$

Classroom Discussion 2.2.4: Limits and Intersecting Streets

This investigation's goal is to provide a geometric interpretation of the definition of the limit of a function at a point.

1. Return to the function h from Problem 1 in Classroom Discussion 2.2.1. Do you agree with the following statement?

> If we want a person's height $h(x)$ to be within 0.5 inches of 63.44, it suffices to have that person's femur length x within 0.21551724 inches of 17.

2. Does it suffice to choose x within 0.01 of 17 in order to get $h(x)$ within 0.001 of 63.44?

3. Determine $\delta > 0$ so that if $|x - 17| < \delta$ with $x \neq 17$, then $|h(x) - 63.44| < 0.001$.

4. In general, let $\varepsilon > 0$ be a small number. Determine $\delta > 0$ so that if $|x - 17| < \delta$ with $x \neq 17$, then $|h(x) - 63.44| < \varepsilon$.

The computation in Problem 4 shows that we can get $h(x)$ arbitrarily close to 63.44 by choosing x sufficiently close to 17; each time we decide how close to 63.44 we want $h(x)$ to be, we can find an open interval containing 17 such that if $x \neq 17$ is in that interval, then $h(x)$ is as close as we want to 63.44. Geometrically, this is how we can understand this statement.

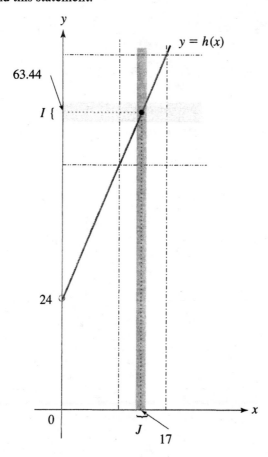

a. If $h(x)$ is close to 63.44, that means $h(x)$ is contained in an open interval I centered at 63.44. The length of this interval is determined by how close $h(x)$ is to 63.44.

b. The horizontal band, whose intersection with the y-axis is I, cuts a portion of the graph of h.

c. The portion of the graph of h cut by the horizontal band is projected onto the x-axis.

d. Now the goal is to select an open interval J on the x-axis containing 17, $J \setminus \{17\}$ being contained in both the domain of h and the projection obtained in c, with the following property: for all $x \in J$, with $x \neq 17$, the points $(x, h(x))$ are in the portion of the graph of h cut by the horizontal band. Observe that this last condition guarantees that $h(x)$ will be in I for all x in J.

If this construction can be completed for all intervals I of arbitrarily small length, then "$h(x)$ can be made arbitrarily close to 63.44 by taking x in the domain of h sufficiently close but not equal to 17."

5. Use this geometric interpretation to show that for the function g given by the following graph, $\lim\limits_{x \to 11} g(x) = 9$ while $\lim\limits_{x \to -3} g(x) \neq 6$.

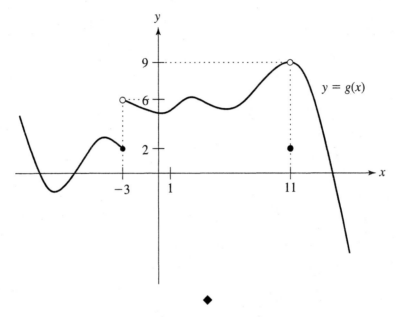

Definition 2.2.1 captures the essence of the notion of the limit of a function at a point. However, the phrases "as close as we want" and "sufficiently close" are not mathematically precise. A more rigorous definition follows.

Definition 2.2.3 Let f be a real valued function whose domain is a subset of the real line. We say that a real number L is the limit of f as x approaches a; we write $\lim\limits_{x \to a} f(x) = L$ if for every open interval I centered around L there exists an open interval J containing a, with $J \setminus \{a\}$ contained in the domain of f, such that for every $x \in J \setminus \{a\}$, we have $f(x) \in I$.

EXAMPLE Use Definition 2.2.3 to show that $\lim\limits_{x \to 2} f(x) = 8$ where

$$f(x) = \begin{cases} x^3 & \text{if } x \neq 2 \\ 100 & \text{if } x = 2. \end{cases}$$

Solution Fix an open interval I centered at 8. We want to find an open interval J containing 2, with $J \setminus \{2\}$ contained in the domain of f, such that for every $x \in J \setminus \{2\}$, we have $f(x) \in I$. Let $2l$ be the length of the interval I; that is, $I = (8 - l, 8 + l)$. Select $J = (\sqrt[3]{8 - l}, \sqrt[3]{8 + l})$. Then, $2 \in J$, $J \setminus \{2\}$ is contained in the domain of f, and for all $x \in J \setminus \{2\}$, the inequality $|x^3 - 8| < l$ is satisfied. Hence, $f(x) \in I$ for all $x \in J \setminus \{2\}$, which completes the proof. ∎

Practice Problem

Use Definition 2.2.3 to show that $\lim\limits_{x \to -\frac{1}{2}} f(x) = -\frac{1}{32}$ where

$$f(x) = \begin{cases} x^5 & \text{if } x \neq -\dfrac{1}{2} \\ 3 & \text{if } x = -\dfrac{1}{2}. \end{cases}$$

EXERCISES 2.3

In Exercises 1–4, determine if $\lim\limits_{x \to a^-} f(x)$, $\lim\limits_{x \to a^+} f(x)$, and $\lim\limits_{x \to a} f(x)$ exist. In case a limit exists, compute it.

1.
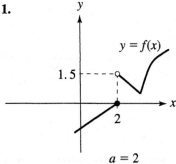
$a = 2$

2.
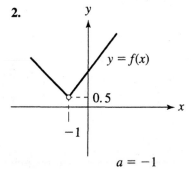
$a = -1$

3.
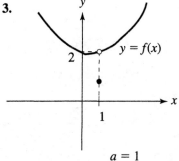
$a = 1$

4.
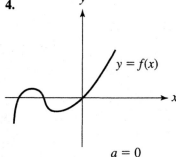
$a = 0$

In Exercises 5–10, by analyzing the graph $y = f(x)$ evaluate f and the limits of f at the given point, provided the corresponding limits exist.

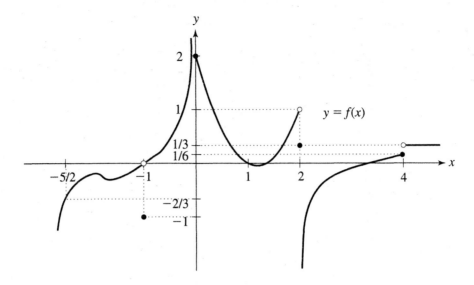

5. $f(-1)$, $\lim\limits_{x \to -1^-} f(x)$, $\lim\limits_{x \to -1^+} f(x)$, $\lim\limits_{x \to -1} f(x)$

6. $f(4)$, $\lim\limits_{x \to 4^-} f(x)$, $\lim\limits_{x \to 4^+} f(x)$, $\lim\limits_{x \to 4} f(x)$

7. $f(0)$, $\lim\limits_{x \to 0^-} f(x)$, $\lim\limits_{x \to 0^+} f(x)$, $\lim\limits_{x \to 0} f(x)$

8. $f(2)$, $\lim\limits_{x \to 2^-} f(x)$, $\lim\limits_{x \to 2^+} f(x)$, $\lim\limits_{x \to 2} f(x)$

9. $f(1)$, $\lim\limits_{x \to 1^-} f(x)$, $\lim\limits_{x \to 1^+} f(x)$, $\lim\limits_{x \to 1} f(x)$

10. $f(-\frac{5}{2})$, $\lim\limits_{x \to -\frac{5}{2}^-} f(x)$, $\lim\limits_{x \to -\frac{5}{2}^+} f(x)$, $\lim\limits_{x \to -\frac{5}{2}} f(x)$

In Exercises 11–28, evaluate the limits using P1–P6.

11. $\lim\limits_{x \to 1} (x^3 - 6x^2 - 4)$

12. $\lim\limits_{x \to -1} (3x^4 + 5x^3 + 2x - 1)$

13. $\lim\limits_{x \to 0} \sqrt{2x + 3}$

14. $\lim\limits_{x \to 4} \sqrt{x^2 - 3}$

15. $\lim\limits_{x \to -1} \frac{2x+1}{x-3}$

16. $\lim\limits_{x \to 0} \frac{x-6}{3x+1}$

17. $\lim\limits_{x \to 2} (4 - 2x)$

18. $\lim\limits_{x \to 3} (5 - x^2)$

19. $\displaystyle\lim_{x\to 0} \frac{-1}{x^2+x+1}$

20. $\displaystyle\lim_{x\to 0} \frac{x+1}{x^2-x-1}$

21. $\displaystyle\lim_{x\to 1} \sqrt[3]{x^3 - 5x + 1}$

22. $\displaystyle\lim_{x\to -2} \sqrt[3]{2x^3 - x^2 + 3x - 4}$

23. $\displaystyle\lim_{x\to 2} \left(\frac{1}{x} - \frac{1}{2} \right)$

24. $\displaystyle\lim_{x\to 4} \frac{2}{x+1}$

25. $\displaystyle\lim_{x\to 6} \sqrt[4]{x^2 - 4x + 4}$

26. $\displaystyle\lim_{x\to 2} \sqrt[4]{x^3 + x - 1}$

27. $\displaystyle\lim_{x\to -2} \frac{x^3-2x-3}{x+3}$

28. $\displaystyle\lim_{x\to 0} \frac{x-x^4}{2x-1}$

29. Let $n \geq 1$ be an arbitrary integer and define the function $f(x) = \frac{x^{n+1}-1}{x-1}$.

  a. Determine the domain of f. Can you compute $\displaystyle\lim_{x\to 1} f(x)$ using P4?

  b. Using sums of consecutive terms of a geometric sequence, prove that if $x \neq 1$, then

$$x^n + x^{n-1} + \cdots + x + 1 = \frac{x^{n+1} - 1}{x - 1}.$$

  c. Compute $\displaystyle\lim_{x\to 1} \frac{x^{n+1}-1}{x-1}$.

30. Use Definition 2.2.3 to show that $\displaystyle\lim_{x\to 1}(x^3 + 2) = 3$.

31. Use Definition 2.2.3 to show that $\displaystyle\lim_{x\to 1}(3x^3 - 1) = 2$.

PROJECTS AND EXTENSIONS 2.2

More on Limits of Rational Functions

Let $p(x)$ and $q(x)$ be polynomial functions. You have seen that if $q(b) \neq 0$ for some real number b, then $\displaystyle\lim_{x\to b} \frac{p(x)}{q(x)} = \frac{p(b)}{q(b)}$. This project's goal is to investigate the case when $q(b) = 0$. You have seen one example in this direction: the function $\frac{1}{x}$ that corresponds to $p(x) = 1$, $q(x) = x$, and $b = 0$. In this particular example, you have determined that $\displaystyle\lim_{x\to 0} \frac{1}{x}$ does not exist. More precisely, $\displaystyle\lim_{x\to 0^-} \frac{1}{x} = -\infty$ and $\displaystyle\lim_{x\to 0^+} \frac{1}{x} = +\infty$.

1. Consider the function $f(x) = \frac{x-2}{x^2-4}$. What is the domain of f?

2. Can you compute $\displaystyle\lim_{x\to 2} f(x)$ using P4? Explain.

3. Generate a table of values for x and $f(x)$ with the values of x approaching but never equal to 2. What do you observe?

4. What can you say about $\displaystyle\lim_{x\to 2^+} f(x)$, $\displaystyle\lim_{x\to 2^-} f(x)$, and $\displaystyle\lim_{x\to 2} f(x)$? Is this surprising? Explain.

5. Elain did the following computation:

$$\lim_{x \to 2} \frac{x - 2}{x^2 - 4} = \lim_{x \to 2} \frac{x - 2}{(x - 2)(x + 2)} = \lim_{x \to 2} \frac{1}{x + 2} = \frac{1}{4}.$$

Jessica argued that Elain's computation is incorrect since she canceled $x - 2$ from the numerator and the denominator, and this is not allowed when computing the limit as $x \to 2$. With whom do you agree?

6. Elain looks at the function $g : \mathbb{R} \backslash \{-2, 2\} \to \mathbb{R}$, $g(x) = \frac{1}{x+2}$ and claims that the functions f and g are not equal. Do you agree with her? Explain.

7. Compute $\lim_{x \to 0} \frac{x^3 - x^2}{3x^6 + x^2}$ and $\lim_{x \to 1} \frac{x^3 - 2x^2 + x}{x^4 - 2x^2 + 1}$. Hint: Simplify first the given fractions.

8. Can you compute $\lim_{x \to -2} f(x)$ using P4? Is the cancellation technique applicable here? Explain.

9. Generate a table of values for x and $f(x)$ with the values of x approaching but never equal to -2. What do you observe?

10. Conjecture a value for $\lim_{x \to -2^-} f(x)$, $\lim_{x \to -2^+} f(x)$, and $\lim_{x \to -2} f(x)$.

11. Without generating a table of values for $\frac{x^2}{x^2 - 4}$, find $\lim_{x \to -2^-} \frac{x^2}{x^2 - 4}$, $\lim_{x \to -2^+} \frac{x^2}{x^2 - 4}$, $\lim_{x \to -2} \frac{x^2}{x^2 - 4}$, $\lim_{x \to 2^-} \frac{x^2}{x^2 - 4}$, $\lim_{x \to 2^+} \frac{x^2}{x^2 - 4}$, and $\lim_{x \to 2} \frac{x^2}{x^2 - 4}$. Your reasoning should take into account the behavior of x^2 and $x^2 - 4$ as x approaches -2 and 2 from the left, right, or both sides.

12. Compute the limit at 1 of the functions $g_1(x) = \frac{x^2 + 3}{x^2 - 2}$, $g_2(x) = \frac{x^2 + 3}{x^2 - 1}$, $g_3(x) = \frac{x^2 - 1}{x^2 - 2}$, $g_4(x) = \frac{x^2 + x - 2}{x^2 - 1}$.

2.3   CONTINUITY

Continuous functions

You have seen in Section 2.2 that for some functions f, including polynomial functions, $\lim_{x \to x_0} f(x)$ is simply $f(x_0)$ for any x_0 in the domain of f. Functions satisfying this property are referred to as *continuous functions*. In order to understand the notion of continuity, you will analyze the equality $\lim_{x \to x_0} f(x) = f(x_0)$ from a geometric point of view.

Classroom Discussion 2.3.1: Setting the Stage

1. The following are the graphs of four functions. Can you trace any of them without lifting your pen?

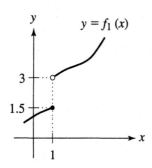

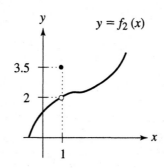

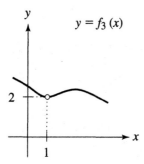

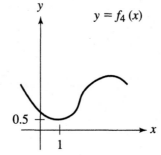

Describe what happens to the graph of each function at $x = 1$. Also, in each case, analyze the existence of the limit at 1, the value of the function at 1 (if defined), and how the two compare.

2. Recall the two functions h and f from Classroom Discussion 2.2.1, Problems 1 and 3.

   a. Can you trace the graph of h without lifting the pen? How about the graph of f?

   b. Are there any "jumps" in the graphs of the two functions?

   c. Does either function model a process without gaps or sudden changes? ◆

In everyday speech, a *continuous* process is one that proceeds without interruption or sudden change. You can trace the graph of a function modeling a continuous process without lifting the pen. In Euler's words, it is "a curve described by freely leading the hand." Using the preceding observations, it is natural to make the following definition.

Definition 2.3.1 Let f be a real valued function and let c be a point in its domain. We say that f is **continuous at** c if

$$\lim_{x \to c} f(x) = f(c). \tag{1}$$

The function f is said to be **continuous on the interval** (a, b) if it is continuous at every point in (a, b). The function f is said to be **discontinuous at** c whenever f is not continuous at c.

Important Remark.   For the condition (1) to be satisfied, it is implicitly understood that c is in the domain of f and that $\lim_{x \to c} f(x)$ exists (so that both sides of the equality (1) are well defined). Thus, when checking the continuity of a function f at a point c, we must make sure that the following hold:

    a.   the function f, is defined at c;

    b. $\lim_{x \to c} f(x)$ exists;

    c. $\lim_{x \to c} f(x) = f(c)$.

If either of the conditions **a**, **b**, or **c** fails, the function f is discontinuous at c.

EXAMPLES

1. In Section 2.2, you saw that if $p(x)$ is a polynomial function, then $\lim_{x \to c} p(x) = p(c)$ for all real numbers c. Thus, all polynomial functions are continuous on $\mathbb{R}$.

2. You also saw in Section 2.2 that if $p(x)$ and $q(x)$ are polynomial functions, then $\lim_{x \to a} \frac{p(x)}{q(x)} = \frac{p(a)}{q(a)}$ for all $a \in \mathbb{R}$ satisfying $q(a) \neq 0$. Thus, all rational functions are continuous on their domains.

3. Using the properties of limits P1–P6 from Section 2.2, we see that if f and g are functions that are both continuous at some real number a, and $r \in \mathbb{Q} \backslash \{0\}$, $c \in \mathbb{R}$, then

$$f + g, \quad f - g, \quad c \cdot f, \quad f \cdot g, \quad \frac{f}{g}, \text{ and } \sqrt[r]{f}$$

are also continuous at a, provided $g(a) \neq 0$ in the case of $\frac{f}{g}$, and provided $\sqrt[r]{f(a)}$ is defined in the case of $\sqrt[r]{f}$.

4. Consider the function

$$g(x) = \begin{cases} x^3 + 2 & \text{if } x \leq -1 \\ 2x + 3 & \text{if } x > -1. \end{cases}$$

Discuss the continuity of g on the real line.

5. Study the continuity of the function

$$g(x) = \begin{cases} x - 3 & \text{if } x \leq 2 \\ x^2 - 1 & \text{if } x > 2. \end{cases}$$

6. Determine the value of the real number a for which the function

$$g(x) = \begin{cases} 2x - a & \text{if } x \leq 0 \\ 3x - 4 & \text{if } x > 0 \end{cases}$$

is continuous on the real line.

Solutions 4. On the interval $(-\infty, -1)$, the law for the function g is given by the polynomial expression $x^3 + 2$. Since polynomial functions are continuous on the real line, we conclude that g is continuous on $(-\infty, -1)$. By a similar argument, g is continuous on the interval $(-1, \infty)$. We are left with checking the continuity of g at $x = -1$; that is, checking the validity of the equality $\lim_{x \to -1} g(x) = g(-1)$. This last equality is true if g satisfies conditions a–c in the remark after Definition 2.3.1. Condition a holds since the function g is defined at -1: $g(-1) = 1$. Using the continuity of the polynomial function $x^3 + 2$ at $x = -1$, we have

$$\lim_{x \to -1^-} (x^3 + 2) = (x^3 + 2)|_{x=-1} = 1,$$

thus, $\lim_{x \to -1^-} g(x) = 1$. In this book, given a function f, $f(x)|_{x=x_0}$ is just another notation for $f(x_0)$. Similarly,

$$\lim_{x \to -1^+} g(x) = \lim_{x \to -1^+} (2x + 3) = (2x + 3)|_{x=-1} = 1.$$

Hence, $\lim_{x \to -1} g(x) = 1$, so b holds. In addition, $\lim_{x \to -1} g(x) = 1 = g(-1)$, thus c holds. Therefore, g is continuous at $x = -1$. Now we can conclude that g is continuous on the real line.

5. Since $g(x)$ coincides with polynomial functions on the intervals $(-\infty, 2)$ and $(2, \infty)$, we conclude that g is continuous on these intervals. We are left with analyzing the continuity of g at $x = 2$. Because

$$\lim_{x \to 2^-} g(x) = \lim_{x \to 2^-} (x - 3) = (x - 3)|_{x=2} = -1$$

and

$$\lim_{x \to 2^+} g(x) = \lim_{x \to 2^+} (x^2 - 1) = (x^2 - 1)|_{x=2} = 3,$$

$\lim_{x \to 2} g(x)$ does not exist. Thus, condition b fails and g is discontinuous at $x = 2$. We conclude that g is continuous on $(-\infty, 2) \cup (2, \infty)$.

6. The function g is continuous on $(-\infty, 0) \cup (0, \infty)$ since it coincides with polynomial functions on these intervals. For g to be continuous on the whole real line, g must also be continuous at 0. The function g is defined at 0, so a holds. For b to hold, we must have $\lim_{x \to 0^-} g(x) = \lim_{x \to 0^+} g(x)$. Moreover,

$$\lim_{x \to 0^-} g(x) = \lim_{x \to 0^-} (2x - a) = (2x - a)|_{x=0} = -a$$

and

$$\lim_{x \to 0^+} g(x) = \lim_{x \to 0^+} (3x - 4) = (3x - 4)|_{x=0} = -4.$$

For b to hold, we must have $a = 4$. Hence, setting $a = 4$, we have $\lim_{x \to 0} g(x) = -4 = g(0)$, which also satisfies condition c. For the choice $a = 4$, the function g becomes

$$g(x) = \begin{cases} 2x - 4 & \text{if } x \le 0 \\ 3x - 4 & \text{if } x > 0, \end{cases}$$

and it is continuous on the entire real line. ∎

Practice Problem

Study the continuity of the function

$$g(x) = \begin{cases} 3x + 2 & \text{if } x \leq 0 \\ \dfrac{1}{x + 1} & \text{if } x > 0. \end{cases}$$

EXERCISES 2.4

In Exercises 1–4, study the continuity of f on the real line.

1. $f(x) = \begin{cases} -x + 3 & \text{if } x \leq -1 \\ -x^2 + 2 & \text{if } x > -1 \end{cases}$

2. $f(x) = \begin{cases} x^3 - 2 & \text{if } x \leq 2 \\ x - 2x^2 - 1 & \text{if } x > 2 \end{cases}$

3. $f(x) = \begin{cases} x^2 + 2x + 3 & \text{if } x \leq 0 \\ \dfrac{5x + 3}{2x + 1} & \text{if } x > 0 \end{cases}$

4. $f(x) = \begin{cases} x^2 - \frac{1}{2}x + 1 & \text{if } x \leq 1 \\ \dfrac{3x}{x + 1} & \text{if } x > 1 \end{cases}$

In Exercises 5–8, find the values of a for which the function f is continuous on $\mathbb{R}$.

5. $f(x) = \begin{cases} 2x^2 - 3 & \text{if } x \leq 1 \\ 3x - a & \text{if } x > 1 \end{cases}$

6. $f(x) = \begin{cases} x - a & \text{if } x \leq 0 \\ x^2 + 1 & \text{if } x > 0 \end{cases}$

7. $f(x) = \begin{cases} |x| & \text{if } x \leq 2 \\ ax^2 - 6 & \text{if } x > 2 \end{cases}$

8. $f(x) = \begin{cases} -2x - a^2 & \text{if } x \leq 1 \\ 1 - 4ax & \text{if } x > 1 \end{cases}$

In Exercises 9–14, give an example of a function f as specified.

9. f is continuous at $x = 4$.

10. f is continuous at $x = 2$.

11. f is discontinuous at $x = 4$, but $\lim\limits_{x \to 4^-} f(x)$ exists.

12. $\lim\limits_{x \to 2} f(x)$ exists, and f is discontinuous at $x = 2$.

13. $\lim\limits_{x \to 4^-} f(x) \neq \lim\limits_{x \to 4^+} f(x)$.

14. $\lim\limits_{x \to 2^-} f(x) = f(2)$, and f is discontinuous at $x = 2$.

15. How many continuous functions $f : \mathbb{R} \to \mathbb{R}$ are there that satisfy $[f(x)]^2 = x^2$ for all $x \in \mathbb{R}$?

PROJECTS AND EXTENSIONS 2.3

I. Linear Functions with Graphs Passing through the Origin

Recall that a function f is linear if and only if $f(x) = mx + b$ for some real numbers m and b. In this investigation, we focus only on those linear functions whose graphs pass through the origin. This latter condition is equivalent to $f(0) = 0$, thus $b = 0$; consequently,

$$f(x) = mx. \tag{2}$$

1. Show that if a function f is of type (2), then

$$f \text{ is continuous and } f(x + y) = f(x) + f(y) \text{ for all real numbers } x, y. \quad (3)$$

Functions f satisfying $f(x + y) = f(x) + f(y)$ for all real numbers x, y are called *additive*. It is natural to wonder whether description (2) contains more information than description (3). You have seen that (2) implies (3), but the intriguing aspect is the possibility that the converse could also hold! It is therefore natural to ask,

Is it true that if $f : \mathbb{R} \to \mathbb{R}$ satisfies (3), then there exists $m \in \mathbb{R}$ such that $f(x) = mx$ for all $x \in \mathbb{R}$?

For an answer, let f be as in (3) and complete the following tasks.

2. Compute $f(0)$ by making a suitable choice for x and y in (3).
3. Using (3) and Problem 2, show that f is odd, i.e., $f(-x) = -f(x)$ for each real number x.
4. What does (3) yield for the choice $x = y$? Build on this idea to prove that $f(nx) = nf(x)$ for each integer n.
5. Write $f(x) = f(n \frac{x}{n})$ $(n \neq 0)$ and apply Problem 4 to show that

$$f\left(\frac{x}{n}\right) = \frac{1}{n}f(x).$$

6. In light of Problems 4 and 5, justify the fact that, for each rational number q, $f(qx) = qf(x)$ for all $x \in \mathbb{R}$.
7. Using Problem 6, determine m for which $f(q) = mq$ whenever q is a rational number.
8. Now, of course, you want to conclude that $f(x) = mx$ for each *real* number x and not only when $x = q$, where q is a rational number. Up to this point, we've used purely algebraic reasoning. It's time to put the hypothesis of continuity to good use, but how?

   Hint: For each real number x, let $\{q_n\}_n$ be a sequence of rational numbers converging to x. Such a sequence always exists (see Problem V in Projects and Extensions 1.1). Take the limit as $n \to \infty$ of both sides of the identity $f(q_n) = mq_n$.
9. Summarize this investigation's conclusion in the form of a theorem.

II. The Intermediate Value Theorem

As pointed out before, the graph of a continuous function defined on an interval can be traced without lifting the pen from the paper. Consequently, if f is a continuous real valued function defined on an interval I, and if there exist two points c and d in I with $f(c) < 0$ and $f(d) > 0$, then there should be a point between c and d where f takes the value zero. In other words, to join a point on the graph that is below the x-axis with a point on the graph that is above the x-axis, one must cross the x-axis. More generally, *a continuous function defined on an interval has the property that it takes all values between any two points in its image.* In particular, it *transforms intervals into intervals.* This is an important fact that has many applications. Here are a few.

1. Show that if $f : [0,1] \to [0,1]$ is a continuous function, then there exists at least one number $x \in [0,1]$ satisfying $f(x) = x$. Such a number is called a *fixed point for f*.

a. Solution from a Geometric Perspective

Geometrically, the existence of a solution in $[0,1]$ for the equation $f(x) = x$ is equivalent to the existence of at least a point of intersection between the graph of the function f and the line segment l joining the points $(0,0)$ and $(1,1)$. Here is a generic example.

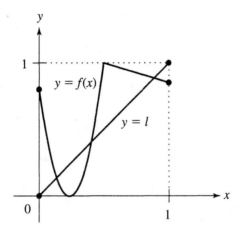

Use the geometric interpretation of graphs of continuous functions to explain why the graph of f must cross the line segment l at least once, yielding an intersection point.

b. The Intermediate Value Theorem at Work

- First assume $f(0) = 0$ or $f(1) = 1$. Does f have a fixed point in these cases?
- Assume that $f(0) \neq 0$ and $f(1) \neq 1$. Consider the function $g : [0,1] \to \mathbb{R}$, defined by $g(x) = x - f(x)$. Determine the sign of $g(0)$ and the sign of $g(1)$. Show that g is continuous, and use the intermediate value theorem for g to conclude that there exists a point x_0 in $(0,1)$ for which $g(x_0) = 0$. Finish the proof of the fact that f has a fixed point.

2. Let a, b be arbitrary real numbers with $a < b$, and let $f : [a,b] \to [a,b]$ be a given continuous function. Does there exist a number $x \in [a,b]$ satisfying $f(x) = x$? Explain.

3. Let $f : [0,1] \to [0,1]$ be an arbitrary continuous function. Prove that for each positive integer n, there exists a value $x \in [0,1]$ satisfying $f(x) = x^n$.

CHAPTER 2 REVIEW

Let A and B be two subsets of the real line. A real valued function $f : A \to B$ associates to each element $x \in A$ a unique element $y \in B$. A is called the *domain of the function*, B is called the *range of the function*, and f is called the *law of correspondence* (see Definition 2.1.1). If the law of correspondence f associates to x the value y, we write $y = f(x)$. Whenever an algebraic expression $f(x)$ is referred to as being a function without further specifications regarding its domain or range, it is implicitly understood that this is the function consisting of

- domain = the set of all real numbers x for which the expression $f(x)$ is meaningful;
- range = the whole real line;
- law = x goes into $f(x)$.

The image of a function f is the set

$$\text{Image}(f) = \{f(x) : x \in A\},$$

while its graph is the set of points $\{(x, f(x)) : x \in \text{Domain}(f)\}$.

Vertical line test: a curve in the plane is the graph of a function if and only if any vertical line intersects the curve at most once.

The composition of two functions f and g, denoted by $f \circ g$, is the function given by the law $(f \circ g)(x) = f(g(x))$ defined for all real values x in the domain of g for which $g(x)$ is in the domain of f.

We say that a real number L is the limit of f as x approaches a from the left, we write $\lim_{x \to a^-} f(x) = L$, provided $f(x)$ can be made arbitrarily close to L by choosing $x < a$ sufficiently close to a within the domain of f. Similarly, one defines $\lim_{x \to a^+} f(x) = L$ (see Definition 2.2.1). Also, if $\lim_{x \to a^-} f(x) = L = \lim_{x \to a^+} f(x)$, then we say that L is the limit of f as x approaches a within the domain of f, and we write $\lim_{x \to a} f(x) = L$. If $\lim_{x \to a^-} f(x)$ or $\lim_{x \to a^+} f(x)$ does not exist or if they both exist but are different, we say that $\lim_{x \to a} f(x)$ does not exist.

The study of the limit of a function f at a point can be done based on Definition 2.2.1 (or Definition 2.2.3), graphically, numerically, and/or employing P1–P6 below. If f, g are real valued functions for which $\lim_{x \to a} f(x)$ and $\lim_{x \to a} g(x)$ exist, and $n \in \mathbf{N}$, a, $c \in \mathbb{R}$, then the following are true:

P1 $\lim_{x \to a} [f(x) \pm g(x)] = \lim_{x \to a} f(x) \pm \lim_{x \to a} g(x)$.

P2 $\lim_{x \to a} [c \cdot f(x)] = c \cdot \lim_{x \to a} f(x)$.

P3 $\lim_{x \to a} [f(x) \cdot g(x)] = \lim_{x \to a} f(x) \cdot \lim_{x \to a} g(x)$.

P4 $\lim\limits_{x \to a} \dfrac{f(x)}{g(x)} = \dfrac{\lim\limits_{x \to a} f(x)}{\lim\limits_{x \to a} g(x)}$ whenever $\lim\limits_{x \to a} g(x) \neq 0$.

P5 $\lim\limits_{x \to a} \sqrt[n]{f(x)} = \sqrt[n]{\lim\limits_{x \to a} f(x)}$ with the additional assumption that $\lim\limits_{x \to a} f(x) \geq 0$ whenever n is even.

P6 If $f(x) \leq g(x)$ for all x in an open interval containing a, then $\lim\limits_{x \to a} f(x) \leq \lim\limits_{x \to a} g(x)$.

Given a function f and a point c in its domain, we say that f is continuous at c provided $\lim\limits_{x \to c} f(x) = f(c)$ (see Definition 2.3.1). As such, if a function g is not defined at c or if $\lim\limits_{x \to c} g(x)$ does not exist or if the latter exists but $\lim\limits_{x \to c} g(x) \neq g(c)$, then g is not continuous at c, and we say that g is discontinuous at c. A function f is said to be continuous on an interval (a, b) provided f is continuous at all the points $c \in (a, b)$.

In general, if f and g are functions that are continuous at some real number a, and $r \in \mathbb{Q} \setminus \{0\}, c \in \mathbb{R}$, then

$$f + g, \quad f - g, \quad c \cdot f, \quad f \cdot g, \quad \frac{f}{g}, \text{ and } \sqrt[r]{f}$$

are continuous at a, provided $g(a) \neq 0$ when considering the continuity of $\frac{f}{g}$, and provided $\sqrt[r]{f(a)}$ is defined when considering the continuity of $\sqrt[r]{f}$.

Important examples of continuous functions are polynomial functions that are continuous on the real line, that is if $p(x)$ is an arbitrary polynomial function, then $\lim\limits_{x \to a} p(x) = p(a)$ for all real numbers a. In addition, $\lim\limits_{x \to a} \sqrt[r]{p(x)} = \sqrt[r]{p(a)}$ provided $r \in \mathbb{Q} \setminus \{0\}$, and $\sqrt[r]{p(a)}$ is defined. Rational functions are also continuous at all the points in their domains, that is, if $p(x)$ and $q(x)$ are polynomial functions, then $\lim\limits_{x \to a} \dfrac{p(x)}{q(x)} = \dfrac{p(a)}{q(a)}$ for all $a \in \mathbb{R}$ with $q(a) \neq 0$.

Informally, from a geometric point of view, a function is continuous if and only if its graph can be traced without lifting the pen.

CHAPTER 2 REVIEW EXERCISES

In Exercises 1–4, determine if the given equation describes y as a function of x.

1. $x^3 - y^2 = 2$ **2.** $y^3 - x^2 = 2$

3. $2y - 5x = 1$ **4.** $xy = -2$

5. Let $f(x)$ be equal to $x^2 + x$ if $x \leq 0$, and to $x^2 - x$ if $x > 0$. Write this function in a condensed form, sketch its graph, and determine its image.

6. Let $f(x)$ be equal to $2 + x - x^2$ if $x \leq 1$, and to $2x + 4$ if $x > 1$. Write this function in a condensed form, sketch its graph, and determine its image.

In Exercises 7–16, determine the domain for the given function and its corresponding image.

7. $f(x) = x^2 - x + 2$

8. $f(x) = \frac{3}{x^2+x+1}$

9. $f(x) = \sqrt{x^3 + x}$

10. $f(x) = \sqrt{x^2 - 2x - 3}$

11. $f(x) = \sqrt[3]{x^5 - 1}$

12. $f(x) = \sqrt[5]{1 - 2x}$

13. $f(x) = \frac{x+2}{\sqrt{x^2+4x+4}}$

14. $f(x) = \frac{(x-1)\sqrt{x-1}}{(x-1)^{\frac{3}{2}}}$

15. $f(x) = \frac{x+3}{x^2-9}$

16. $f(x) = \frac{x^2-2x}{x-2}$

In Exercises 17–20, find the law and the domain of the specified composite function.

17. $f \circ g$ if $f(x) = \frac{x}{x-3}$ and $g(x) = x^2 + 2x$

18. $f \circ g$ if $f(x) = \frac{\sqrt{x+1}}{x}$ and $g(x) = x^2 - 4x - 1$

19. $g \circ f$ if $f(x) = x - 3$ and $g(x) = \sqrt{x}$

20. $g \circ f$ if $f(x) = x^2 + x$ and $g(x) = \sqrt{x}$

21. Let

$$f : \mathbb{R} \to \mathbb{R}, \ f(x) = \begin{cases} 2x + 3 & \text{if } x < 0 \\ x^2 + 3 & \text{if } x \geq 0, \end{cases}$$

and let

$$g : \mathbb{R} \to \mathbb{R}, \ g(x) = x + 1.$$

Compute $f \circ g$ and $g \circ f$.

22. Let

$$f : \mathbb{R} \to \mathbb{R}, \ f(x) = \begin{cases} 3x + 1 & \text{if } x \leq -1 \\ -2 & \text{if } x > -1, \end{cases}$$

and let

$$g : \mathbb{R} \to \mathbb{R}, \ g(x) = \begin{cases} -3 & \text{if } x \leq -2 \\ x - 1 & \text{if } x > -2. \end{cases}$$

Compute $f \circ g$ and $g \circ f$.

In Exercises 23–28, by analyzing the graph $y = f(x)$, evaluate f and the limits of f at the given point, provided they exist.

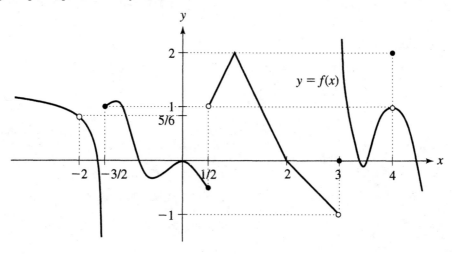

23. $f(-2)$, $\lim\limits_{x \to -2^-} f(x)$, $\lim\limits_{x \to -2^+} f(x)$, $\lim\limits_{x \to -2} f(x)$

24. $f(\frac{1}{2})$, $\lim\limits_{x \to \frac{1}{2}^-} f(x)$, $\lim\limits_{x \to \frac{1}{2}^+} f(x)$, $\lim\limits_{x \to \frac{1}{2}} f(x)$

25. $f(-\frac{3}{2})$, $\lim\limits_{x \to -\frac{3}{2}^-} f(x)$, $\lim\limits_{x \to -\frac{3}{2}^+} f(x)$, $\lim\limits_{x \to -\frac{3}{2}} f(x)$

26. $f(3)$, $\lim\limits_{x \to 3^-} f(x)$, $\lim\limits_{x \to 3^+} f(x)$, $\lim\limits_{x \to 3} f(x)$

27. $f(2)$, $\lim\limits_{x \to 2^-} f(x)$, $\lim\limits_{x \to 2^+} f(x)$, $\lim\limits_{x \to 2} f(x)$

28. $f(4)$, $\lim\limits_{x \to 4^-} f(x)$, $\lim\limits_{x \to 4^+} f(x)$, $\lim\limits_{x \to 4} f(x)$

In Exercises 29–36, evaluate the limits.

29. $\lim\limits_{x \to 2} (3x^5 - 2x^4 + x - 10)$

30. $\lim\limits_{x \to -1} (100 - 3x^2 + x^3 - 7x^5)$

31. $\lim\limits_{x \to -2} \sqrt[3]{x^2 + 3x + 1}$

32. $\lim\limits_{x \to 1} \sqrt{3x^5 - 2x + 2}$

33. $\lim\limits_{x \to 0} \frac{x^2 - x}{2x - 1}$

34. $\lim\limits_{x \to 3} \frac{x + 2}{x^2 - 6}$

35. $\lim\limits_{x \to -1} \left(\frac{1}{x} - \sqrt{3 - x^3} \right)$

36. $\lim\limits_{x \to 0} \left(\frac{3x - 2}{x^3 - 1} + \sqrt[3]{2 - x^2 - x} \right)$

In Exercises 37–42, give an example of a function f as specified.

37. $\lim\limits_{x \to 0^-} f(x) \neq \lim\limits_{x \to 0^+} f(x)$.

38. $\lim\limits_{x \to -1^+} f(x) = f(-1)$, and f is discontinuous at $x = -1$.

39. $\lim\limits_{x \to 1} f(x)$ exists, and f is discontinuous at $x = 1$.

40. f is discontinuous at $x = 0$, but $\lim\limits_{x \to 0} f(x)$ exists.

41. f is continuous at $x = 0$.

42. f is continuous at $x = -1$.

In Exercises 43–46, sketch the graph of f and study its continuity on the real line.

43. $f(x) = \begin{cases} 2 & \text{if } x \leq 0 \\ x - 1 & \text{if } x > 0 \end{cases}$

44. $f(x) = \begin{cases} x^2 - 2x & \text{if } x \leq 1 \\ -1 & \text{if } x > 1 \end{cases}$

45. $f(x) = \begin{cases} x - 2 & \text{if } x \leq 1 \\ 2 - 3x & \text{if } x > 1 \end{cases}$

46. $f(x) = \begin{cases} 3x + 1 & \text{if } x \leq 0 \\ 2x^2 - 1 & \text{if } x > 0 \end{cases}$

In Exercises 47–48, find the values of a for which the function f is continuous on $\mathbb{R}$.

47. $f(x) = \begin{cases} x^2 + x - a & \text{if } x \leq -1 \\ -2x + a & \text{if } x > -1 \end{cases}$

48. $f(x) = \begin{cases} a^2x + 2 & \text{if } x \leq 1 \\ x^2 - x + 6 & \text{if } x > 1 \end{cases}$

49. Determine all the linear functions f with the property that $(f \circ f)(x) = x$ for all $x \in \mathbb{R}$.

CHAPTER 3

Differentiation

3.1 **AVERAGE RATES OF CHANGE**
3.2 **INSTANTANEOUS RATES OF CHANGE AND SLOPES OF GRAPHS**
3.3 **MOTION AND DERIVATIVES**
3.4 **RULES FOR COMPUTING DERIVATIVES**
3.5 **THE CHAIN RULE**

3.1 AVERAGE RATES OF CHANGE

Average rate of change of a function

What is a **rate**? How do we use rates in daily life? To address these issues, work in groups and answer the following questions.

Classroom Discussion 3.1.1: Rates of Change in Real Life

1. What is a person's heart rate? How do you compute it? What units of measurement do you use for heart rate? Fill in the blanks:

$$\text{heart rate} = \frac{\text{number of} \underline{\qquad}}{\text{time, measured in} \underline{\qquad}}$$

2. A person's painting rate is the ratio of the area the person paints to the time it takes to paint that area. Two painters must paint a rectangular wooden fence that is 15 yards long and 5 feet tall. The first painter paints at a rate of 25 square feet per minute. The second painter paints at a rate of 20 square feet per minute. How long will it take for the two workers to finish painting both sides of the wooden fence?

3. A person's walking rate is the distance the person walks divided by the time it takes to walk that distance. In other words,

$$\text{walking rate} = \frac{\text{distance}}{\text{time}}.$$

What is Adrian's walking rate if he covers 7 kilometers in 1.5 hours? Express your answer in meters per second.

These examples describe how a given entity changes in relationship to another entity. Consider Problem 3 again. To compute a person's walking rate at different times during a walk, look at the change in the distance the person has walked with respect to the change in time: $\frac{\text{change in distance}}{\text{change in time}}$. Suppose that after 1 hour, Adrian has walked 5 kilometers and after 2 hours he has walked 11 kilometers. His walking rate for the second hour is $\frac{11-5}{2-1} = 6$ kilometers per hour.

In order to formalize rates of change, we use the function notation. In each case, you computed the ratio between the change of a given function $f(x)$ and the change of x. This is why it is appropriate to define

the **average rate of change** of $f(x)$ with respect to x as the ratio, $\dfrac{\text{change in } f(x)}{\text{change in } x}$.

In particular, if $a < b$ and x changes from a to b, then $f(x)$ changes from $f(a)$ to $f(b)$. This change of f is also referred to as the change of f over the interval $[a, b]$. ◆

Definition 3.1.1 The average rate of change of a function $f(x)$ with respect to x over the interval $[a, b]$ is

$$\frac{f(b) - f(a)}{b - a}.$$

Classroom Discussion 3.1.2: Rates of Change of Linear and Nonlinear Functions

1. Water Levels

The following graph represents a lake's water level over a 30-day period.

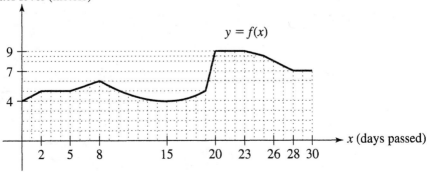

Denote the water level by $f(x)$, where x represents the number of days that have passed. In a–c, compute the water level's average rate of change over the given time period. First write the formula for the rate of change using the function notation, and then use the values of f given by the graph.

a. From the end of day 1 to the end of day 2
b. From the beginning of day 3 to the end of day 5
c. From the end of day 8 to the end of day 15, and from the end of day 15 to the end of day 30

How do these average rates of change compare?

2. **Walking Rates**

Diana walks from home to school in 25 minutes. The distance d in meters she has traveled after a t minute walk is given by $d(t) = 60t$.

a. Compute Diana's average walking rate over the following intervals of time: $[1, 4]$, $[6, 7]$, $[0, 25]$. What do you observe? Interpret the results.
b. Make a conjecture about how the average rates of change of d computed over two arbitrary intervals compare. Prove your conjecture.

3. **Linear Functions versus Nonlinear Functions**

a. Let m and c be arbitrary real numbers, and consider the linear function $g(x) = mx + c$. Compute the average rate of change of g over the interval $[a, b]$. What do you observe?
b. Consider a function g for which the average rate of change over any interval $[a, b]$ is constant and equal to a real number m. Prove that g is linear.

Hint: Fix a point a and let x be arbitrary. What is the average rate of change of g over the interval with endpoints a and x?

Observation. Combine a and b to obtain the following result: A function is linear if and only if the function's average rate of change over any interval is constant.

c. Recall the example in Problem 1. Is f a linear function? What did you observe about the rates of change of f over different intervals?
d. The following graph shows the change in x and the change in $f(x)$ over the interval $[a, b]$. Use this sketch to fill in the blank:

The average rate of change of a function f over an interval $[a, b]$ is equal to the slope of _____.

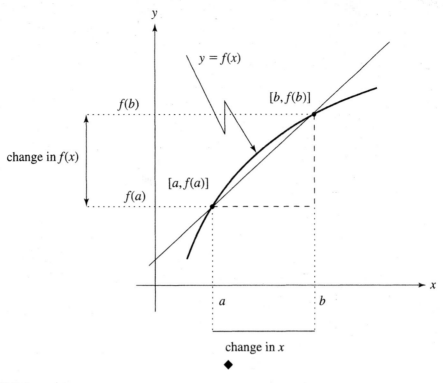

change in $f(x)$

change in x

◆

EXERCISES 3.1

1. One criterion for judging a football team is the team's winning percentage:

$$\text{winning percentage} = \frac{\text{number of games won}}{\text{number of games played}}.$$

The winning percentage is the rate of change of the number of games won with respect to the number of games played. What can you conclude if you know that the winning percentage for a football team in a season was 0.4? Can you find the number of total games the team played that season if the team won eight games?

2. **Altitudes and the Heart Rate.** The human body is optimally equipped for existence at an air pressure close to 760 millimeters of mercury (the air pressure at sea level) with an oxygen concentration of 21%. When the altitude increases, the atmospheric pressure decreases, which in turn leads to a decrease in the number of oxygen molecules per breath. Consequently, the amount of oxygen available in the body's blood and tissue decreases. Such a lack of oxygen can cause potentially life-threatening illness at high altitudes. The following graph (from a report by Catherine M. Quinn) shows the results of studies conducted at the University of Limerick. Subjects breathed into air bags containing concentrations of oxygen that would be found at various altitudes. The relationship between the heart rate

H in beats per minute for a person at rest and the altitude *x* measured in meters is sketched here.

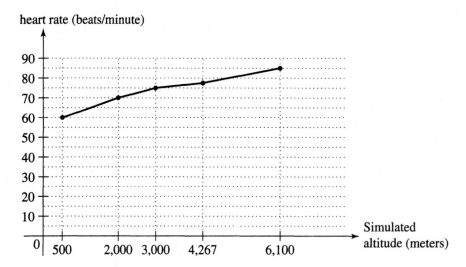

heart rate (beats/minute)

What is the average rate of change of *H* with respect to *x* experienced by a person in a hot air balloon as the balloon rises from 2,000 to 3,000 meters? How about from 4,267 to 6,100 meters? How do the two compare?

3. Jesse filled the gas tank of her Chevy Beretta before beginning a trip from Kansas City to Houston. After driving 199 miles on I-70 and I-35, she stopped in Wichita to pick up three friends. Since she was stopped anyway, she filled the tank again with 6 gallons of gasoline. The four friends made good time on the 363-mile stretch from Wichita to Dallas and filled the tank with 11.7 gallons of gas. The next morning, they picked up yet another friend and drove the 239 miles from Dallas to Houston, making slow progress because of heavy traffic in both cities. When they finally arrived in Houston, they put 8.2 gallons of gas in the car to fill up the tank. The rate at which a car burns fuel is expressed as the number of miles traveled divided by the number of gallons of gasoline used. What was the rate at which Jesse's car burned fuel (a) between Kansas City and Wichita, (b) between Wichita and Dallas, (c) between Dallas and Houston, and (d) on the entire trip?

4. Rachel's 3-cylinder Geo Metro accelerates from 0 to 60 miles per hour in 31 seconds. What is the average acceleration over the 31 seconds that it takes her to reach 60 miles per hour? Express your answer in miles per hour squared.

5. During the summer, Diana earns money by selling lemonade. She charges 50 cents for each cup of lemonade.
   a. Let $f(x)$ denote the amount of money (in dollars) Diana earns by selling x cups of lemonade. Determine a formula for $f(x)$ and sketch the graph of f.
   b. What is the average rate of change of f with respect to x as x increases from a to b, where $a < b$ and a, b are arbitrary whole numbers? Explain.

6. When selling lemonade for 50 cents per cup, Diana can sell 40 cups per day. She observes that the number of cups sold per day decreases by 2 for every 5-cent increase in the price of a cup of lemonade.

a. Determine a formula for the number of cups C of lemonade sold each day as a function of the price p in cents of a cup of lemonade if all price increases occur only in multiples of 5-cent increments. Graph this function.

b. What is the average rate of change in the number of cups sold per day if the price per cup increases from 60 cents to 75 cents? How about from 70 cents to 90 cents? Explain.

3.2 INSTANTANEOUS RATES OF CHANGE AND SLOPES OF GRAPHS

Instantaneous rate of change and the limit process • Slopes of tangent lines • Derivatives • Sketching graphs of derivatives • Increasing and decreasing functions • Differentiability and continuity

You learned in Section 3.1 that the average rate of change of a function f over an interval $[a, b]$ is equal to $\frac{f(b)-f(a)}{b-a}$ and that it is the same as the slope of the line segment joining the points $(a, f(a))$ and $(b, f(b))$. For a linear function f, the average rate of change does not depend on the interval and equals the slope of the line $y = f(x)$. However, this is not the case for a nonlinear function for which different intervals may yield different average rates of change.

In order to understand the behavior of a given function, one must know how the function changes at each point (i.e., one looks at *instantaneous rates of change*). For example, when driving a car, knowing only the average rate of change of the distance traveled with respect to time (which is also called the *average speed*) does not provide sufficiently detailed information. A police officer will ticket a driver if the speedometer, which indicates the instantaneous rate of change of the distance traveled with respect to time, indicates 40 miles per hour in a 30-mile-per-hour zone, even if the car was going at an average speed of 30 miles per hour for the last hour. But how can we define instantaneous rates of change? To answer this question, first recall the notion of a **tangent** line to a graph. In the following figure, the lines passing through the points P_1, P_2, P_3, P_4, and P_5 are examples of tangent lines to the given graph.

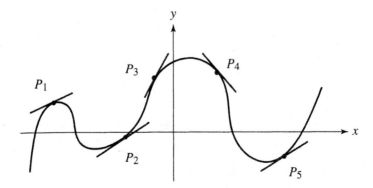

Intuitively, a line is tangent to a curve at a point P on the curve whenever it best approximates the curve *near P*. See what happens as we "zoom in" toward the curve and its tangent line at point P_5.

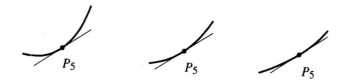

To decide if a graph $y = f(x)$ admits a tangent line at a point $(a, f(a))$, proceed as follows. Consider another point $(b, f(b))$ on the graph and analyze the behavior of the line passing through $(a, f(a))$ and $(b, f(b))$ as $b \to a$. If there exists a line L passing through $(a, f(a))$ such that the slope of the line passing through $(a, f(a))$ and $(b, f(b))$ approaches the slope of L as $b \to a$, then L is by definition called the tangent line to $y = f(x)$ at $(a, f(a))$.

In particular, if we assume that:

i) the slope of the line passing through $(a, f(a))$ and $(b, f(b))$ approaches as $b \to a^-$ the slope of a line L_1 passing through $(a, f(a))$ and

ii) the slope of the line passing through $(a, f(a))$ and $(b, f(b))$ approaches as $b \to a^+$ the slope of a line L_2 passing through $(a, f(a))$,

then a tangent line to $y = f(x)$ at $(a, f(a))$ exists if and only if $L_1 = L_2$.

For example, the function f whose graph is sketched here does not have a tangent line at $(a, f(a))$.

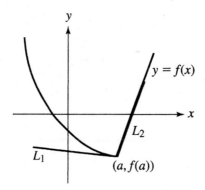

Classroom Discussion 3.2.1: Instantaneous Rate of Change of the Function $f(x) = x^2$

1. Sketch the graph of $f(x) = x^2$ on a sheet of graph paper.
2. Compute the average rate of change of $f(x)$ over the interval $[1, 1.1]$, draw the line passing through the points $(1, f(1))$ and $(1.1, f(1.1))$.
3. Use your calculator to compute the average rates of change of f over the intervals $[1, 1 + h]$ for $h > 0$, taking the values $10^{-1}, 10^{-2}, \ldots, 10^{-6}$. What do you observe? Describe what you expect to happen to the line passing through points $(1, f(1))$ and $(1 + h, f(1 + h))$ as h decreases.
4. Use your calculator to compute the average rates of change of f over the intervals $[1 + h, 1]$ for $h < 0$, taking the values $-10^{-1}, -10^{-2}, \ldots, -10^{-6}$.

What do you observe? Describe what you expect to happen to the line passing through points $(1 + h, f(1 + h))$ and $(1, f(1))$ as the absolute value of h decreases.

5. Assuming that your calculator can be as precise as you want, what do you expect to happen to the average rates of change of f over the intervals with endpoints 1 and $1 + h$ as $h \to 0$? What is the connection with the tangent line to the graph of f at the point $(1, 1)$?

6. Using calculus, you can verify the prediction you made in Problem 5. Write the general formula for the average rate of change of f over the interval with endpoints 1 and $1 + h$, then simplify the expression you obtained under the assumption that $h \neq 0$. Next, take the limit as $h \to 0$ of your simplified expression. What is the geometric interpretation of the value you obtained?

7. Why do you think the value you obtained in Problem 6 is called the *instantaneous rate of change* of f at 1, or simply the rate of change of f at 1?

8. Fill in the blank:

   The slope of the tangent line to the graph $y = x^2$ at the point $(1, 1)$ is equal to _____.

9. Consider now an arbitrary point $(x, f(x))$ on the graph of f. Compute the average rate of change of f over the interval with endpoints x and $x + h$. Simplify your expression and then take the limit as $h \to 0$.

10. Fill in the blanks:

   The instantaneous rate of change of f at x is $\lim\limits_{h \to 0} \frac{f(x+h)-f(x)}{h}$. The value of this limit coincides with the slope of the _____ to the graph $y = f(x)$ at the point _____. ◆

In the preceding example, you constructed a new function, which associates to each x the value of $\lim\limits_{h \to 0} \frac{f(x+h)-f(x)}{h}$. This new function is called the **derivative** of f and is denoted by f'. You have proved that

$$\text{if } f(x) = x^2, \text{ then } f'(x) = 2x. \tag{1}$$

In addition, you have seen that the slope of the tangent line to the graph of f at a point $(x, f(x))$ is equal to $f'(x)$.

EXAMPLE Determine the equation of the tangent line to the graph $y = x^2$ at the point $(-2, 4)$.

Solution   The slope of the tangent line to the graph of the function $f(x) = x^2$ at the point $(-2, 4)$ is $m = f'(-2) = 2(-2) = -4$. Thus, the equation of the tangent line is $y - 4 = -4(x + 2)$, or, equivalently, $y = -4x - 4$. ■

Practice Problem

Determine the equation of the tangent line to the graph $y = x^2$ at the point $(3, 9)$.

Definition 3.2.1 Let f be a real valued function. The **derivative** f' at a point x in the domain of f is defined by

$$f'(x) = \lim_{h \to 0} \frac{f(x + h) - f(x)}{h}, \tag{2}$$

provided this limit exists. If $f'(x)$ exists, we say that the function f is **differentiable** at x. If a function f is differentiable at any point x in an interval (a, b), we say that f is **differentiable on** (a, b).

Notation and Terminology.   The process of computing derivatives of functions is called *differentiation*. We use the notation $(f(x))'$, $f'(x)$ (read "f prime of x"), or simply f', when taking the derivative of a function $f(x)$. Sometimes we use the notation Δx instead of h and use Δf for $f(x + \Delta x) - f(x)$. In that case, $f'(x) = \lim_{\Delta x \to 0} \frac{\Delta f}{\Delta x}$. Whenever we say a function f is differentiable without specifying the points at which f is differentiable, we mean that f is differentiable at all the points x in the domain of f that are approachable from both sides within the domain of f.

Observations

1. At times it is useful to compute a function's derivative by means of the following formula (3), which is equivalent to formula (2).

$$f'(x) = \lim_{y \to x} \frac{f(y) - f(x)}{y - x} = \lim_{y \to x} \frac{f(x) - f(y)}{x - y}. \tag{3}$$

2. The slope of the tangent line to the graph $y = f(x)$ at the point $(x, f(x))$ is equal to the derivative $f'(x)$. You have seen this to be true for $f(x) = x^2$, but the same reasoning holds for arbitrary differentiable functions f.

Classroom Discussion 3.2.2: Derivatives of Linear and Quadratic Functions

1. Let $f(x) = 3$. Determine the slope of any tangent line to the graph of f by analyzing its graph. What is the expression of $f'(x)$? Compute the derivative of $f(x) = 3$ using Definition 3.2.1.
2. Fix a real number c and let $f(x) = c$. By analyzing the graph $y = f(x)$, determine the slope of any tangent line to the graph of f. What is the expression of $f'(x)$? Compute the derivative of $f(x) = c$ using Definition 3.2.1. Fill in the blank:

$$\text{if } f(x) = c \text{ for every } x, \text{ then } f'(x) = \underline{\hspace{2em}}. \tag{4}$$

3. Let $g(x) = 3x + 1$. By analyzing the graph $y = g(x)$, determine the slope of any tangent line to the graph of g. What is the expression of $g'(x)$? Compute the derivative of $g(x)$ using Definition 3.2.1.
4. Consider a linear function $f(x) = mx + b$, where m and b are arbitrary real numbers. By analyzing the graph $y = f(x)$, determine the slope of any tangent

line to the graph of f. What is the expression of $f'(x)$? Compute the derivative of $f(x)$ using Definition 3.2.1. Fill in the blank:

$$\text{if } f(x) = mx + b \text{ for every } x, \text{ then } f'(x) = \underline{\hspace{2cm}}. \qquad (5)$$

5. Let a, b, and c be arbitrary real numbers and consider the function $f(x) = ax^2 + bx + c$. Determine $f'(x)$ using Definition 3.2.1. Fill in the blank:

$$\text{if } f(x) = ax^2 + bx + c \text{ for every } x, \text{ then } f'(x) = \underline{\hspace{2cm}}. \qquad (6) \quad \blacklozenge$$

Definition 3.2.2 Let (a, b) be an open interval and let f be a function defined on (a, b).

    a. f is **increasing** on (a, b) provided $f(x_1) < f(x_2)$ whenever x_1, $x_2 \in (a, b)$ are such that $x_1 < x_2$.

    b. f is **decreasing** on (a, b) provided $f(x_1) > f(x_2)$ whenever x_1, $x_2 \in (a, b)$ are such that $x_1 < x_2$.

    c. f is **nondecreasing** on (a, b) provided $f(x_1) \leq f(x_2)$ whenever x_1, $x_2 \in (a, b)$ are such that $x_1 < x_2$.

    d. f is **nonincreasing** on (a, b) provided $f(x_1) \geq f(x_2)$ whenever x_1, $x_2 \in (a, b)$ are such that $x_1 < x_2$.

Informally, a function is increasing on an interval if its graph is "going up" or "rising" from left to right on that interval. Similarly, a function is decreasing on an interval if its graph is "going down" or "falling" from left to right on that interval.

Classroom Discussion 3.2.3: Rising or Falling Graphs and Derivatives of Functions

This discussion's goal is to establish a link between "rising" or "falling" graphs of functions and the sign of their derivatives.

1. On two separate coordinate systems, sketch the graphs of $g(x) = x^2$ and $g'(x) = 2x$.
2. On which interval is g nonincreasing? What is the sign of $g'(x)$ on this interval?
3. On which interval is g nondecreasing? What is the sign of $g'(x)$ on this interval?
4. Determine the points x_0 at which g' takes the value zero. What can you say about the tangent line to the graph of g at the point $(x_0, g(x_0))$?
5. Suppose that a differentiable function f is nonincreasing on an interval (a, b). Given the geometric interpretation of the derivative, how do the values of f' on (a, b) compare to 0? Use Definition 3.2.1 of the derivative to check your conjecture.
6. Suppose that a differentiable function f is nondecreasing on an interval (a, b). Given the geometric interpretation of the derivative, how do the values of f' on (a, b) compare to 0? Use Definition 3.2.1 to check your conjecture.

7. Explain why the following statement is true:

> For a differentiable function f, the tangent line to the graph of f at a point $(c, f(c))$ is horizontal if and only if $f'(c) = 0$.   ◆

You discovered that for a differentiable function f, being nondecreasing/nonincreasing provides information about the sign of the derivative f'. Should we expect the sign of f' to provide information about the behavior of f? Intuitively, this seems to be the case. For example, if $f' < 0$ on an interval (a, b), then the slopes of the tangent lines to the graph of f corresponding to points $x \in (a, b)$ are negative. This suggests that f is decreasing on the interval (a, b). Similarly, if $f' > 0$ on an interval (a, b), then the slopes of the tangent lines to the graph of f corresponding to points $x \in (a, b)$ are positive, and we expect that f is increasing on the interval (a, b). Transform this informal reasoning into a proof by carefully using Definition 3.2.1. The section "More on the Connection between the Sign of f' and the Behavior of f" in Projects and Extensions 3.2 addresses this issue. In many applications, we determine the sign of f' in order to identify the intervals on which f is increasing/decreasing (as you will see in Chapter 4). The following is a summary of the relationship between the behavior of a function f and the sign of its derivative, including the results you proved in Classroom Discussion 3.2.3.

The Relationship between a Function's Behavior and the Sign of Its Derivative.
Let f be a differentiable function on an interval (a, b). Then the following hold.

1. f is nonincreasing on (a, b) if and only if $f' \leq 0$ on (a, b).
2. f is nondecreasing on (a, b) if and only if $f' \geq 0$ on (a, b).
3. If $f' < 0$ on (a, b), then f is decreasing on (a, b).
4. If $f' > 0$ on (a, b), then f is increasing on (a, b).
5. The tangent line to the graph of f at a point $(c, f(c))$ $(c \in (a, b))$ is horizontal if and only if $f'(c) = 0$.

Classroom Discussion 3.2.4: Sketching Graphs of Derivatives

In each example here, sketch the graph of the derivative based on the graph of the function. (Your sketch cannot be exact in some of these cases, but you can draw a rough approximation for the graph of the derivative.) Describe what happens to the derivative function as x increases. To do so, look at the graph from left to right and answer the following questions:

a. Where is the derivative positive?
b. Where is the derivative negative?
c. Where is the derivative zero?
d. Are there any points $(x, f(x))$ where the graph does not have a tangent line?

1.

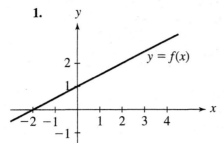

2.

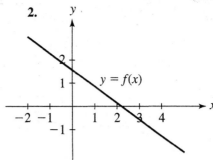

3.

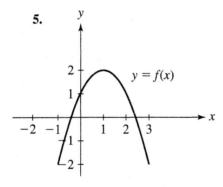

4.

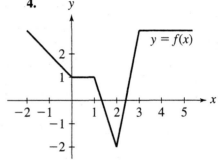

5.

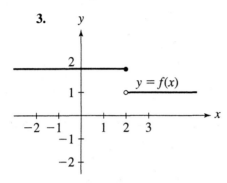

6.

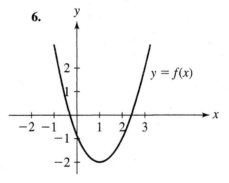

7.

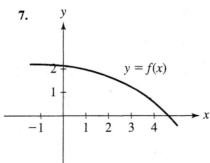

8.

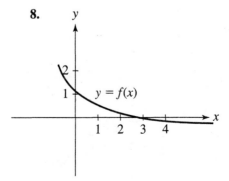

9.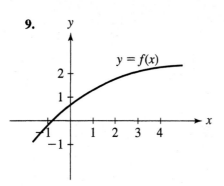

10.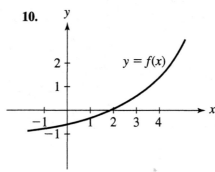

Classroom Discussion 3.2.5: Differentiability and Continuity

From Problem 4 in the previous Classroom Discussion, you can see that not all continuous functions are differentiable. In particular, if the graph of a function contains a corner at some point $(a, f(a))$, then the function does not have a tangent line at that point, thus, the function is not differentiable at $x = a$. Hence, continuity does not imply differentiability. How about the converse? Is it natural to expect that differentiable functions are continuous? From a geometric point of view, the question can be restated as follows: If a graph has a tangent line at every point, does it follow that the graph can be traced without lifting the pen? This discussion's goal is to answer this question.

Suppose f is a function that is differentiable at x. By definition, f is continuous at x provided $\lim_{y \to x} f(y) = f(x)$. In particular, if we denote by h the difference $y - x$, then $y = x + h$ and $y \to x$ if and only if $h \to 0$. Hence, f is continuous at x provided $\lim_{h \to 0} f(x + h) = f(x)$. The following outline is structured around investigating the validity of this equality. First, observe that the algebraic identity

$$f(x + h) - f(x) = \frac{f(x + h) - f(x)}{h} \cdot h$$

holds for any $h \neq 0$.

a. What can you say about $\lim_{h \to 0} \frac{f(x+h) - f(x)}{h}$ and $\lim_{h \to 0} h$?

b. Use a to compute $\lim_{h \to 0} [f(x + h) - f(x)]$.

c. Use b to compute $\lim_{h \to 0} f(x + h)$.

d. Fill in the blanks:

If f is differentiable at x, then $\lim_{h \to 0} f(x + h) = $ _____; that is, f is

_____ at x. ◆

EXERCISES 3.2

In Exercises 1–4, compute the derivative of the given function.

1. $f(x) = x^2 + 2x$

2. $g(x) = x^2 - 3x$

3. $A(s) = 2s^2 + s - 2$

4. $f(t) = -t^2 + 3t - 10$

5. Determine the equation of the tangent line to the curve $y = 2x^2 - x + 10$ at the point $(-1, 13)$.

6. Determine the equation of the tangent line to the curve $y = -3x^2 + 5x - 2$ at the point $(1, 0)$.

7. Scientists have found that radioactive carbon-14 (C14) has a half-life of 5,730 years. This means that if the amount of C14 now is α, then the amount 5,730 years from now will be $\frac{1}{2}\alpha$, the amount 11,460 years from now will be $\frac{1}{2^2}\alpha$, and so on. The amount of C14 remains constant in living organisms due to metabolic processes but decreases once the organism dies. This is the idea behind carbon dating.

  a. Make a sketch of the amount $A(t)$ of C14 in an organism t years after it has died, if the amount of C14 present while the organism was living is α. Is A increasing, decreasing, or neither? Will $A(t)$ ever be zero? As $t \to \infty$, what value does $A(t)$ approach?

  b. Use the graph you have sketched in a to answer the following questions: Is A' positive or negative? Will $A'(t)$ ever be zero? Is A' increasing, decreasing, or neither? As $t \to \infty$, what value does $A'(t)$ approach? Make a sketch of A'.

8. Match each limit a–c with the corresponding derivative from i–iii.

  a. $\lim\limits_{h \to 0} \frac{1}{h}\left(\frac{1}{x+h} - \frac{1}{x}\right)$ **i.** $(x^3)'$

  b. $\lim\limits_{h \to 0} \frac{(x+h)^3 - x^3}{h}$ **ii.** $(\sqrt{x})'$

  c. $\lim\limits_{h \to 0} \frac{\sqrt{x+h} - \sqrt{x}}{h}$ **iii.** $\left(\frac{1}{x}\right)'$

9. Match each function a–c with the graph of its derivative i–iii.

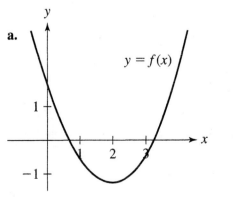

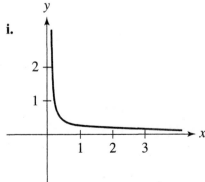

b.

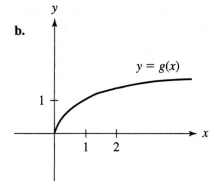

$y = g(x)$

ii.

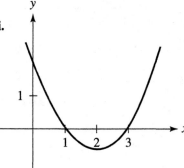

c.

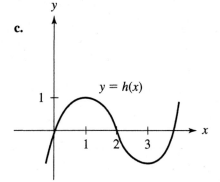

$y = h(x)$

iii.

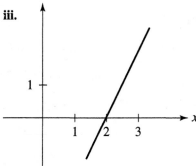

10. Sketch the graph of the derivative of the function $f(x)$.

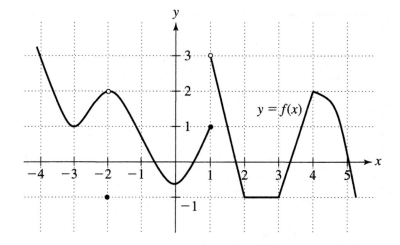

$y = f(x)$

11. Sketch the graph of the derivative of the function $f(x)$.

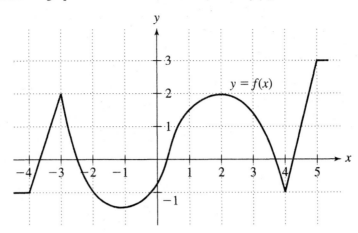

12. Is it true that a function $f : \mathbb{R} \to \mathbb{R}$ that is increasing on $(-\infty, 0)$ and on $(0, \infty)$ is increasing on the entire real line? If yes, explain why, if no, give a counterexample.

13. Let $f, g, h : \mathbb{R} \to \mathbb{R}$ be nondecreasing functions such that $f(x) \le g(x) \le h(x)$ for all $x \in \mathbb{R}$. Is it true or false that $f(f(x)) \le g(g(x)) \le h(h(x))$ for all $x \in \mathbb{R}$? To answer this question, complete the following outline. Start by fixing an arbitrary real number x.

  a. How does $f(f(x))$ compare to $f(g(x))$? Explain.

  b. How does $f(g(x))$ compare to $g(g(x))$? Explain.

  c. Combine a and b to determine the relationship between $f(f(x))$ and $g(g(x))$.

  d. Reason as in a–c to determine the relationship between $g(g(x))$ and $h(h(x))$.

  e. What is the correct answer to the true-false question?

14. Let $f : \mathbb{R} \to \mathbb{R}$ be such that $|f(x)| \le |x|^2$ for all $x \in \mathbb{R}$. Prove that f is differentiable at $x = 0$ and that $f'(0) = 0$ by using the following outline.

  a. Determine the value of $f(0)$.

  b. Write out the definition of $f'(0)$ and use the inequality verified by f to estimate the absolute value of the fraction appearing in the definition of $f'(0)$.

  c. Use b and the properties of limits to determine $f'(0)$.

15. Let $f : \mathbb{R} \to \mathbb{R}$ and let $\alpha > 0$ be a real number.

  a. If $\alpha > 1$ and $|f(x)| \le |x|^\alpha$ for all $x \in \mathbb{R}$, prove that f is differentiable at $x = 0$ and that $f'(0) = 0$.

  b. If $0 < \alpha < 1$, $|f(x)| \ge |x|^\alpha$ for all $x \in \mathbb{R}$ and if $f(0) = 0$, prove that f is not differentiable at $x = 0$.

  c. What can you conclude if $|f(x)| \le |x|$ for all $x \in \mathbb{R}$?

Projects and Extensions 3.2

More on the Connection between the Sign of f' and the Behavior of f

Let f be a function differentiable on an interval (a, b). The goal of this project is to prove the following:

  a. If $f' > 0$ on (a, b), then f is increasing on (a, b).

b. If $f' < 0$ on (a,b), then f is decreasing on (a,b).

c. If $f' \geq 0$ on (a,b), then f is nondecreasing on (a,b).

d. If $f' \leq 0$ on (a,b), then f is nonincreasing on (a,b).

I. Complete steps 1–7 to prove a. Start by assuming that $f' > 0$ on (a,b).

1. Write the condition that $f(x_1)$ and $f(x_2)$ should satisfy, for any two real numbers x_1, x_2 such that $a < x_1 < x_2 < b$, for the function f to be increasing on the interval (a,b).

2. Fix x_1 and x_2 with $a < x_1 < x_2 < b$. Reason by contradiction to show that there exists $c \in (x_1,b]$ with the property that $f(x_1) < f(x)$ for all $x \in (x_1,c)$.

   Hint: $\lim\limits_{x \to x_1^+} \dfrac{f(x)-f(x_1)}{x-x_1} = f'(x_1) > 0$.

3. Let I be the largest interval of the form (x_1,c), where c satisfies the condition in Step 2. Denote by $c*$ the right endpoint of I; that is, $I = (x_1,c*)$. (Observe that the interval I can be constructed by taking the union of all intervals (x_1,c), where c satisfies the condition in Step 2.) Prove that if $x_2 < c*$, then $x_2 \in I$ and $f(x_1) < f(x_2)$, and we are done in this case.

4. Prove that if $c* \leq x_2$, then there exists $x_0 \in I = (x_1,c*)$ such that $f(x_0) < f(c*)$.

   Hint: Reason by contradiction, much as in Step 2, this time using

   $$\lim_{x \to c*^-} \frac{f(c*) - f(x)}{c* - x} = f'(c*) > 0.$$

5. Run the same type of argument as in Step 2 to show that there exists $h > 0$ such that $f(c*) < f(x)$ for all $x \in (c*,c* + h) \subset (a,b)$.

6. Combine Steps 4 and 5 in order to show that $f(x_1) < f(x)$ for all $x \in (x_1,c* + h)$. How does this contradict the definition of I?

7. Which assumption (made along the way) has generated the contradiction in Step 6? Is the proof of a complete at this point?

II. In order to prove b, follow the same outline as in Problem I, making the necessary changes due to the fact that $f' < 0$ this time. An alternative, faster way, to prove b is to proceed in two steps. First, apply a to the function $g(x) = -f(x)$ for all $x \in (a,b)$ to show that g is increasing on (a,b). Second, use the fact that g is increasing on (a,b) to show that f is decreasing on (a,b).

III. Complete Steps 1–4 (following) to prove c. Start by assuming that $f' \geq 0$ on (a,b).

1. Fix an arbitrary number $\varepsilon > 0$ and define the function $g(x) = f(x) + \varepsilon x$ for $x \in (a,b)$. Use Definition 3.2.1 to show that g is differentiable on (a,b) and compute g'.

2. How do the values of g' compare to 0 on (a,b)?

3. Let $x_1, x_2 \in (a, b)$ be such that $x_1 < x_2$. Based on a and b, which of the following inequalities is true?

$$\underbrace{f(x_1) + \varepsilon x_1}_{g(x_1)} < \underbrace{f(x_2) + \varepsilon x_2}_{g(x_2)} \quad \text{or} \quad \underbrace{f(x_1) + \varepsilon x_1}_{g(x_1)} > \underbrace{f(x_2) + \varepsilon x_2}_{g(x_2)}$$

4. Pass to the limit as $\varepsilon \to 0$ in the preceding appropriate inequality in order to complete the proof of c.

IV. In order to prove d, follow the same outline as in Problem III, making the necessary changes entailed by the fact that $f' \le 0$ this time. Alternatively, you may apply c to the function $h(x) = -f(x), x \in (a, b)$.

3.3 MOTION AND DERIVATIVES

Position function • Velocity • Acceleration

In this section we analyze the motion of an object moving along a straight line. The object can be a toy rocket launched vertically, a ball dropped from a tower, or a person walking along a straight path. In each case, the function describing the motion is called the **position function**, and it represents the displacement of the object from a certain reference point. Thus, the values of the position function can be positive or negative, depending on which side of the reference point the object is located.

It is very important to distinguish between the position at time t of the object and the distance traveled during the time interval $[0, t]$. Suppose Diana walks to the board, writes a formula, returns to her desk, and after a few seconds joins her classmate at the desk behind hers. The distance traveled will not fully describe Diana's motion. Look at the following picture and decide which graph describes the displacement of Diana from her desk and which one describes the distance Diana traveled.

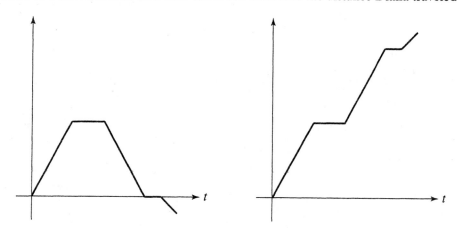

Imagine that you are driving a car on a straight road. You are interested in your displacement from the starting point, and you want to know how fast the car is moving at various times and if the car is decelerating, accelerating, or moving at a constant rate. These are various ways to measure the change of the position function over time. Denote by $h(t)$ the position function of an object moving along a straight line; then consider

- the **velocity** of the object at time t, denoted by $v(t)$, which is the instantaneous rate of change of h with respect to time, i.e., $v(t) = h'(t)$;
- the **speed** of the object at time t, denoted by $s(t)$, which is the absolute value of the velocity $v(t)$, i.e., $s(t) = |v(t)|$;
- the **acceleration** of the object at time t, denoted by $a(t)$, which is the instantaneous rate of change of the velocity function with respect to time, i.e., $a(t) = v'(t)$.

From these definitions, you can see that the unit for measuring the velocity equals the unit for measuring the position, divided by the unit for measuring time. Similarly, the unit for measuring the acceleration equals the unit for measuring the velocity, divided by the unit for measuring time; this is the same as the unit for measuring the position divided by the square of the unit for measuring time.

Classroom Discussion 3.3.1: Velocity, Speed, and Acceleration

1. Adrian received a model rocket for his birthday. The launcher is powered by an air pump, and the height attained by the rocket depends on the number of times Adrian pumps air. Adrian knows that five pumps will project the rocket upward according to the formula $h(t) = -16t^2 + 128t$, where $h(t)$ is the rocket's height, measured in feet from the ground, t seconds after take off.

   a. Sketch the graph of $h(t)$ for $t \geq 0$. Is this the trajectory of the rocket? Explain.

   b. Compute the velocity $v(t)$ of the rocket.

   c. Sketch the graph of $v(t)$ and interpret it based on the rocket's actual motion.

   d. Compute the rocket's velocity at $t = 2$, $t = 3$, and $t = 5$. How do you interpret the difference in signs of the values you obtained?

   e. Sketch the graph of $s(t)$. What is the speed of the rocket at times $t = 2$, $t = 3$, and $t = 5$?

   f. After how many seconds will the rocket start falling? What will be the maximum height attained by the rocket?

   g. After how many seconds will the rocket hit the ground?

   h. Determine the rocket's acceleration.

2. A rock is thrown up from a cliff. Measured from the cliff's base, the height (in feet) of the rock after t seconds is given by the expression $h(t) = -16t^2 + 64t + 80$.

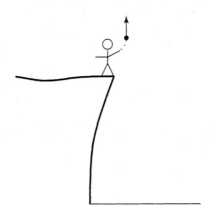

a. Sketch the graph of $h(t)$.

b. Compute the rock's velocity $v(t)$. Sketch the graph of $v(t)$.

c. What does 64 represent in the expression of $v(t)$?

d. What is the rock's velocity when it hits the ground?

e. What does 80 represent in the expression of $h(t)$? Explain your reasoning.

f. During which intervals of time is the speed decreasing? During which intervals of time is the speed increasing?

g. Determine the acceleration of the rock.   ◆

Remark.   The fact that the rocket and the rock have the same constant acceleration is not a coincidence. We provide an explanation in Chapter 5.

EXERCISES 3.3

1. A train, moving with a constant speed, makes daily trips between Columbia, Saint Louis, and Kansas City.

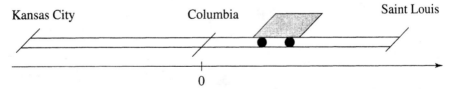

The motion of the train traveling from Columbia to Saint Louis is given by the position function $d(t) = 110t$. Here d is measured in miles, t is the time in hours measured from the instant the train leaves Columbia, and the displacement of the train is measured with respect to Columbia.

a. What is the train's velocity?

b. It takes 70 minutes for the train to reach Saint Louis. The train waits in Saint Louis for 20 minutes, after which it returns to Columbia. What is the expression of the position function for the trip from Saint Louis to Columbia?

c. After reaching Columbia, the train stays in the station for 30 minutes and then leaves for Kansas City. What is the expression of the position function for the trip from Columbia to Kansas City if Columbia is halfway between Kansas City and Saint Louis?

d. After a 1-hour wait in Kansas City, the train returns to Columbia. Write the expression of the position function for the Kansas City–Columbia route.

e. Sketch the graphs of the displacement $d(t)$ and the velocity $v(t)$ from the moment the train began its trip to Saint Louis until it returned to Columbia from Kansas City.

2. The Washington Monument is 555 feet tall. The height h of a coin dropped from the top of the monument is $h(t) = -16t^2 + 555$. Here h is measured in feet from the ground, and t is measured in seconds from the moment the coin was dropped.

a. Sketch the graph of $h(t)$.

b. How long will it take for the coin to hit the ground?

c. Determine the coin's velocity when it hits the ground.

d. Sketch the graph of the velocity $v(t)$. Interpret the graph.

e. Determine the coin's acceleration $a(t)$ at time t.

3. The height h in feet of an object fired straight up is given by $h(t) = -16t^2 + 100t$, where time t is measured in seconds from the instant the object was fired.

a. From where was the object fired, and what was its initial velocity?

b. What is the velocity of the object 1.2 seconds after it was fired?

c. What is the maximum height attained by the object?

d. During which intervals of time is the speed decreasing?

e. During which intervals of time is the speed increasing?

f. Compute the acceleration of the object at $t = 1$ seconds and $t = 2$ seconds.

4. A diver jumps from a diving board. The diver's height (measured in feet) at time t is $h(t) = -16t^2 + 32t + 48$, where t is the time in seconds $h(t)$.

a. How high is the diving board?

b. When does the diver hit the water?

c. What is the diver's velocity at impact?

5. A car pulls out of a driveway, accelerates, and then stops for a red light. This motion takes place on a straight road. The car's displacement, in miles, from the front of the driveway is given by the function

$$d(t) = \begin{cases} 0.17t^3, & 0 \le t \le 1 \\ -0.17(t - 2)^2 + 0.34, & 1 \le t \le 2, \end{cases}$$

where t is the time in minutes.

a. Find the car's velocity $v(t)$ at time t. Graph the function $v(t)$. What can you say about $v(1)$?

b. Find the car's acceleration $a(t)$ at time t. Graph the function $a(t)$. What can you say about $a(1)$? How do you interpret the sign of $a(t)$?

3.4 RULES FOR COMPUTING DERIVATIVES

The power rule • The constant multiple rule • The sum and difference rule • The product rule • The quotient rule

Recall that, by definition, the derivative $f'(x)$ of a given function $f(x)$ is equal to $\lim_{h \to 0} \frac{f(x+h)-f(x)}{h}$, whenever such a limit exists. Using this definition, you proved in

Section 3.2 that for any real numbers a, b, and c,

$$\text{if } f(x) = ax^2 + bx + c, \text{ then } f'(x) = 2ax + b. \tag{7}$$

This formula will save you a lot of time, since you no longer need to use Definition 3.2.1 to compute the derivative of quadratic functions. All you have to do is apply the preceding formula. This is why it is beneficial to continue deriving more rules for computing derivatives. Once such formulas are in place, you can apply them without having to return to Definition 3.2.1.

Classroom Discussion 3.4.1: Rules for Computing Derivatives

1. **Nonnegative Integer Powers of x**

   Our goal is to find the derivative of x^n for a nonnegative integer n by progressively completing the following table.

$f(x)$	$f'(x)$
c	
x	
x^2	
x^3	
x^4	
x^5	
x^6	
x^n	

   a. Apply formula (7) to determine the derivatives of c, x, and x^2.

   b. Compute the derivative of x^3 using Definition 3.2.1. Here you may use the identity $a^3 - b^3 = (a - b)(a^2 + ab + b^2)$.

   c. Compute the derivative of x^4 using Definition 3.2.1. Here you may use the identity $a^2 - b^2 = (a - b)(a + b)$.

   d. Use the pattern you observed to complete the rest of the table.

   e. Prove that your guess in d is correct. To do so, fix an arbitrary natural number n. By multiplying out the right-hand side, verify first that the following formula holds true:

   $$y^n - x^n = (y - x)(y^{n-1} + y^{n-2}x + \cdots + yx^{n-2} + x^{n-1}).$$

   Next, use formula (3) in Section 3.2 to complete the proof.

2. **Constant Multiples of a Function**

   Suppose f is a differentiable function and c is an arbitrary real number. Our goal is to show that the function cf is differentiable and to determine the relationship between $(cf)'$, c, and f'.

a. Compute and compare the derivatives $(3x^2)'$ and $3(x^2)'$. What happens if we replace 3 by any other real number c?

b. Suppose now that $f(x)$ is an arbitrary differentiable function, and fix a real number c. What relationship between $(cf)'$, c, and f' do the calculations in a suggest?

c. Use Definition 3.2.1 to show that cf is differentiable and to express $(cf(x))'$ in terms of c and $f'(x)$. Fill in the blank $(cf)' =$ _____. Compare the result with your guess in b.

3. Sums and Differences of Functions

Let f and g be two differentiable functions. Are the functions $f + g$ and $f - g$ differentiable? If yes, is there a relationship between $(f + g)'$, f', and g'? How about between $(f - g)'$, f', and g'? Find the answers to these questions by completing the following tasks.

a. Compute and compare $(x + x^2)'$ and $(x)' + (x^2)'$. What do you observe?

b. Compute and compare $(x^2 + 3)'$ and $(x^2)' + (3)'$. What do you observe?

c. Suppose now that f and g are arbitrary differentiable functions. The preceding calculations suggest that $f + g$ is differentiable and that there is a relationship between $(f + g)'$, f', and g'. What relationship between $(f + g)'$, f', and g' do the previous calculations suggest?

d. In order to check your guess in c, consider $S(x) = f(x) + g(x)$. What limit should you compute in order to determine if S is differentiable? Rewrite this limit in terms of f and g. Use the hypothesis on f and g to finish the computation of the derivative of S. In the process, you proved that $f + g$ is differentiable. Compare this computation's result with your guess in c and fill in the blank: $(f + g)' =$ _____.

e. Use the formulas you obtained so far in Problems 2 and 3 to compute $(f - g)'$ in terms of f' and g'. Fill in the blank: $(f - g)' =$ _____.

f. Use the rules you have obtained to compute the derivatives of the following functions: $2x^3 - 4x$, $x^{99} + 9$, $5x^6 + 2x^4 - 4x^3$, $(-x + 2)^2$, $(x - 1)(2x^2 + 3)$.

g. Recall that a polynomial function is a function of the form $f(x) = a_n x^n + a_{n-1}x^{n-1} + \cdots + a_1 x + a_0$, where n is a nonnegative integer and $a_n, a_{n-1}, \ldots, a_1, a_0$ are real numbers. What is the derivative of f?

4. Product of Two Functions

Consider two differentiable functions f and g. The goal is to show that $f \cdot g$ is differentiable and to find a formula for $(f \cdot g)'$.

a. Is it true that $(f \cdot g)' = f' \cdot g'$ for any differentiable functions f and g? Check to see if the identity holds for the following choice of f and g: $f(x) = x$ and $g(x) = x^2$.

b. Denote by $P(x)$ the product of $f(x)$ and $g(x)$, that is, $P(x) = f(x) \cdot g(x)$. What limit should you compute in order to prove that P is differentiable? Rewrite this limit in terms of f and g.

c. Check that for $h \neq 0$, the following identity holds:

$$\frac{f(x + h)g(x + h) - f(x)g(x)}{h} \tag{8}$$

$$= \frac{f(x + h) - f(x)}{h} g(x + h) + f(x) \frac{g(x + h) - g(x)}{h}.$$

d. What does it mean for f and g to be differentiable? What is $\lim\limits_{h \to 0} g(x + h)$?

e. Use d and (8) to compute P'. Fill in the blank: $(f \cdot g)' = $ _____.

f. Verify your formula for the choice of functions $f(x) = x^3$ and $g(x) = x^6$. When taking the derivative of the product $f(x) \cdot g(x) = x^9$, use the rule determined in Problem 1.

5. The Function $\frac{1}{x}$

The objective here is to prove that the function $f(x) = \frac{1}{x}$ is differentiable on $(-\infty, 0) \cup (0, \infty)$ and to compute its derivative. The graph of f is sketched here.

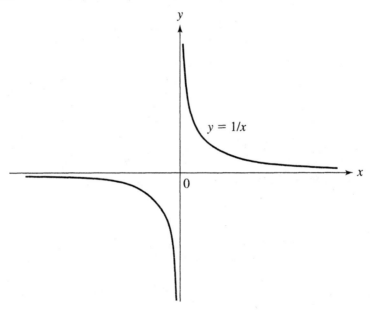

$y = 1/x$

a. Starting from the graph of $f(x)$, sketch roughly the graph of the derivative $f'(x)$.

b. Write the difference $f(x + h) - f(x)$ as a single quotient and simplify as much as possible.

c. Use Definition 3.2.1 and the simplified expression from b to show that f is differentiable on $(-\infty, 0) \cup (0, \infty)$ and to compute its derivative. Fill in the blank: $\left(\dfrac{1}{x}\right)' = $ _____.

d. Use a graphing calculator to graph $f'(x)$. Compare it with the graph of $f'(x)$ you sketched earlier.

6. More General Reciprocals

You have seen that the function $\frac{1}{x}$ is differentiable on $(-\infty, 0) \cup (0, \infty)$ and that its derivative is $-\frac{1}{x^2}$. Are the functions $\frac{1}{x^2}$ and $\frac{1}{x^2+3}$ differentiable on $(-\infty, 0) \cup (0, \infty)$? Or more generally, if f is a differentiable function satisfying $f(x) \neq 0$ for all x in its domain, can we conclude that $\frac{1}{f}$ is differentiable? Moreover, if these functions are differentiable, what are their derivatives? To answer these questions, do the following:

a. Denote by g the reciprocal of f; that is, $g(x) = \frac{1}{f(x)}$ for all $x \in \text{Domain}(f)$. Show that for $h \neq 0$,

$$\frac{g(x+h) - g(x)}{h} = \frac{-1}{f(x+h)f(x)} \cdot \frac{f(x+h) - f(x)}{h}.$$

b. Compute $\lim_{h \to 0} f(x+h)$ and $\lim_{h \to 0} \frac{1}{f(x+h)f(x)}$.

c. Use a and b to prove that g is differentiable and to fill in the blank:

$$\left(\frac{1}{f}\right)' = \underline{\hspace{3cm}}.$$

d. Use the rules you have obtained so far to compute $\left(\frac{1}{x^2}\right)'$ and $\left(\frac{1}{x^2+3}\right)'$.

Remark. If one knows in advance that $\frac{1}{f}$ is differentiable, an alternative computation of its derivative is as follows. Set again $g(x) = \frac{1}{f(x)}$. Take the derivative of the identity $f(x) \cdot g(x) = 1$ and use Problem 4 to obtain $f'(x) \cdot g(x) + f(x) \cdot g'(x) = 0$. Next, solve for $g'(x)$ in the last identity and then replace $g(x)$ by $\frac{1}{f(x)}$ to conclude that

$$g'(x) = -\frac{f'(x) \cdot g(x)}{f(x)} = -\frac{f'(x) \cdot \frac{1}{f(x)}}{f(x)} = -\frac{f'(x)}{[f(x)]^2}.$$

7. Negative Powers of x

In Problem 1, you obtained a formula for the derivative of x^n when n is a nonnegative integer. Now you treat the case when the integer n is negative. Fix such an n and recall that, by definition, $x^n = \frac{1}{x^{-n}}$ for $x \neq 0$. Use the formulas obtained in Problems 1 and 6 to prove that x^n is differentiable and that $(x^n)' = nx^{n-1}$.

8. Quotient of Two Functions

Consider two differentiable functions f and g. Suppose that $g(x) \neq 0$ and let $Q(x) = \frac{f(x)}{g(x)}$. The goal is to prove that the function Q is differentiable and to find the relationship between Q', f, g, f', and g'.

a. Start with $Q(x) = f(x) \cdot \frac{1}{g(x)}$ and use Problems 4 and 6 to prove that Q is differentiable.

b. Use a and the formulas from Problems 4 and 6 to compute Q'. Fill in the blank: $\left(\frac{f}{g}\right)' = $ _____.

c. Compute the derivatives of $\frac{1}{x+1}$, $\frac{x}{x^3+1}$, and $\frac{x^3-2x}{x^2-1}$ using the rules you have obtained so far.

9. The Square Root of a Function

Suppose that f is a positive, differentiable function. The goal of this discussion is to prove that $\sqrt{f}$ is differentiable and to find a formula for $(\sqrt{f})'$.

a. Define the function $g(x) = \sqrt{f(x)}$. Show that for $h \neq 0$,

$$\frac{g(x+h) - g(x)}{h} = \frac{f(x+h) - f(x)}{h} \cdot \frac{1}{\sqrt{f(x+h)} + \sqrt{f(x)}}.$$

b. Compute $\lim\limits_{h \to 0} \dfrac{1}{\sqrt{f(x+h)} + \sqrt{f(x)}}$.

c. Use a and b to show that g is differentiable and to fill in the blank: $(\sqrt{f})' = $ _____.

d. Evaluate the derivatives of $\sqrt{x}$ and $\sqrt{x^4 - 3x^2 + 100}$.

Remark.   If one knows in advance that $\sqrt{f}$ is differentiable, an alternative computation of its derivative is as follows. Set again $g(x) = \sqrt{f(x)}$. Take the derivative of the identity $f(x) = g(x) \cdot g(x)$ to obtain $f'(x) = g'(x)g(x) + g(x)g'(x) = 2g(x)g'(x)$. Next, solve for $g'(x)$ in the last identity; then replace $g(x)$ by $\sqrt{f(x)}$ to conclude that $g'(x) = \dfrac{f'(x)}{2g(x)} = \dfrac{f'(x)}{2\sqrt{f(x)}}$.

10. Rational Powers of x

You have shown in Problem 9 that $(\sqrt{x})' = \frac{1}{2\sqrt{x}}$. Write this identity as $(x^{\frac{1}{2}})' = \frac{1}{2}x^{-\frac{1}{2}}$. Is it just a coincidence that the rules proved in Problems 1 and 7 also hold for the fractional power $\frac{1}{2}$? It turns out that the formula $(x^a)' = ax^{a-1}$ holds for any real number a. We will refer to this rule as **the power rule** and will use it without a proof. Since we have not yet discussed irrational powers, we restrict our attention to the case where a is rational. Use the power rule to compute the derivatives of $x^{\frac{2}{3}}$, $x^{-\frac{1}{3}}$, and $x^{\frac{6}{5}}$. ◆

The formulas for finding derivatives proved in this section are summarized here. The functions f and g are assumed to have derivatives. All functions appearing in the denominators are assumed to be nonzero.

Rule	Function	Derivative
Power rule	x^a, a rational	ax^{a-1}
Constant multiple rule	$cf(x)$, c real	$cf'(x)$
Sum rule	$f(x) + g(x)$	$f'(x) + g'(x)$
Reciprocal rule	$\dfrac{1}{f(x)}$	$-\dfrac{f'(x)}{f^2(x)}$
Product rule	$f(x)g(x)$	$f'(x)g(x) + f(x)g'(x)$
Quotient rule	$\dfrac{f(x)}{g(x)}$	$\dfrac{f'(x)g(x)-f(x)g'(x)}{g^2(x)}$
Square-root rule	$\sqrt{f(x)}$	$\dfrac{f'(x)}{2\sqrt{f(x)}}$

Historical Note: Gottfried Wilhelm von Leibniz (1646–1716; from Leipzig, Saxony [now Germany])

Gottfried Leibniz entered the university in Leipzig at the age of fifteen; at seventeen, he earned his bachelor's degree. He studied theology, law, philosophy, and mathematics and is sometimes regarded as the last scholar to achieve universal knowledge. In 1667, he received a doctorate in law from the University of Altdorf after which he entered the diplomatic service. During his travels to Paris (1672) and London (1673) as a diplomat, Leibniz made contacts with mathematicians and philosophers there. Amongst them, Christian Huygens (1629–1695) had the most influence. By the time Leibniz visited London again in 1676, he had the main ideas of differential calculus. He had also come up with the notation dx for the differential in x; he later used the symbol $\int y\,dx$, the integral sum being an enlarged letter s (for sum). He published the first account of differential calculus in 1684 under the title "Nova Methodus pro Maximis et Minimis, Itemaque Tangentibus, qua nec Irrationales Quantitates Moratur" (A new method for maxima and minima, and also for tangents, which is not obstructed by irrational quantities). Here Leibniz gave some of the most basic rules for differentiation, written in the form $d(xy) = x\,dy + y\,dx$, $dx^n = nx^{n-1}dx$, and $d(\frac{x}{y}) = \frac{y\,dx - x\,dy}{y^2}$, together with geometric applications. Leibniz's other mathematical contributions include work on infinite series, the multinomial theorem, and logic.

EXERCISES 3.4

In Exercises 1–20, compute the derivative of the given function by using the formulas obtained in this section. At each step specify the formula you applied.

1. $f(x) = 3x^2 - 4x + 2$

2. $f(x) = -4x^6 + 3x^2 + x$

3. $g(t) = 3 - 6t + 10t^{2,004}$

4. $g(t) = 3t^{-301} + 5t^5 + 7$

5. $f(s) = (s - 2)(2s + 3)$

6. $f(x) = (x - 1)^3$

7. $g(t) = -4 - t(5 - t)$

8. $h(x) = (x - 1)(x + 1)(x^2 + 1)$

9. $h(t) = -\frac{2}{t}$

10. $A(l) = 3l + \frac{5}{l}$

11. $h(t) = -\frac{1}{t^3}$

12. $f(x) = \frac{1}{x-1}\left(3 - \frac{1}{x+1}\right)$

13. $f(x) = \frac{1}{x^2+3x}$

14. $h(x) = -\frac{1}{(2x-1)^2}$

15. $f(x) = \frac{x+2}{x+1}$

16. $f(x) = \frac{x^2}{x+1}$

17. $f(x) = \sqrt{x^2 + 3x}$

18. $g(x) = 2\sqrt{x^3} - 2x$

19. $h(x) = \sqrt{6}$

20. $g(t) = t\sqrt{t}$

In Exercises 21–32, compute the derivative of the given function.

21. $f(s) = \sqrt{s(s - 1)}$

22. $h(x) = x^2\sqrt{1 - x}$

23. $A(x) = x\sqrt{4 - x^2}$

24. $f(t) = t\sqrt{t^3 + t}$

25. $f(s) = \frac{s}{\sqrt{s+1}}$

26. $g(x) = \sqrt{\frac{x+1}{x-1}}$

27. $f(x) = (x^3 + x^2 - 3x)(2x + 1)$

28. $h(x) = x\left(2 - \frac{2}{x-1}\right)$

29. $g(x) = \frac{3x^{\frac{1}{3}}}{x^2}$

30. $g(t) = \sqrt[3]{t}(\sqrt{t} - 2t)$

31. $f(x) = \frac{x^2+3x}{2\sqrt{x}}$

32. $h(t) = \frac{\sqrt{t}-1}{t+2}$

33. Determine the equation of the line tangent to the curve $y = \sqrt{x^2 + 9}$ at the point $(4, 5)$.

34. Determine the equation of the line tangent to the curve $y = \frac{x^2-3x+4}{4x^3-2x-1}$ at the point corresponding to $x = 1$.

Determine the intervals where each of the functions in 35–42 is increasing, and determine the intervals where it is decreasing.

35. $f(x) = \frac{1}{x+2}$

36. $f(x) = x - \frac{1}{x}$

37. $g(x) = x + \frac{1}{x^2}$

38. $g(x) = x^2 + \frac{2}{x}$

39. $f(x) = \frac{x-1}{x+1}$

40. $f(x) = \frac{x+3}{x-2}$

41. $f(x) = \frac{x+1}{x^2+2}$

42. $f(x) = \frac{x^2+1}{x-1}$

43. Let f and g be differentiable functions such that $g(x) \neq 0$. You have proved that the function $Q(x) = \frac{f(x)}{g(x)}$ is differentiable and that $Q' = \frac{f'g - fg'}{g^2}$. Under the assumption that Q is differentiable, use the identity $g(x) \cdot Q(x) = f(x)$ and the product rule to give an alternative proof for the quotient rule.

44. Compute the derivative of $f(x) = (x - 1)(x + 1)(x^2 + 1)(x^4 + 1) \cdots (x^{1,024} + 1)$.

Projects and Extensions 3.4

Polynomial Functions with Double Roots

Let $f(x)$ be a polynomial function. A real number a is called a *root of f* if $f(x) = 0$. It is known that if a is a root of f, then $f(x) = (x - a)g(x)$ for some polynomial function $g(x)$. In this spirit, we say that a is a double root of f if there exists a polynomial function $g(x)$ such that $f(x) = (x - a)^2 g(x)$ for all real numbers x. This project's goal is to investigate polynomial functions with double roots.

1. Prove that if the real number a is a double root of f, then a is a root of both f and f'.

2. Prove that if the real number a is a root of both f and f', then a is a double root of f.

3. Let $a_1, a_2, \ldots, a_n$ be distinct real numbers. Prove that there exists a polynomial function f of degree $2n$ such that $f(a_i) = f'(a_i) = 0$ for $i = 1, 2, \ldots, n$.

4. Prove that a polynomial function f of degree $2n$ satisfying the conditions in Problem 3 is unique if we also set $f(0) = c$ for some given real number c.

3.5 THE CHAIN RULE

The chain rule • The general power rule • Applications

Consider the function $k(x) = (x^3 - 2x + 7)^{100}$. To compute $k'(x)$ using only the properties of derivatives obtained up to this point, you need to first expand $(x^3 - 2x + 7)^{100}$ and then take the derivative of the polynomial function you obtain. This is a time-consuming computation. On the other hand, k is a composite function: $k(x) = f(g(x))$, where $f(y) = y^{100}$ and $g(x) = x^3 - 2x + 7$. The derivatives of f and g can be computed quickly: $f'(y) = 100y^{99}$, $g'(x) = 3x^2 - 2$. If the derivatives f' and g' could be combined in an easy way to get k', then we could save a lot of time.

This brings us to the central question of this section: Is there a formula that relates the derivative of the composition of two functions to the derivatives of the functions themselves?

Classroom Discussion 3.5.1: Derivatives of Composite Functions

Start your investigation with simpler examples.

1. Let $k(x) = (2x + 1)^2$

   a. Expand $k(x)$ and compute $k'(x)$.
   b. If $g(x) = 2x + 1$, determine $f(y)$ such that $k(x) = f(g(x))$.
   c. Compute $f'(y)$ and $g'(x)$.
   d. Is it true that $k'(x) = f'(y) \cdot g'(x)$? How about $k'(x) = f'(g(x)) \cdot g'(x)$? Explain which equality is not true and why.

2. Let $k(x) = (x^2 + x)^2$.

   a. Expand $(x^2 + x)^2$ and use the resulting expression to compute $k'(x)$.
   b. Determine $g(x)$ and $f(y)$ so that $k(x) = f(g(x))$.
   c. Compute $f'(y)$ and $g'(x)$.
   d. What is the expression for $f'(g(x))$?
   e. Evaluate $f'(g(x)) \cdot g'(x)$.
   f. Compare $k'(x)$ with $f'(g(x)) \cdot g'(x)$.

3. In Examples 1–2, you saw that $(f(g(x)))' = f'(g(x)) \cdot g'(x)$. Actually this formula works for any differentiable functions f and g for which the $f \circ g$ and $f' \circ g$ make sense. For a proof, fix two such functions and use the following outline:

   a. Fix a real number x contained in the domain of $f \circ g$ and of $f' \circ g$ and let $k(x) = f(g(x))$. By definition,

   $$k'(x) = \lim_{t \to 0} \frac{k(x + t) - k(x)}{t} = \lim_{t \to 0} \frac{f(g(x + t)) - f(g(x))}{t}.$$

   b. The last fraction in a can be written as

   $$\frac{f(g(x + t)) - f(g(x))}{t} = \frac{f(g(x + t)) - f(g(x))}{g(x + t) - g(x)} \cdot \frac{g(x + t) - g(x)}{t}.$$
   (9)

   To show that k' exists, it suffices to prove that the limit as $t \to 0$ of each fraction on the right-hand side exists. We will treat one fraction at a time.

   c. The limit as $t \to 0$ of the second fraction is

   $$\lim_{t \to 0} \frac{g(x + t) - g(x)}{t} = g'(x).$$

   d. To find the limit as $t \to 0$ of the first fraction, observe that since g is continuous, $\lim_{t \to 0} g(x + t) = g(x)$. Hence, if we define h as

$h = g(x + t) - g(x)$, then $h \to 0$ as $t \to 0$. Now $g(x + t) = g(x) + h$, and

$$\lim_{t \to 0} \frac{f(g(x + t)) - f(g(x))}{g(x + t) - g(x)} = \lim_{h \to 0} \frac{f(g(x) + h) - f(g(x))}{h} = f'(g(x)).$$

e. The conclusion is that

$$k'(x) = \lim_{t \to 0} \frac{f(g(x + t)) - f(g(x))}{g(x + t) - g(x)} \cdot \lim_{t \to 0} \frac{g(x + t) - g(x)}{t}$$

$$= f'(g(x)) \cdot g'(x).$$

Observation.   There is one problem with the reasoning in Problem 3: the identity (9) is true only when $g(x + t) \neq g(x)$ for t close to zero. To remedy this problem, proceed as follows.

Case I: Suppose there exists some (possibly small) number $\delta > 0$ such that $g(x + t) \neq g(x)$ for all $|t| < \delta$. In this case, repeat the reasoning in Problem 3 under the assumption that $|t| < \delta$. The restriction on t does not affect the validity of the formula obtained in e since t gets arbitrarily small as $t \to 0$.

Case II: Suppose that for each $\delta > 0$ (no matter how small), we can always find some t, such that $|t| < \delta$ and $g(x + t) = g(x)$. In this case, define two sets;

$$A = \{t : g(x + t) = g(x)\} \quad \text{and} \quad B = \{t : g(x + t) \neq g(x)\}.$$

Claim 3.5.1: Assume that $\mathbb{R}$ is partitioned into two sets A and B, each of which contain arbitrarily small numbers. Then, given an arbitrary function $F(t)$, $\lim_{t \to 0} F(t)$ exists and is equal to a real number L if and only if $\lim_{t \to 0, t \in A} F(t)$ and $\lim_{t \to 0, t \in B} F(t)$ exist and are equal to L. As expected, we use the notation $\lim_{t \to 0, t \in A}$ whenever we compute the limit $\lim_{t \to 0}$ and in addition we require the points t to belong to a given set A.

While we accept this claim without a proof, it is worth pointing out that, informally speaking, these limits, with restrictions on t, can be thought of as generalizations of the left-hand and right-hand limits we defined in Section 2.2. Then, since g is differentiable at x, we can write

$$g'(x) = \lim_{t \to 0} \frac{g(x + t) - g(x)}{t} = \lim_{t \to 0, t \in A} \frac{g(x + t) - g(x)}{t} = 0. \qquad (10)$$

The second equality in (10) follows from Claim 3.5.1, while the last equality follows from the definition of A. One more use of the definition of A implies that

$$\lim_{t \to 0, t \in A} \frac{k(x + t) - k(x)}{t} = \lim_{t \to 0, t \in A} \frac{f(g(x + t)) - f(g(x))}{t} = 0. \qquad (11)$$

In addition, since by Claim 3.5.1 we have $\displaystyle\lim_{t\to 0,\,t\in B}\frac{g(x+t)-g(x)}{t}=\lim_{t\to 0,\,t\in A}\frac{g(x+t)-g(x)}{t}$,

we conclude that $\displaystyle\lim_{t\to 0,\,t\in B}\frac{g(x+t)-g(x)}{t}=0$. Hence, if we reason as in Case I for $t\in B$, we have

$$\lim_{t\to 0,\,t\in B}\frac{k(x+t)-k(x)}{t}=\lim_{t\to 0,\,t\in B}\frac{f(g(x+t))-f(g(x))}{g(x+t)-g(x)}$$

$$\cdot\ \lim_{t\to 0,\,t\in B}\frac{g(x+t)-g(x)}{t}=0. \qquad (12)$$

Combining (11) and (12), we can conclude that

$$\lim_{t\to 0,\,t\in A}\frac{k(x+t)-k(x)}{t}=0=\lim_{t\to 0,\,t\in B}\frac{k(x+t)-k(x)}{t}.$$

Thus, one more use of Claim 3.5.1 gives that $k'(x)=\displaystyle\lim_{t\to 0}\frac{k(x+t)-k(x)}{t}$ exists and is equal to 0. In particular, $(f(g(x)))'=0=f'(g(x))\cdot 0=f'(g(x))\cdot g'(x)$, so the identity we want also holds true in this case. ◆

Proposition 3.5.1. Chain Rule

Suppose that $f(y)$ and $g(x)$ are two differentiable functions for which the composition $f(g(x))$ makes sense. Then the formula

$$(f(g(x)))'=f'(g(x))\cdot g'(x) \qquad (13)$$

holds for all x for which $f'(g(x))$ is defined.

Observation.   Using the composition notation, (13) can be written as $(f\circ g)'=(f'\circ g)\cdot g'$.

EXAMPLES

1. In a–c, use the chain rule to compute the derivative of the given function.

   a. $k(x)=(x^5+2x^2+3)^8$ **b.** $k(x)=\left(\dfrac{x-1}{x+2}\right)^9$ **c.** $k(x)=\sqrt[5]{x^2+2}$

Solutions

   a. Let $f(y)=y^8$, $g(x)=x^5+2x^2+3$. Then, $k(x)=f(g(x))$; since $f'(y)=8y^7$ and $g'(x)=5x^4+4x$, by the chain rule we have

$$k'(x)=f'(g(x))\cdot g'(x)=f'(x^5+2x^2+3)\cdot g'(x)$$
$$=8(x^5+2x^2+3)^7(5x^4+4x).$$

   b. Let $f(y)=y^9$ and $g(x)=\dfrac{x-1}{x+2}$. Then, $k(x)=f(g(x))$ and $f'(y)=9y^8$,

$$g'(x)=\frac{(x-1)'(x+2)-(x-1)(x+2)'}{(x-2)^2}=\frac{3}{(x-2)^2}.$$

The chain rule implies that

$$k'(x) = f'(g(x)) \cdot g'(x) = f'\left(\frac{x-1}{x+2}\right) \cdot g'(x) = 9\left(\frac{x-1}{x+2}\right)^8 \cdot \frac{3}{(x-2)^2}.$$

c. Let $f(y) = \sqrt[5]{y}$ and $g(x) = x^2 + 2$. Then, $k(x) = f(g(x))$ and $f'(y) = \frac{1}{5} \cdot y^{\frac{1}{5}-1} = \frac{1}{5} \cdot y^{-\frac{4}{5}}, g'(x) = 2x$. Hence, by the chain rule,

$$k'(x) = f'(g(x)) \cdot g'(x) = f'(x^2 + 2) \cdot g'(x) = \frac{1}{5} \cdot (x^2 + 2)^{-\frac{4}{5}} \cdot 2x$$

$$= \frac{2x}{5\sqrt[5]{(x^2 + 2)^4}}. \qquad \blacksquare$$

2. Melting Snowballs

Adrian loves ice cream. On a hot summer day, he put a big scoop of ice cream in his bowl and went outside to eat it. When Adrian had to answer the phone, he left his ice cream on the porch, and it started to melt. When the radius of the ice cream ball was 2 centimeters, the radius was changing at a rate of -1 centimeter per minute. What was the rate of change of the ice cream's volume with respect to time at the moment when the radius was 2 centimeters? Here, we assume that the ice cream is sphere shaped at any time.

Solution   The first step is to express in mathematical terms the information given. To do so, denote by $r(t)$ the radius of the ice cream ball at time t; r is a function of time. The formula for the ice cream's volume V in terms of the radius r is $V(r) = \frac{4\pi r^3}{3}$. Thus, we have a chain of functions. The volume V depends on the radius r, while the radius r depends on the time t.

Denote by t_0 the time when $r = 2$ centimeters. What we know is that $r'(t_0) = -1$ centimeter per minute; the minus sign shows that r is decreasing in time. We have to find the rate of change of the volume of the ice cream with respect to time at the moment t_0 when the radius was 2 centimeters; that is, $(V(r(t)))'$ evaluated at $t = t_0$. From the chain rule, it follows that $(V(r(t)))' = V'(r(t)) \cdot r'(t)$ for t in the domain of $V \circ r$. In particular,

$$(V(r(t)))'|_{t=t_0} = V'(r(t_0)) \cdot r'(t_0) = V'(2) \cdot (-1).$$

Since $V'(r) = 4\pi r^2$, we have that $V'(2) = 4\pi \cdot 4 = 16\pi$ square centimeters. Therefore, $(V(r(t)))'|_{t=t_0} = -16\pi$ cubic centimeters per minute.

Practice Problem

Use the chain rule to compute the derivatives of $k(x) = (x^3 - 2x + 7)^{100}$ and $l(x) = (x^2 - 2x + 4)^{\frac{2}{3}}$.

Classroom Discussion 3.5.2: General Power Rule

Let g be a differentiable function, and fix an arbitrary rational number a, such that g^a and g^{a-1} are well-defined functions.

1. Use the chain rule to show that for each rational number a,

$$((g(x))^a)' = a(g(x))^{a-1} \cdot g'(x).$$

This is a generalization of the power rule from Section 3.4 and is true for any real number a for which g^a and g^{a-1} are well-defined functions.

2. Use the general power rule to give an alternative proof to the square-root rule. ◆

EXERCISES 3.5

In Exercises 1–6, use the chain rule to compute $(f(g(x)))'$.

1. $f(y) = 3y^{51} - 4y^{39}, g(x) = 2 - 3x + x^3$
2. $f(y) = y^{35} + 2y^{16}, g(x) = x^2 + 2x - 1$
3. $f(y) = \frac{1}{y^2}, g(x) = -x^2 + 3$
4. $f(y) = \frac{1}{y}, g(x) = x^3 + 3x$
5. $f(y) = \sqrt[3]{y^4 + 6}, g(x) = 10 - 4x^3 + 5x$
6. $f(y) = \sqrt{y^2 + 6y}, g(x) = 1 + 4x - x^2$

In Exercises 7–18, use the chain rule to compute the derivative of the given function.

7. $(x^3 - 4x + 7)^{2,004}$
8. $(4x^2 + 5x + 3)^{1,001}$
9. $\sqrt[5]{x^2 - x - 1}$
10. $\sqrt[7]{x^{10} + x^9}$
11. $\frac{2}{(2x-1)^{1,001}}$
12. $\frac{3}{(2x^4+x)^{99}}$
13. $h(t) = \sqrt[5]{x^4 - 3x^2 + 4x}$
14. $f(x) = \sqrt{\frac{x}{x+1}}$
15. $f(t) = -\frac{5}{(t^2-t^3)^3}$
16. $f(x) = \left(\frac{4-2x}{x^2+x}\right)^4$
17. $f(x) = \sqrt[3]{\frac{x+2}{x-2}}$
18. $g(x) = \sqrt{\frac{x^3}{x^3+1}}$

19. Find an equation of the tangent line to the graph of $f(x) = \left(\frac{3x}{x+4}\right)^3$ at the point $(-1, -1)$.

20. Find an equation of the tangent line to the graph of $f(x) = \left(\frac{x}{x-2}\right)^4$ at the point $(1, 1)$.

In Exercises 21–24, compute the value of $(f \circ g)'(t)$ at the given value of t.

21. $f(y) = y^2 - 2, g(t) = \sqrt{t}, t = 9$
22. $f(y) = y^3 - 5y^2 + 1, g(t) = \sqrt{t - 1}, t = 5$
23. $f(y) = \frac{y-2}{y}, g(t) = t^4 - t^2, t = 2$

24. $f(y) = \frac{y+1}{y^2+1}, g(t) = \frac{1}{t}, t = -1$

25. A person standing on a riverbank observes an empty boat floating down the middle of the river. The river is 100 meters wide, and the water flows at a rate of 1 meter per second. Let D be the distance between the person and the boat at time t. Determine the rate of change of D with respect to time: 2 seconds prior to, 3 seconds after, and 10 minutes after the moment the boat passes in front of the person.

Hint: You must compute $(D(s(t)))'$ at the specified times knowing that $s'(t) = 1$ meter per second.

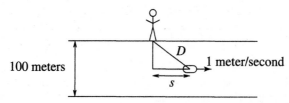

26. On a hot summer day, a spherically shaped scoop of ice cream is left outside in a bowl. The ice cream starts melting. What is the rate of change with respect to time of this ice cream ball's radius at the moment when the radius is 3 centimeters, if the ice cream's volume is changing at a rate of -5 cubic centimeters per minute?

27. In Section 3.3, you studied the motion of the rocket launched by Adrian. Recall that the rocket's height in feet is $h(t) = -16t^2 + 128t$ when measured from the ground t seconds after launching. Diana watches the launching of the rocket from a distance of 20 feet from the launcher. Set D as the distance from Diana to the rocket at time t. Compute the rate of change of D with respect to time at $t = 2$ seconds and $t = 5$ seconds after launching.

28. A function f is even if for any x in the domain of f, $-x$ also belongs to the domain of f and if $f(-x) = f(x)$. Similarly, a function g is odd if for any x in the domain of g, $-x$ also belongs to the domain of g and if $g(-x) = -g(x)$. Use the chain rule to show that the derivative of a differentiable even function is an odd function, while the derivative of a differentiable odd function is an even function.

29. Suppose $f(x) = x^3 + \sqrt{x^2 + 1}$ and g, h are two functions such that $g'(x) = f(x - 1), g(1) = 2, h'(x) = x^5 - 5x^3 + 7,$ and $h(1) = -1$. Compute $(f \circ g)'(1)$ and $(h \circ f)'(0)$.

30. Let $f(x) = x^4 + 5x^3 - 6x$ and let $k(x) = f(x^3)$. Compute $k'(x^3)$.
Warning: $k'(x^3) \neq (k(x^3))'$.

CHAPTER 3 REVIEW

The average rate of change of a function f over an interval $[a, b]$ is by definition the value of $\frac{f(b)-f(a)}{b-a}$ (see Definition 3.1.1).

The derivative of a function f at a point x, denoted by $f'(x)$, is the limit $\lim\limits_{h \to 0} \frac{f(x+h)-f(x)}{h}$ (also called the *instantaneous rate of change of f at x*), provided it exists. A function for which $f'(x)$ exists is differentiable at x. Alternatively, f is differentiable at x provided $\lim\limits_{y \to x} \frac{f(y)-f(x)}{y-x}$. In addition, if f is differentiable at x, then $f'(x)$ equals the slope of the tangent line to the graph of f at the point $(x, f(x))$. The sign of the derivative $f'(x)$ provides information about the behavior of the function f, while the behavior of the function f provides information about the sign of $f'(x)$. More precisely, if f is a differentiable function on (a, b), then the following are true.

1. f is nonincreasing on (a, b) if and only if $f' \leq 0$ on (a, b).
2. f is nondecreasing on (a, b) if and only if $f' \geq 0$ on (a, b).
3. If $f' < 0$ on (a, b), then f is decreasing on (a, b).
4. If $f' > 0$ on (a, b), then f is increasing on (a, b).
5. The tangent line to the graph of f at $(c, f(c))$ is horizontal if and only if $f'(c) = 0$.

While a function that is continuous at a point c is not necessarily differentiable at c, the converse is always true: a function that is differentiable at a point c is also continuous at c.

If $h(t)$ denotes the position function at time t of an object moving along a straight line, then $v(t) = h'(t)$ is the velocity of the object at time t, $s(t) = |v(t)|$ is the object's speed at time t, while $a(t) = v'(t)[= h''(t)]$ is the object's acceleration at time t.

The following rules are very useful for computing derivatives of various functions. Here the functions f and g are supposed to be differentiable, the functions appearing in the denominators are assumed to be nonzero, and f is assumed to be positive when considering the $(\sqrt{f})'$.

- Power rule: $(x^a)' = ax^{a-1}$ for any rational number a
- Constant multiple rule: $(cf)' = c(f)'$ for any real number c
- Sum and difference rule: $(f \pm g)' = f' \pm g'$
- Product rule: $(f \cdot g)' = f' \cdot g + f \cdot g'$
- Reciprocal rule: $\left(\dfrac{1}{f}\right)' = -\dfrac{f'}{f^2}$
- Quotient rule: $\left(\dfrac{f}{g}\right)' = \dfrac{f' \cdot g - f \cdot g'}{g^2}$
- Square-root rule: $(\sqrt{f})' = \dfrac{f'}{2\sqrt{f}}$

An important rule proved in this chapter is the chain rule (see Proposition 3.5.1), which says that for any differentiable functions f and g,

$$(f(g(x)))' = f'(g(x)) \cdot g'(x)$$

at any point x where both $f(g(x))$ and $f'(g(x))$ are defined. Applying this formula, we obtain the general power rule that says that for any rational number a and any differentiable function g,

$$(g^a(x))' = a \cdot g^{a-1}(x) \cdot g'(x)$$

at any point x where both $g^a(x)$ and $g^{a-1}(x)$ are defined.

CHAPTER 3 REVIEW EXERCISES

In Exercises 1–2, compute the average rate of change of f over the interval $[a, b]$. Write the equation of the line containing the points $(a, f(a))$ and $(b, f(b))$.

1. $f(x) = \sqrt{x + 2}, a = 2, b = 7$ **2.** $f(x) = \frac{2}{x}, a = 2, b = 3$

3. Use the graph of f to sketch the graph of its derivative.

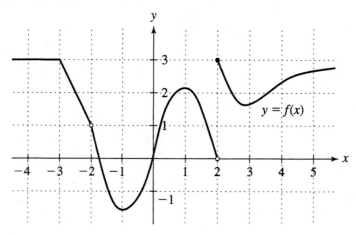

4. Use the graph of f to sketch the graph of its derivative.

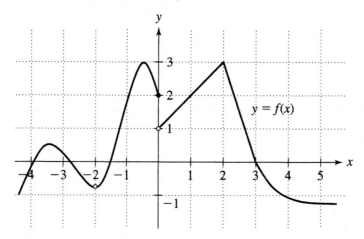

5. The velocity of an ice skater moving along a horizontal straight line is $v(t) = 2(t - 3)^2 - 2$ feet per second, where t is the time in seconds.

 a. Plot the velocity of the skater for t between 0 and 8 seconds, and determine his initial speed.

 b. Determine the acceleration of the skater as a function of time, and calculate his initial acceleration.

 c. When does the skater reverse his motion?

 d. Based on the results just obtained, describe the skater's motion.

 e. At what moment is the skater's velocity minimal, and what is its velocity then?

6. The velocity of an ice skater moving along a horizontal straight line is $v(t) = t^2 - 6t + 5$, where t is the time in seconds from the starting point.
   a. Plot the velocity of the skater for t between 0 and 10 seconds, and determine his initial speed.
   b. Determine the skater's acceleration as a function of time, and calculate his initial acceleration.
   c. Based on the results obtained in a, describe the skater's motion. In particular, determine the times when the skater reverses his motion.
   d. At what moment is the skater's velocity minimal, and what is his velocity then?

7. The height (in feet) above the ground of a coin dropped from the top of the Empire State Building is $h(t) = 1{,}250 - 16t^2$, t seconds into the fall.
   a. What is the coin's velocity, speed, and acceleration at time t?
   b. How long will it take for the coin to hit the ground?
   c. Determine the velocity of the coin when it hits the ground.

8. The height of a coin dropped from a hot air balloon is $h(t) = 4{,}000 - 16t^2$ feet above the ground t seconds into the fall.
   a. How long will it take for the coin to hit the ground?
   b. Determine the velocity and acceleration of the coin when it hits the ground.

9. A rock thrown vertically upward from the Earth's surface reaches a height of $h(t) = 72t - 16t^2$ feet in t seconds.
   a. How high does the rock go?
   b. How long is the rock aloft?
   c. What is the rock's velocity when it hits the ground?

10. A rock thrown vertically upward from the surface of the moon reaches a height of $h(t) = 72t - 2.6t^2$ feet in t seconds.
    a. How high does the rock go?
    b. How long is the rock aloft?
    c. What is the rock's velocity when it hits the ground?

In Exercises 11–32, compute the derivative of the given function.

11. $g(x) = 3x^4 + 2x - 6x^{-2}$

12. $f(x) = 3x^{11} + \sqrt{2}x^2 - \frac{\pi}{\pi+1}$

13. $f(x) = 5 - 0.2x^5 - 2(x^3 - \pi^2)$

14. $h(x) = (3 - x^2)(2x + 3)$

15. $f(x) = (2x^4 - x + 2)(x^3 - x)$

16. $g(x) = (x^2 + x - 3)(5x^2 + 3x + 6)$

17. $f(x) = \frac{x^2+2}{x^3+2x-1}$

18. $f(x) = \frac{3x^4-4x^2-3}{1+x+x^2}$

19. $g(t) = (t - 1)^{-1}(t^2 + 2)$

20. $g(x) = (2x + 3)^3(x - 1)^{-4}$

21. $f(t) = (t^2 - 3t)^{10}$

22. $g(s) = (s^4 - 3s^2 + 10)^{-21}$

23. $f(t) = 3\left(\sqrt{t - 1} - \frac{1}{\sqrt{t}}\right)$

24. $g(x) = x^2\sqrt{3x - \sqrt{x}}$

25. $h(x) = \sqrt{1 + (x^3 + x - 2)^5}$

26. $f(x) = \sqrt{\frac{x^2-x}{x^2}}$

27. $f(x) = \frac{\sqrt{x}+1}{\sqrt{x}-1}$

28. $h(x) = \frac{\sqrt{x}}{1+\sqrt{x}}$

29. $f(x) = (\sqrt{x} + 2)^{-1}$

30. $f(t) = \left(\dfrac{t}{2 + \sqrt{t}}\right)^2$

31. $f(x) = \left(\dfrac{x^3}{3} - x^2 + \dfrac{2}{x}\right)^5$

32. $h(x) = \left(\dfrac{3\sqrt{x+1}}{2 + 3\sqrt{x+1}}\right)^2$

In Exercises 33–36, compute the value of $(f \circ g)'(t)$ at the given value of t.

33. $f(y) = y^4 - 3y, g(t) = \sqrt{t}, t = 4$

34. $f(y) = -y^5 + 6y^2 + 2, g(t) = \sqrt{t + 3}, t = 6$

35. $f(y) = \dfrac{y}{y+1}, g(t) = t^3 - 3t + 1, t = 0$

36. $f(y) = \dfrac{2}{y^2 - 3}, g(t) = \dfrac{1}{t}, t = 1$

37. A spherical balloon is inflated with gas. The balloon's volume is changing at a rate of 10 cubic feet per minute. How fast is the balloon's radius changing at the instant when the radius is 1.5 feet?

38. Andrew is dropping pebbles into a calm pool of water, creating ripples in the form of concentric circles. The radius of the outer ripple is increasing at a rate of 0.5 feet per second. How fast is the area inside the outer ripple changing at the instant the outer ripple's radius is 3 feet?

More on Differentiation

4.1 OPTIMIZATION
4.2 CURVE SKETCHING
4.3 EXPONENTIAL CHANGE

4.1 OPTIMIZATION

Critical points • Local and absolute minima • Local and absolute maxima • Solving optimization problems

So far, you have seen that derivatives are very useful for solving problems involving motion. In this section, you will see that derivatives also play an important role in *optimization* problems. These are problems in which we try to maximize or minimize some quantity. For example, we might want to determine the least amount of material, the greatest profit, or the largest enclosed area. Such tasks usually involve finding the smallest or largest value of some function defined on a certain domain.

Classroom Discussion 4.1.1: Largest Area

The following exploration is from pages 39–40 in the eighth-grade textbook *Mathematics in Context*, *Get the Most Out of It*. Work through tasks 1–5.

F. HYPERBOLAS

In Section C, you worked with a feasible region that has a curved border.

That is not uncommon in real-world problems. In this section, you will look at another problem involving a curve.

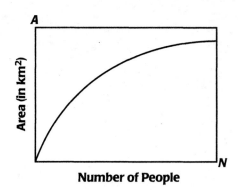

Number of People

In the 19th century, many adventurers traveled to North America to search for gold. A man named Dan Jackson owned some land where gold had been found. Instead of digging for the gold himself, he rented plots of land to the adventurers. The "rent" was to give Dan 50% of any gold found on the plot of land.

Dan gave each adventurer four stakes and a rope that was exactly 100 meters long. Each adventurer had to use the stakes and rope to mark off a rectangle with north-south and east-west sides.

1. Did everyone get the same area to dig for gold? Explain your answer.

Get the Most Out of It

There are many different rectangles you can make that have a perimeter of 100 meters.

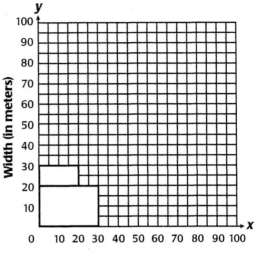

Length (in meters)

2. **a.** On graph paper, draw at least five different rectangles with perimeters of 100 meters. Draw your rectangles to scale, using one grid unit to represent five meters.

   b. Cut out the rectangles. Tape them on a graph so the lower-left corners lie on the origin. Use the same grid size as in part a, as shown at the right.

For each rectangle, only the upper-right vertex will not touch an axis.

3. You should notice that all the upper-right vertices of your rectangles seem to lie on a line. What is the equation of that line? Use *x* and *y* as the dimensions of the rectangles.

4. Calculate the area of each rectangle, and organize all of your information in a table.

One of the diggers discovered that one kind of rectangle always had the greatest area. He decided to sell the secret to other diggers.

5. What was the secret?

1. Does the reasoning you used for answering Question 5 constitute a complete proof? How would you show that no other rectangle with a perimeter of 100 meters will have an area larger than the rectangle you discovered when answering Question 5? Do this in two ways.

    a. Give a geometric proof using the following figure.

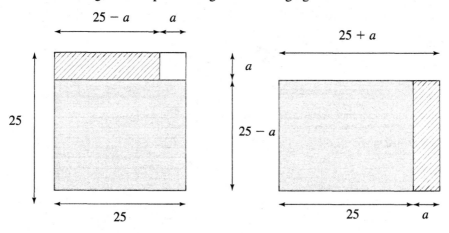

    b. Give an algebraic proof by showing that $(25 - a)(25 + a) \le 625$.

2. Complete the following tasks to tackle Question 5 by other means.

    a. Denote by x and y the dimensions of a plot of land measured in meters. Use the information about the rope's length to write an identity involving x and y. We refer to this identity as the **constraint** for x and y.

    b. Let A denote the area, measured in square meters, of such a rectangle. Write A first in terms of x and y, and then use the constraint for x and y to express A as a function of x only. Call this new expression $A(x)$. What is the domain of $A(x)$?

    c. Use a calculator to graph $A(x)$. What do you see? Use your calculator's zoom and trace features to find the value of x for which $A(x)$ is the largest. How does your answer compare to the answer you obtained in Problem 1?

3. **Derivatives at Work**

Return to the function $A(x)$ from Problem 2. You should have seen that the domain of $A(x)$ is the interval $(0, 50)$ and that the graph of A is a parabola. Use calculus to determine the coordinates of the highest point on this parabola, which corresponds to the value of x that gives the largest area.

    a. What slope does the tangent line have at the highest point on the graph of A? What does this information tell you about the derivative $A'(x)$?

    b. Compute $A'(x)$ and graph it. Find the value x_0 in the interval $(0, 50)$ with the property that $A'(x_0) = 0$.

    c. What is the sign of $A'(x)$ to the left of x_0? How about to the right of x_0? What sort of information do you obtain about $A(x)$ from the sign of $A'(x)$? What can you say about the value of $A(x_0)$? How does this relate to Question 5 in Classroom Discussion 4.1.1? ◆

You have just seen an example of how you can use calculus to solve an optimization problem. You essentially used the connection between a function's behavior and the sign of its derivative. The idea is that if there exists a point x_0 in the domain of a function f such that $f'(x_0) = 0$, f' is positive to the left of x_0, and f' is negative to the right of x_0, then $f(x_0)$ is the largest value taken by f. In this case, f has a **maximum** at x_0, and $f(x_0)$ is the maximum value of f.

value of x	$x < x_0$	x_0	$x > x_0$
sign of $f'(x)$	$+$	$f'(x_0) = 0$	$-$
behavior of f	↗	$f(x_0)$ is the maximum value of f	↘

Similarly, if $f'(x_0) = 0$, f' is negative to the left of x_0, and f' is positive to the right of x_0, then $f(x_0)$ is the smallest value taken by f. In this case, f has a **minimum** at x_0, and $f(x_0)$ is the minimum value of f.

value of x	$x < x_0$	x_0	$x > x_0$
sign of $f'(x)$	$-$	$f'(x_0) = 0$	$+$
behavior of f	↘	$f(x_0)$ is the maximum value of f	↗

The values of x in the domain of f for which $f'(x) = 0$ are called **critical points**. In Classroom Discussion 4.1.1, the function $A(x) = 50x - x^2$, defined on the interval $(0, 50)$, had one critical point at $x = 25$.

EXAMPLES   Determine the critical points for the following functions:

1. $f(x) = x^2 - 5x + 6$ with Domain$(f) = (-2, 10)$.
2. $g(x) = x^3 - 3x$ with Domain$(g) = [0, 4]$.

Solutions

1. The derivative of $f(x)$ is $f'(x) = 2x - 5$. The condition $f'(x) = 0$ implies $2x - 5 = 0$. This equation's solution is $x = \frac{5}{2}$, which is in the domain of f. Thus, f has one critical point, $x_0 = \frac{5}{2}$.
2. The derivative of $g(x)$ is $g'(x) = 3x^2 - 3$, and $g'(x) = 0$ implies $3x^2 - 3 = 0$. This last equation's solutions are -1 and 1. Since 1 is in the domain of g and -1 is not, it follows that g has only one critical point, $x_0 = 1$. ■

Practice Problem

Determine the critical points for the function $f(x) = 2x^2 - 3x + 1$ with Domain$(f) = (-1, 1)$.

Classroom Discussion 4.1.2: Smallest Perimeter

The following exploration is from page 48 in the eighth-grade textbook *Mathematics in Context, Get the Most Out of It*. Work through tasks 6–9.

A New Plot

Once the secret was out, Dan changed his rental agreement. He still gave the adventurers four stakes, and they still marked off rectangles with north-south and east-west sides, but with a new constraint:

> The rectangle had to have an *area* of 400 square meters.

Since the adventurers had to use their *own* rope to mark off the plot, they had to decide how long their rope should be.

6. a. Using graph paper, draw at least five different rectangles with areas of 400 square meters. Again, use the scale that one grid unit represents five meters. How long is the rope needed for each rectangle?

b. As you did before, cut out the rectangles and tape them on grid paper so two sides lie on the axes.

You should notice that the upper-right vertices no longer lie on a straight line, but instead on a curve called a *hyperbola*.

7. Connect the points of the hyperbola on your grid paper. What equation using *x* and *y* corresponds to the hyperbola?

8. One adventurer made a rectangle with a perimeter of 208 meters (and an area of 400 square meters, of course). Find the dimensions of the rectangle.

9. Using as little rope as possible, how would you mark off a rectangle with an area of 400 square meters?

---

1. Does the reasoning you used for answering Question 9 constitute a complete proof? How do you show that no other rectangle with an area of 400 square meters will have a perimeter smaller than the rectangle you discovered when answering Question 9?

   Hint: Use algebra and show that $2\left(20 + a + \dfrac{400}{20+a}\right) < 80$ leads to a contradiction.

2. Complete the following tasks to tackle Question 9 by other means.

   a. Denote by x and y the dimensions of a plot of land measured in meters. What is the constraint for x and y in this case?
   b. Let P denote the perimeter, measured in meters, of such a rectangle. Write P first in terms of x and y, and then use the constraint you found for x and y to express P as a function of x. Call this new expression $P(x)$. What is the domain of $P(x)$?
   c. Use a calculator to graph $P(x)$. You need to choose an appropriate window scale. What do you see? Use your calculator's zoom and trace features to find the value of x for which $P(x)$ is the smallest. How does your answer compare to the answer you obtained in Problem 1?

3. Return to the function $P(x)$ from Problem 2. Use calculus to determine the smallest value attained by $P(x)$.

   a. Compute $P'(x)$ and determine the critical points of $P(x)$.
   b. Determine the sign of $P'(x)$ on the intervals in the domain to the left and right of the critical points.
   c. Use your answer from b to determine if $P(x)$ is increasing or decreasing on each interval.
   d. Determine the smallest value of $P(x)$ and determine the point x_0 for which this value is attained. Compare $P(x_0)$ to the solutions you obtained in Problems 1 and 2. ◆

   In the Classroom Discussions 4.1.1 and 4.1.2 you have solved two optimization problems using calculus. In the process, you completed several steps that are typical of any optimization problem. Let's look back and outline these steps.

Step 1: Read the problem carefully and decide which variables to use, the function involving your variables that you need to optimize, and the constraint that you have for the variables. In both examples in Classroom Discussions 4.1.1 and 4.1.2 the variables were x and y, the length and width of a plot of land. In the problem from Classroom Discussion 4.1.1, you had to maximize the area xy under the constraint $2(x + y) = 100$. In the one from Classroom Discussion 4.1.2, you had to minimize the perimeter $2x + 2y$ under the constraint $xy = 400$.

Step 2: Use the constraint to write the function to be optimized in terms of only one variable. Let us call this function f for the purpose of our discussion. In the Classroom Discussion 4.1.1, the resulting function was $f(x) = x(50 - x)$, in the Classroom Discussion 4.1.2 the resulting function was $f(x) = 2x + \frac{800}{x}$.

Step 3: Determine the domain of f as dictated by the physical assumptions of the problem. In the examples discussed, the domains were $(0, 50)$ for Classroom Discussion 4.1.1 and $(0, \infty)$ for Classroom Discussion 4.1.2.

Step 4: Compute the derivative of f and find the critical points, i.e., the values of x in the domain of f where $f'(x) = 0$. Be careful here since the equation $f'(x) = 0$ might yield solutions that are not in the domain of f. In the first and second example, the critical points were $x = 25$ and $x = 20$, respectively.

Step 5: Determine the intervals where f' is negative and where it is positive. Use this information to decide whether f increases or decreases on these intervals. The function in the first example was increasing for $x < 25$ and decreasing for $x > 25$. In the second example, the function was decreasing for $x < 20$ and increasing for $x > 20$.

Step 6: Decide if there are any points where your function attains its maximum or minimum, as required by the problem. Evaluate the function at any such points. In the first example, the function attained its maximum at $x = 25$, with the value of this maximum being 625. In the second example, the function attained its minimum at $x = 20$, with the value of this minimum being 80.

Step 7: Interpret and check your solution. In the first example, the conclusion was that, among all the rectangles with perimeters equal to 100 meters, the one with the largest area is the square with side length 25 meters. In the second example, you proved that among all the rectangles having an area of 400 square meters, the one with the smallest perimeter is the square with side length 20 meters.

Practice Problems

1. Use calculus to prove generalizations of the examples discussed in the Classroom Discussions 4.1.1 and 4.1.2.

   a. Among all the rectangles with a fixed perimeter P, the one with the largest area is the square with side length $\frac{P}{4}$.

   b. Among all the rectangles with fixed area A, the one with the smallest perimeter is the square with side length $\sqrt{A}$.

2. Denote by x and y the dimensions of a rectangle, by A its area, and by P its perimeter. The following identity then holds:

$$\frac{P^2}{4} = (x - y)^2 + 4A.$$

Give an algebraic and a geometric proof of this identity; for the geometric proof, use the following picture. Then, utilize the preceding identity to prove, without using calculus, the statements a and b in Problem 1.

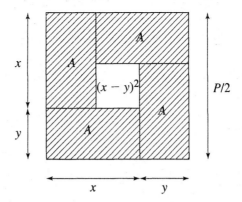

Classroom Discussion 4.1.3: The Behavior of Quadratic Functions

Fix arbitrary real numbers a, b, c with $a \neq 0$, and consider the quadratic function $f(x) = ax^2 + bx + c$. Since f is a polynomial function, the domain of f is the whole real line $\mathbb{R}$. This discussion's goal is to analyze the behavior of $f(x)$, first using algebra and then using calculus. This will be done first for some particular quadratic functions and then for a general quadratic function.

1. **The Case** $f(x) = 2x^2 + 4x + 1$

   a. **Algebraic Approach**

      i. Why do you think the computation

      $$2x^2 + 4x + 1 = 2\left(x^2 + 2x + \tfrac{1}{2}\right) = 2(x^2 + 2x + 1) - 1$$
      $$= 2(x + 1)^2 - 1$$

      is called *completing the square*?

      ii. How small can $2(x + 1)^2$ be? For what value of x is $2(x + 1)^2$ smallest?

      iii. How small can $f(x)$ be? For what value of x is $f(x)$ smallest?

      iv. How large can $f(x)$ be? Explain.

   b. **Calculus Approach**

      Using calculus, compute the minimum value of $f(x) = 2x^2 + 4x + 1$ on the real line.

2. **The Case** $f(x) = -3x^2 + 6x - 2$

   a. **Algebraic Approach**

      i. Complete the square for the polynomial $-3x^2 + 6x + 2$.

      ii. How large can $f(x) = -3x^2 + 6x - 2$ be if x is allowed to take any real value?

b. Calculus Approach

Using calculus, compute the maximum value of $f(x) = -3x^2 + 6x + 2$ when x takes values on the real line.

3. **The General Case $f(x) = ax^2 + bx + c$**

Let a, b, c be arbitrary fixed real numbers, with $a \neq 0$.

a. Algebraic Approach

i. Show that after completing the square, the expression for $f(x) = ax^2 + bx + c$ becomes

$$f(x) = a\left(x + \frac{b}{2a}\right)^2 - \frac{b^2 - 4ac}{4a}.$$

ii. What can you say about the values of $\left(x + \dfrac{b}{2a}\right)^2$ for values of x on the real line?

iii. Assume $a > 0$. Use the conclusion in ii to decide if $f(x)$ has a maximum or minimum value for some value of x. How about if $a < 0$?

iv. For what value of x does f attain the corresponding minimum or maximum? What is the value of that minimum or maximum?

b. Calculus Approach

Use calculus to optimize $f(x) = ax^2 + bx + c$ on the real line. Make sure you distinguish between the cases $a < 0$ and $a > 0$.   ◆

Classroom Discussion 4.1.4: Local Maximum/Minimum versus Absolute Maximum/Minimum

The goal in this discussion is to determine the largest and smallest values attained by the function $f(x) = 3x^4 - 4x^3 - 12x^2$ on the interval $[-2, 3]$. Follow the outline here.

1. Compute the critical points of $f(x)$ and find the subintervals of $[-2, 3]$ determined by the critical points.

2. Determine the sign of $f'(x)$ on each subinterval from Problem 1. Explain why the sign of f' cannot change within these subintervals.

3. Decide whether f is increasing or decreasing on each subinterval from Problem 1. Compare your results with the following table.

x	-2		-1		0		2		3
$f'(x)$		$-$	0	$+$	0	$-$	0	$+$	
f	32	↘	-5	↗	0	↘	-32	↗	27

Observe that the function $f(x)$ is decreasing to the left and increasing to the right of both $x = -1$ and $x = 2$. In this case we say that $f(x)$ has **local minima** at $x = -1$ and $x = 2$. Since $f(-1) = -5 > -32 = f(2)$, and the pattern is decreasing-increasing-decreasing-increasing, the smallest value $f(x)$ that takes on the interval $[-2, 3]$ is -32. This is why -32 is called the **absolute minimum** for $f(x)$.

Similarly, because $f(x)$ is increasing to the left of $x = 0$ and decreasing to the right of $x = 0$, we say that $f(x)$ has a **local maximum** at $x = 0$. The local maximum value is $f(0) = 0$. Is this also the **absolute maximum** for f (i.e., is this the largest value taken by f on the interval $[-2, 3]$)? To answer this question, we look at the behavior of f. The decreasing-increasing-decreasing-increasing pattern suggests that the values of f at the endpoints $x = -2$ and $x = 3$ must be taken into account. A direct computation gives $f(-2) = 32$ and $f(3) = 27$, so the absolute maximum of f is $32 = f(-2)$, not $0 = f(0)$.

4. Do critical points always yield local minima or local maxima? To answer this question, analyze the critical points of the function $f(x) = x^3$ defined on the real line. ◆

Summary of Main Steps in Optimization Problems

1. Select the variables and write the expression for the function f to be optimized.
2. Write the constraint and use it to express the function f in terms of one variable.
3. Determine the domain of f.
4. Find all critical points of f.
5. Determine the intervals where f' is positive and the intervals where f' is negative. Determine the intervals where f is increasing and the intervals where f is decreasing.
6. Determine maxima and/or minima for f.
7. Interpret and check your solution.

EXERCISES 4.1

In Exercises 1–4, for each given function, determine the critical points, the subintervals in the domain determined by the critical points, the sign of the derivative on each subinterval, and whether the function is increasing or decreasing on each subinterval.

1. $f(x) = 3x - 2$, Domain $= [-4, 12]$
2. $f(x) = x^2 - 3x + 2$, Domain $= [-1, 8]$
3. $f(x) = \frac{1}{3}x^3 - 9x$, Domain $= (-\infty, \infty)$
4. $f(x) = -x^3 + 12x$, Domain $= (0, \infty)$
5. For each function from 1–4, determine local maxima, local minima, absolute maxima, and absolute minima, provided they exist.
6. **Classroom Connection 4.1.1: Fenced In**
   The following exploration is from page 26 in the eighth-grade textbook *MathScape, Family Portraits*.

a: How does the last question in this exploration relate to the practice problems in Section 4.1?

Fenced In

SOLVING A
PROBLEM THAT
INVOLVES A
PARABOLA

Suppose you have 32 m of fencing material. What is the largest rectangle you can fence off? And what does this problem have to do with parabolas? You will explore this situation and make a graph to describe it. This will help you see the connection between perimeters, areas, and parabolas.

Explore Rectangular Pens

How can you find a rectangle with the greatest area for a given perimeter?

A farmer has 32 m of fencing material and wants to fence off a rectangular pen for animals. One side of the pen must lie along a creek. What length along the creek results in a pen with the greatest area for the animals?

1 What is the perimeter of any pen the farmer can make?

2 Use a sheet of graph paper to help sketch all of the possible pens that have whole-number lengths. One possible pen is shown here.

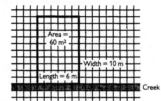

Area = 60 m²
Width = 10 m
Length = 6 m
Creek

3 Find the area of each pen.

Which length along the creek results in the pen with the greatest area?

b: Now suppose there exists a wall along the creek, so no fencing is needed for that side of the pen. Use calculus to determine the dimensions of the rectangular pen with the greatest area if the wall is used for one side of the pen. ◆

7. Chocolate Boxes: A candy store is closed for remodeling. The owner has purchased square pieces of thin plastic of different colors in order to make new candy display boxes. The plastic squares have a 10-inch side length. He decides to cut out four equal squares from the corners of each piece and then form the boxes by turning up the sides. What should be the side length of the squares he cuts so that the resulting boxes will have the largest volume?

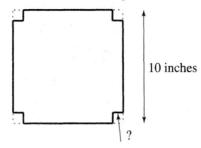

10 inches

?

8. **Pillows:** Tracy has a home-based business in which she sews decorative pillows and sells them on eBay. Each pillow requires $2 of materials and 2 hours of sewing. Tracy finds that if she charges $12 per pillow, she sells 30 pillows per week. For each $1 increase in price, she sells 5 fewer pillows per week; for each $1 decrease in price, she sells 5 more pillows per week. Use calculus to determine what price Tracy should charge to maximize her profit.

9. **A Rollerblading Track:** A city plans to construct a new park that will have a rollerblading track that is 400π meters long. The portion enclosed by the track will consist of a rectangular playground with two semicircular ends in which to plant flowers. At what dimensions will the rectangular area in the center be the largest? For these dimensions, how much space will be available for planting flowers?

10. **Wooden Beams:** In the construction of houses, beams with rectangular cross sections are often used. These beams are cut from wooden logs. A beam's strength is directly proportional to the product of its width and the square of its height. Suppose you have a log with a circular cross section of diameter 12 inches. What are the dimensions of the cross section of the strongest beam you can cut from this log?

12 in

11. **Watermelon Crop:** A gardener must decide when to harvest and sell his watermelon crop. He estimates that there are approximately 100 pounds of ripe watermelon in his garden. Each week an additional 25 pounds of watermelon ripens, and 5 pounds go to waste. The watermelon's market price is $0.90 per pound, but it drops $0.10 per pound each week that passes. When should the gardener sell his watermelon crop in order to make the maximum revenue? What is the maximum revenue he can make?

12. A gardener is trying to decide what is the best time to harvest and sell his cantaloupe crop. He estimates that there are approximately 200 pounds of ripe cantaloupe in his garden. Each week an additional 60 pounds of cantaloupe ripens, and 10 pounds go to waste. The cantaloupe's market price is $1 per pound, but it drops $0.10 per pound each week that passes. When should the gardener sell his cantaloupe crop in order to maximize revenue? What is the maximum revenue he can make?

13. **Cable Lines:** A house is built along a 1-mile wide river. On the river's other side, 10 miles downstream, there is another building from which cable line will be run to the new house. The underwater cable costs twice as much per foot as the underground cable. How long should the cable line along the river be in order to minimize the cost?

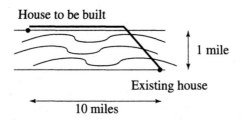

House to be built

1 mile

Existing house

10 miles

14. Tod runs a chocolate store. He invests $4 for each pound of chocolate he makes. He sells 500 pounds of chocolate each month for the price of $12 per pound. Tod discovers that for every 10 cents he takes off the price, he sells 10 more pounds of chocolate each month. What should he charge for 1 pound of chocolate in order to maximize his profit? How many pounds of chocolate will Tod sell at that price?

15. **Inventory Costs:** A retail appliance store sells 500 refrigerators each year. It costs $50 to store one refrigerator for a year. At each reordering, there is a fixed $20 fee for the truck rental and an additional $5 handling fee for each refrigerator ordered. How many times per year should the store place an order to minimize the storage, truck rental, and handling costs, if the number of refrigerators per order is constant? When modeling this problem, assume that at any time during the year, the average number of refrigerators in stock is half the number x of refrigerators ordered each time.

16. The following figure shows the graph of the derivative of a function f. Find all the points where f has a local maximum or a local minimum.

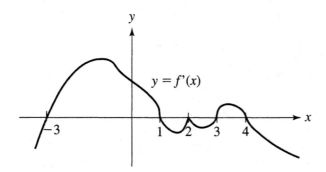

$y = f'(x)$

4.2 CURVE SKETCHING

Second derivative • Concave up and concave down • Inflection points • Sketching graphs

You saw in Section 3.2 that the sign of a function's derivative yields substantial information about the shape of the graph of the function. Take this principle one step further and look at the derivative of $f'(x)$ to get information about $f'(x)$. The derivative of $f'(x)$, also called the **second derivative** of $f(x)$, is denoted by $f''(x)$. Thus, $f''(x) = (f'(x))'$, whenever this makes sense. For example, if we consider the functions $f(x) = 2x^2 - 4x + 7$ and $g(x) = x^5 - 3x^4 + 100$, then $f'(x) = 4x - 4$, $f''(x) = 4$, $g'(x) = 5x^4 - 12x^3$, and $g''(x) = 20x^3 - 36x^2$. The goal is to determine what information the second derivative f'' provides about the shape of the graph of f.

Classroom Discussion 4.2.1: Concavity and the Second Derivative

Let f be a function that is twice differentiable on an interval (a, b); that is, f is differentiable on (a, b), and, in turn, f' is also differentiable on (a, b).

1. Recall the relationship between a function's behavior and the function's derivative. Use this relationship to fill in the blanks:

   f' is nondecreasing on (a, b) if and only if f'' _____ on (a, b).

   f' is nonincreasing on (a, b) if and only if f'' _____ on (a, b).

2. The goal is to understand how a function's behavior relates to the sign of the function's second derivative. Look at the following graphs. Decide in each case whether the derivative is nondecreasing or nonincreasing.

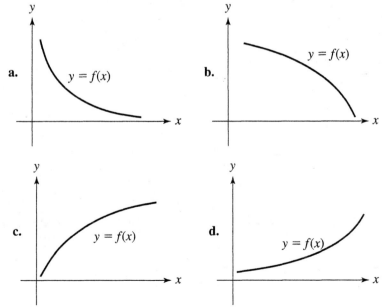

3. Which are the graphs for which $f'' \geq 0$?
4. Which are the graphs for which $f'' \leq 0$? ◆

A portion of a graph that looks like the curves in a or d is said to be *concave up*. Imagine these graphs as taking the shape either one of your arms will make as you casually raise them. Similarly, a portion of a graph that looks like the curves in b or c is said to be *concave down*. Imagine these graphs as taking the shape either one of your arms will make as you casually lower them.

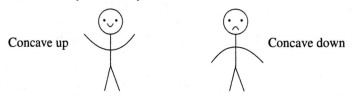

Concave up   Concave down

Definition 4.2.1 A function f defined on an interval (a, b) is **concave up** if for any $x_1, x_2 \in (a, b)$, the line segment joining the points $(x_1, f(x_1))$ and $(x_2, f(x_2))$ lies above the portion of the graph of f in between x_1 and x_2. A function f defined on an interval (a, b) is **concave down** if for any $x_1, x_2 \in (a, b)$, the line segment joining the points $(x_1, f(x_1))$ and $(x_2, f(x_2))$ lies below the portion of the graph of f in between x_1 and x_2.

This definition reinforces the intuitive idea that the graphs a and d are concave up, while the graphs b and c are concave down. See the following pictures.

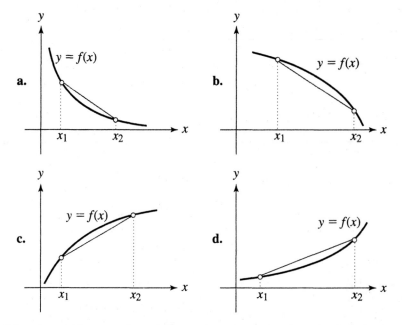

Often in problems we have information about the sign of a function's second derivative and from that we want to deduce information about the shape of the graph of the function's graph. For example, if we know that $f'' \geq 0$ on an interval (a, b), then we can conclude that f' is nondecreasing on (a, b). Furthermore, if f' is nondecreasing on (a, b), then the graph of f is concave up on (a, b). In light of

Classroom Discussion 4.2.1, this is not too surprising, and we will use this result without proving it. Similarly, if $f'' \leq 0$ on an interval (a, b), then f' is nonincreasing on (a, b); this in turn implies that the graph of f is concave down on (a, b).

The Relationship between a Function's Behavior and the Sign of Its Second Derivative.   Let f be a function for which f'' exists on an interval (a, b). Then the following hold:

1. $f'' \geq 0$ on (a, b) if and only if the graph $y = f(x)$, $x \in (a, b)$, is concave up.
2. $f'' \leq 0$ on (a, b), if and only if the graph $y = f(x)$, $x \in (a, b)$, is concave down.

Classroom Discussion 4.2.2: Inflection Points

Here is the graph of a function $f(x)$ defined on the real line.

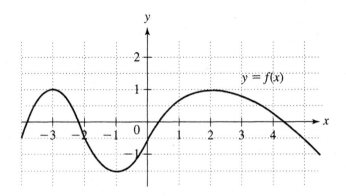

a. Find all the intervals where the graph is concave up and where the graph is concave down.
b. Find the points where the graph changes concavity (i.e., it changes from concave up to concave down or vice versa). Such points are called **inflection points**.
c. What is the value of f'' at an inflection point?
d. Suppose $f''(x_0) = 0$. Can you conclude that x_0 is an inflection point? Consider the function $f(x) = x^4$.   ◆

Classroom Discussion 4.2.3: Sketching Graphs

1. For each case here, sketch a portion of the graph of a function $f(x)$ having the specified features:

   a. increasing, changing from concave up to concave down at the point $(1, 2)$;
   b. $f(3) = 1.5, f' > 0, f''(3) = 0, f''$ negative to the left of $x = 3$ and positive to the right of $x = 3$;
   c. decreasing, changing from concave down to concave up at the point $(1, -1)$;

d. $f(2) = 1.5, f' < 0, f''(2) = 0, f''$ positive to the left of $x = 2$ and negative to the right of $x = 2$;

e. concave up, changing from decreasing to increasing at the point $(3,1)$;

f. $f(2) = 3, f'' < 0, f'$ positive at the left of $x = 2$ and negative at the right of $x = 2$.

2. Limits at Infinity

This discussion's goal is to investigate the behavior of polynomial functions $p(x)$ as $x \to \infty$ and $x \to -\infty$; that is, as x increases without bound and as x decreases without bound, respectively.

a. Let $p(x) = 1$. What can you say about the values of $p(x)$ as $x \to \infty$? Why do you think a condensed way of writing this behavior is $\lim_{x \to \infty} p(x) = 1$? Fill in the blank: $\lim_{x \to -\infty} p(x) = $ _____.

b. Let c be a fixed real number, and suppose $p(x) = c$ for all real numbers x. Fill in the blanks: $\lim_{x \to \infty} p(x) = $ _____, $\lim_{x \to -\infty} p(x) = $ _____.

c. Now take $p(x) = x$. What happens to the values of $p(x)$ as $x \to \infty$? Why do you think a condensed way of writing this behavior is $\lim_{x \to \infty} x = \infty$? Fill in the blank: $\lim_{x \to -\infty} x = $ _____.

d. Fill in the blanks: $\lim_{x \to \infty} x^2 = $ _____, $\lim_{x \to -\infty} x^2 = $ _____. Explain.

e. Fill in the blanks: $\lim_{x \to \infty} x^3 = $ _____, $\lim_{x \to -\infty} x^3 = $ _____. Explain.

f. Let $n \in \mathbb{N}$. Fill in the blanks:

$$\lim_{x \to \infty} x^n = \underline{\quad}, \qquad \lim_{x \to -\infty} x^n = \begin{cases} \underline{\quad} & \text{if } n \text{ is even,} \\ \underline{\quad} & \text{if } n \text{ is odd.} \end{cases}$$

Explain.

g. What happens to the values of $\frac{1}{x}$ as $x \to \infty$ and as $x \to -\infty$? Fill in the blanks: $\lim_{x \to \infty} \frac{1}{x} = $ _____, $\lim_{x \to -\infty} \frac{1}{x} = $ _____.

h. Fill in the blanks: $\lim_{x \to \infty} \frac{1}{x^2} = $ _____, $\lim_{x \to -\infty} \frac{1}{x^2} = $ _____.

i. Let $n \in \mathbb{N}$. Fill in the blanks: $\lim_{x \to \infty} \frac{1}{x^n} = $ _____, $\lim_{x \to -\infty} \frac{1}{x^n} = $ _____.

j. Consider $p(x) = x^2 - 2x - 3$. We want to determine the behavior of $p(x)$ as $x \to \infty$ and as $x \to -\infty$. If $x \neq 0$, we can write

$$x^2 - 2x - 3 = x^2\left(1 - 2 \cdot \frac{1}{x} - 3 \cdot \frac{1}{x^2}\right).$$

Use this to fill in the blanks:

$$\lim_{x \to \infty} (x^2 - 2x - 3) = \underline{\quad},$$

$$\lim_{x \to -\infty} (x^2 - 2x - 3) = \underline{\quad}.$$

Explain your reasoning.

k. Use i and the ideas used in j to fill in the blanks:

$$\lim_{x \to \infty} (5x^3 + 2x^2 - 4x + 12) = \underline{\hspace{1.5cm}},$$

$$\lim_{x \to -\infty} (5x^3 + 2x^2 - 4x + 12) = \underline{\hspace{1.5cm}},$$

$$\lim_{x \to \infty} (-5x^3 + 2x^2 - 4x + 12) = \underline{\hspace{1.5cm}},$$

$$\lim_{x \to -\infty} (-5x^3 + 2x^2 - 4x + 12) = \underline{\hspace{1.5cm}},$$

$$\lim_{x \to \infty} (3x^4 + x - 2) = \underline{\hspace{1.5cm}},$$

$$\lim_{x \to -\infty} (3x^4 + x - 2) = \underline{\hspace{1.5cm}},$$

$$\lim_{x \to \infty} (-3x^4 + x - 2) = \underline{\hspace{1.5cm}},$$

$$\lim_{x \to -\infty} (-3x^4 + x - 2) = \underline{\hspace{1.5cm}}.$$

Explain.

l. Let $n \in \mathbb{N}$, and $a_0, a_1, \ldots, a_n \in \mathbb{R}$, $a_n \neq 0$. Use all the preceding results to fill in the blanks:

$$\lim_{x \to \infty} (a_n x^n + a_{n-1} x^{n-1} + \cdots + a_1 x + a_0)$$

$$= \begin{cases} \underline{\hspace{1.5cm}} & \text{if } a_n > 0, \\ \underline{\hspace{1.5cm}} & \text{if } a_n < 0; \end{cases}$$

$$\lim_{x \to -\infty} (a_n x^n + a_{n-1} x^{n-1} + \cdots + a_1 x + a_0)$$

$$= \begin{cases} \underline{\hspace{1.5cm}} & \text{if } a_n > 0 \text{ and } n \text{ is even,} \\ \underline{\hspace{1.5cm}} & \text{if } a_n < 0 \text{ and } n \text{ is even,} \\ \underline{\hspace{1.5cm}} & \text{if } a_n > 0 \text{ and } n \text{ is odd,} \\ \underline{\hspace{1.5cm}} & \text{if } a_n < 0 \text{ and } n \text{ is odd.} \end{cases}$$

Explain.

Remark.   The results you obtained in Problem 2l show that the limit of a polynomial function as $x \to \infty$, or as $x \to -\infty$, depends only on the leading term of the polynomial. More precisely,

$$\lim_{x \to \infty} (a_n x^n + a_{n-1} x^{n-1} + \cdots + a_1 x + a_0) = \lim_{x \to \infty} (a_n x^n)$$

and

$$\lim_{x \to -\infty} (a_n x^n + a_{n-1} x^{n-1} + \cdots + a_1 x + a_0) = \lim_{x \to -\infty} (a_n x^n).$$

This observation is useful for applications since the limits in the right-hand sides of the equalities above are easier to compute.

3. Consider the function $f(x) = x^5 - 5x^4$ defined on the real line. Use the following outline to sketch the graph of f.

    a. Compute the derivative of f. Find the critical points of f and the sign of f' on each subinterval determined by the critical points.

    b. Decide on which intervals f is increasing and on which intervals f is decreasing.

    c. Compute $f''(x)$ and determine the points where $f''(x) = 0$.

    d. Analyze the sign of f'' and the concavity of the graph of f. What are the inflection points for f?

    e. Determine the values of x where the graph of f crosses the x-axis. Observe that these are the values of x for which $f(x) = 0$. Compute $f(0)$.

    f. Determine $\lim\limits_{x \to \infty} f(x)$ and $\lim\limits_{x \to -\infty} f(x)$.

    g. Use all the information you have gathered to complete the following table and to sketch the graph of f.

x	$-\infty$		0		3		4		∞
$f'(x)$		$+$	0	$-$					
f	$-\infty$	$\nearrow$		$\searrow$					
$f''(x)$		$-$	0	$-$					
f		$\cap$		$\cap$					

    h. Does this function have an absolute maximum?

    i. Does this function have an absolute minimum?

    j. Consider now the function $g(x) = x^5 - 5x^4$ with domain $[-1, 6]$. Use the graph of f to sketch on another system of coordinates the graph of g. Does g have an absolute maximum or an absolute minimum? ◆

EXERCISES 4.2

In Exercises 1–4, decide if the graph of the function is concave up or concave down on the specified interval.

    1. $f(x) = x^2 - x$ on $(0.5, 0.75)$

    2. $f(x) = 2x^3 - 4x - 1$ on $(0, 1)$

    3. $f(x) = x^4 - 6x^2 + 1{,}000$ on $(-0.5, 0.5)$

    4. $f(x) = \frac{1}{3}x^3 - x^2 + x + 1$ on $(4, 5)$

    5. Consider an arbitrary quadratic function $f(x) = ax^2 + bx + c$ with a, b, and c fixed real numbers. How many points of inflection does this function have? What is the relationship between the value of a and the concavity of the graph of f?

    6. Little Red Riding Hood visits her grandmother every weekend. To get from her house to her grandmother's, she follows a straight path that crosses a forest. One day, it took her 40 minutes to complete the trip from her house to grandmother's house. Twice she had to backtrack—once to find the apple she dropped from her basket and once to pick a flower to bring to her grandmother. Here is the graph of Little Red Riding Hood's velocity during the trip.

v (feet/sec)

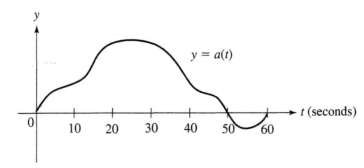

a. Let $d(t)$ denote Little Red Riding Hood's distance (in feet) from her house after t seconds. Determine the intervals of time when d is increasing.

b. At what time t in the interval $[21, 27]$ is Little Red Riding Hood farthest from her house? Explain your reasoning.

c. On which intervals is the graph of d concave up? Explain.

In Exercises 7–18, sketch the graph of the function f. To do so, follow the outline from Problem 3 in Classroom Discussion 4.2.3 and complete a similar table.

7. $f(x) = 4 - x^3$

8. $f(x) = x^3 - 8$

9. $f(x) = x^3 + 3x$

10. $f(x) = x^3 - \frac{1}{3}x + 1$

11. $f(x) = \frac{4}{3}x^3 - 4x$

12. $f(x) = x^3 + x^2$

13. $f(x) = (x - 2)^6 + 1$

14. $f(x) = (x + 1)^5 + 2$

15. $f(x) = 2x^2 - x^4 - 1$

16. $f(x) = \frac{1}{4}x^4 + 2x^2 + 1$

17. $f(x) = x^2(x^2 - 3x + 6)$

18. $f(x) = x^4 + 2x^3 + 6x^2$

19. Steven swims one pool lap in 1 minute. The graph of his acceleration is depicted here.

Let $v(t)$ be Steven's velocity at time t.

a. Is $v(t)$ increasing, decreasing, or neither on the interval $[35, 45]$? Explain.

b. Is the graph of $v(t)$ concave up or concave down on the interval $[15, 20]$? Explain.

c. Is the graph of $v(t)$ concave up or concave down on the interval $[30, 40]$? Explain.

d. Is $v(40) < v(20)$? Explain.

e. Is $v(55) < v(59)$? Explain.

PROJECTS AND EXTENSIONS 4.2

I. The Graph of $f(x) = x^n$

1. Follow the outline from Problem 3 in Classroom Discussion 4.2.3 and complete a similar table for each of the functions here.

   a. $f(x) = x^3$
   b. $f(x) = x^4$
   c. $f(x) = x^5$
   d. $f(x) = x^6$

2. Use the information from Problem 1 to predict the shape of the graph of $f(x) = x^n$ when n is a whole number.

3. Use calculus to show that your prediction for the graph of $f(x) = x^n$ is correct.

II. Limits at Infinity of Rational Functions

The goal of this project is to investigate the behavior of rational functions as $x \to \infty$ and $x \to -\infty$. The following approach builds on the ideas used in Problem 2 of Classroom Discussion 4.2.3.

1. Consider the rational function $f(x) = \frac{x^2+2x+3}{x^3-3x-1}$. For $x \neq 0$, we can write

$$\frac{x^2 + 2x + 3}{x^3 - 3x - 1} = \frac{x^2\left(1 + 2 \cdot \frac{1}{x} + 3 \cdot \frac{1}{x^2}\right)}{x^3\left(1 - 3 \cdot \frac{1}{x^2} - \frac{1}{x^3}\right)} = \frac{1}{x} \cdot \frac{1 + 2 \cdot \frac{1}{x} + 3 \cdot \frac{1}{x^2}}{1 - 3 \cdot \frac{1}{x^2} - \frac{1}{x^3}}.$$

Use this expression of $f(x)$ to determine $\lim\limits_{x \to \infty} f(x)$ and $\lim\limits_{x \to -\infty} f(x)$.

2. Let $g(x) = \frac{x^2+2x+3}{-x^3-3x-1}$. Determine $\lim\limits_{x \to \infty} g(x)$ and $\lim\limits_{x \to -\infty} g(x)$.

3. Let $h(x) = \frac{3x^2-4x-2}{-5x^2+x+1}$. Determine $\lim\limits_{x \to \infty} h(x)$ and $\lim\limits_{x \to -\infty} h(x)$. Explain your reasoning.

4. Let $k(x) = \frac{-x^4-x^3-2}{x-4}$. Determine $\lim\limits_{x \to \infty} k(x)$ and $\lim\limits_{x \to -\infty} k(x)$.

5. Consider now the general case of a rational function $f(x)$; that is, $f(x) = \frac{p(x)}{q(x)}$,

$$p(x) = a_n x^n + a_{n-1} x^{n-1} + \cdots + a_1 x + a_0,$$
$$q(x) = b_m x^m + b_{m-1} x^{m-1} + \cdots + b_1 x + b_0$$

for some $n, m \in \{0, 1, 2, \ldots\}$, and $a_0, a_1, \ldots, a_n, b_0, b_1, \ldots, b_m \in \mathbb{R}, a_n \neq 0$, $b_m \neq 0$.

Determine $\lim\limits_{x \to \infty} f(x)$ and $\lim\limits_{x \to -\infty} f(x)$ in each case here and fill in the blanks.

Case 1: $n < m$ $\lim\limits_{x \to \infty} f(x) =$ _____, $\lim\limits_{x \to -\infty} f(x) =$ _____.

Case 2: $n = m$ $\lim\limits_{x \to \infty} f(x) =$ _____, $\lim\limits_{x \to -\infty} f(x) =$ _____.

Case 3: $n > m$

$$\lim_{x \to \infty} f(x) = \begin{cases} \rule{1.5cm}{0.4pt} & \text{if } \frac{a_n}{b_m} > 0, \\[2mm] \rule{1.5cm}{0.4pt} & \text{if } \frac{a_n}{b_m} < 0; \end{cases}$$

$$\lim_{x \to -\infty} f(x) = \begin{cases} \rule{1.5cm}{0.4pt} & \text{if } n - m \text{ is even and } \frac{a_n}{b_m} > 0, \\[2mm] \rule{1.5cm}{0.4pt} & \text{if } n - m \text{ is odd and } \frac{a_n}{b_m} > 0, \\[2mm] \rule{1.5cm}{0.4pt} & \text{if } n - m \text{ is even and } \frac{a_n}{b_m} < 0, \\[2mm] \rule{1.5cm}{0.4pt} & \text{if } n - m \text{ is odd and } \frac{a_n}{b_m} < 0. \end{cases}$$

Remark. The results you obtained in Problem 5 show that the limit of a rational function as $x \to \infty$, or as $x \to -\infty$, depends only on the leading terms of the polynomials in the numerator and in the denominator of the rational function. More precisely,

$$\lim_{x \to \infty} \frac{a_n x^n + a_{n-1} x^{n-1} + \cdots + a_1 x + a_0}{b_m x^m + b_{m-1} x^{m-1} + \cdots + b_1 x + b_0} = \lim_{x \to \infty} \frac{a_n x^n}{b_m x^m}$$

and

$$\lim_{x \to -\infty} \frac{a_n x^n + a_{n-1} x^{n-1} + \cdots + a_1 x + a_0}{b_m x^m + b_{m-1} x^{m-1} + \cdots + b_1 x + b_0} = \lim_{x \to -\infty} \frac{a_n x^n}{b_m x^m}.$$

This observation is useful for applications since the limits in the right-hand sides of the equalities above are easier to compute (due to the fact that the fractions simplify further).

6. Using the remark just made, compute $\lim\limits_{x \to \infty} f(x)$ and $\lim\limits_{x \to -\infty} f(x)$ for the following choices of f.

a. $f(x) = \dfrac{-3x^5 + 4x^3 - x + 2}{-4x^5 - x^4 - 3x^2 - 2x}$

b. $f(x) = \dfrac{x^6 - 3x^5 + 2x - 7}{-4x^9 + 5x^6 - 3x + 2}$

c. $f(x) = \dfrac{-x^2 - x - 3}{4x^2 + x + 5}$

d. $f(x) = \dfrac{99x^{100} - 33}{-3x^{101} + 1}$

e. $f(x) = \dfrac{x^4 + 5}{2x - 3}$

f. $f(x) = \dfrac{-3x^7 - 1}{2x^5 + 3}$

g. $f(x) = \dfrac{x^5 + 2x^3 + 4}{-x^2 - x + 2}$

h. $f(x) = \dfrac{3x^8 - 5x^4 + 2}{2x^4 - 2x^5 + x + 3}$

4.3 EXPONENTIAL CHANGE

Irrational exponents • Exponential functions • Derivatives of exponential functions

Classroom Connection 4.3.1: Paper Folding

The following exploration is from pages 499–500 in the eighth-grade textbook *Math Thematics*, *Book 3*. Work through Problems 3–8.

Exploration ┄┄┄┄┄┄┄┄┄┄┄┄┄┄┄┄┄┄┄┄┄

exponential Change

GOAL

LEARN HOW TO...
• model exponential change

AS YOU...
• use paper folding to model eating a chocolate bar

SET UP   *You will need: • a rectangular sheet of paper • graph paper*
• graphing calculator (optional)

▶ In Question 2 on page 498, you estimated how long Charlie Bucket's chocolate bar would last if he ate half of the remaining chocolate each day. In this exploration, you will use paper folding to model this situation.

3 a. Fold a sheet of paper in half. With no folds there is one layer. After one fold there are two layers.

0 folds
1 layer
area = 1

1 fold
2 layers
area = ?

b. If the area of the unfolded paper is one square unit, what is the area of each layer after you fold the paper once?

c. Record the number of folds, the number of layers, and the area of each layer in a table like the one shown.

d. Fold your paper as many times as possible. Extend and fill in your table each time you fold the paper.

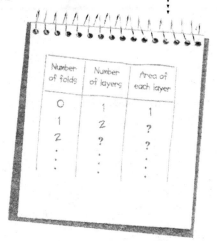

Number of folds	Number of layers	Area of each layer
0	1	1
1	2	?
2	?	?
⋮	⋮	⋮

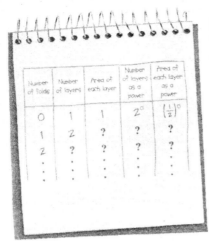

Number of folds	Number of layers	Area of each layer	Number of layers as a power	Area of each layer as a power
0	1	1	2^0	$\left(\frac{1}{2}\right)^0$
1	2	?	?	?
2	?	?	?	?
⋮	⋮	⋮	⋮	⋮

For Questions 4–7, use your completed table from Question 3.

4 Suppose you could continue folding the paper. What would happen to the number of layers as the number of folds increased? What would happen to the area of each layer?

5 Add two columns to your table, as shown. Write the number of layers as a power of 2 and the area of each layer as a power of $\frac{1}{2}$.

6 Suppose you could fold the paper ten times. Predict how many layers there would be. Then predict the area of each layer. Explain your reasoning.

▶ As you folded the paper, the number of layers and the area of each layer changed *exponentially*. You can use equations and graphs to model exponential change.

7 **Try This as a Class** Let x = the number of folds.

   a. Let y = the number of layers. Write an equation for y in terms of x.

   b. Let y = the area of each layer. Write an equation for y in terms of x.

   c. How are your equations in parts (a) and (b) alike? How are they different?

8 **a.** Graph the equation you wrote in Question 7(a) by plotting points for the data in your table from Question 3 and connecting the points with a smooth curve.

   b. Graph the equation you wrote in Question 7(b).

   c. How are the graphs in parts (a) and (b) alike? How are they different? Describe how each graph changes as x increases.

9   Graphing Calculator The *exponential equations* you have seen so far have all had the form $y = b^x$, where $x \geq 0$.

   a. To see how the value of b affects the graph of an equation in the form $y = b^x$, graph the equations $y = 3^x$, $y = 4^x$, and $y = 7^x$ for $x \geq 0$ on the same pair of axes. Describe the differences and the similarities in the curves.

In Problem 8, you had to join the plots corresponding to integer values of x to get a "smooth curve." Is there a particular curve passing through all the points you plotted for integer values of x that would be best for our model? If yes, what function has this "smooth curve" as its graph? To answer these questions, return to Problem 7a in Classroom Connection 4.3.1. The expression you found for the number of layers in terms of x was $y = 2^x$. The function $f(x) = 2^x$ is a great candidate. However, the definition of 2^x when x is an irrational number has not yet been provided. The next discussion's goal is to clarify the issue of how one can raise a given positive number to an irrational power. First, recall how powers with rational exponents were defined.

Rational Exponents. Fix a real number $b > 0$ and take m, n to be whole numbers. Then, rational powers of b are defined as follows.

1. $b^m = b \cdot b \cdots b$ is the product of m copies of b.
2. $b^{-m} = \frac{1}{b^m}$ is the reciprocal of b^m.
3. $b^{\frac{1}{n}} = \sqrt[n]{b}$ is the positive number whose nth power is b. The case $n = 2$ has been discussed in the project "*Computing $\sqrt{x}$ for x a Real Positive Number*" in Section 1.1. The nth root can be defined similarly.
4. $b^{\frac{m}{n}} = \sqrt[n]{b^m}$ is the number whose nth power is b^m.
5. $b^{-\frac{m}{n}} = \frac{1}{b^{\frac{m}{n}}}$ is the reciprocal of $b^{\frac{m}{n}}$.

Thus, for any rational number x, b^x is meaningful. In addition, you can check that for any x, y rational numbers, we have $b^{-x} = \frac{1}{b^x}$, $b^{x+y} = b^x b^y$, and $(b^x)^y = b^{xy}$. The goal is to extend this definition to the case when the exponent x is an irrational number.

Classroom Discussion 4.3.1: Irrational Exponents

Fix an irrational number x. For example, take $x = \pi$. Starting with its decimal representation

$$\pi = 3.14159265358979323846264338327950288419...,$$

you can construct sequences $\{y_n\}_n$ and $\{z_n\}_n$ of rational numbers:

$y_0 = 3$, $y_1 = 3.1$, $y_2 = 3.14$, $y_3 = 3.141$, $y_4 = 3.1415$, $y_5 = 3.14159$,

$y_6 = 3.141592, \ldots$ and

$z_0 = 4$, $z_1 = 3.2$, $z_2 = 3.15$, $z_3 = 3.142$, $z_4 = 3.1416$, $z_5 = 3.1416$,

$z_6 = 3.141593, \ldots$

1. For each n, compare the values of y_n, z_n and π.
2. What can you say about $\lim\limits_{n \to \infty} y_n$ and $\lim\limits_{n \to \infty} z_n$? Explain.
3. According to the definition of rational exponents, the number 2^{y_n} is meaningful for each $n \geq 0$. What happens to the values of 2^{y_n} as $n \to \infty$?
4. Is the sequence $\{2^{y_n}\}_n$ convergent or divergent?
5. Similarly, for each $n \geq 0$, 2^{z_n} is well defined since z_n is rational. What happens to the values of $\{2^{z_n}\}_n$ as $n \to \infty$?
6. Is the sequence $\{2^{z_n}\}_n$ convergent or divergent?
7. Intuitively, it is expected that for $a > 0$, $\lim\limits_{n \to \infty} \sqrt[n]{a} = 1$. Use this fact to compute $\lim\limits_{n \to \infty} (2^{z_n} - 2^{y_n})$.
8. What does the preceding construction suggest as a definition for 2^{π}?

   Take now the general case. Let x be an arbitrary irrational number. Then, there exist two sequences $\{y_n\}_n$ and $\{z_n\}_n$ with the following properties:

   a. y_n and z_n are rational numbers for each n,
   b. $y_1 \leq y_2 \leq y_3 \leq \ldots$ and $z_1 \geq z_2 \geq z_3 \geq \ldots$,
   c. $y_n \leq x \leq z_n$ for each n,
   d. $\lim\limits_{n \to \infty} y_n = \lim\limits_{n \to \infty} z_n = x$

   as seen in the project "*Real Numbers as Limits of Sequences of Rational Numbers*" in Section 1.1.

9. How can you define 2^x?
10. How can you define b^x for $b > 0$, an arbitrary real number, and for x, an arbitrary irrational number?

    For each $b > 0$, the function $f(x) = b^x$ is now well defined on the set of real numbers. This function is the **exponential function with base b**. Observe that, due to the properties of powers with rational exponents, this construction also yields $b^{-x} = \frac{1}{b^x}$, $b^{x+y} = b^x b^y$, and $(b^x)^y = b^{xy}$ for all real numbers x, y. ◆

Classroom Discussion 4.3.2: Exponential versus Linear

Recall that a linear function is a function of the form $g(x) = mx + n$, where m and n are real numbers. Fix $b > 0$ and consider the exponential function $f(x) = b^x$.

1. Compute $g(3) - g(2)$ and $g(70) - g(69)$. What do you observe?
2. What is the value of $g(x + 1) - g(x)$ for an arbitrary x?
3. Compute $\frac{f(5)}{f(4)}$ and $\frac{f(-5)}{f(-6)}$. What do you observe?
4. What is the value of $\frac{f(x+1)}{f(x)}$ for an arbitrary x?
5. Explain the difference between exponential and linear functions.   ◆

Classroom Discussion 4.3.3: Graphs of Exponential Functions

1. Use a graphing calculator to trace the graphs of the functions $f(x) = 2^x$, $g(x) = 3^x$, $h(x) = 4^x$, and $l(x) = 9^x$. Describe the differences and similarities in the curves.

2. Use a graphing calculator to trace the graphs of the functions $F(x) = (\frac{1}{2})^x$, $G(x) = (\frac{1}{3})^x$, $H(x) = (\frac{1}{4})^x$, and $L(x) = (\frac{1}{9})^x$. Describe the differences and similarities in the curves.

3. Make a prediction about the shape of the graph of the exponential function $f(x) = b^x$ for $b > 0$, depending on whether $b < 1$ or $b > 1$.   ◆

Classroom Discussion 4.3.4: Derivatives of Exponential Functions

1. Let $f(x) = 2^x$ be defined for all real numbers x. Here is the graph of $f(x)$:

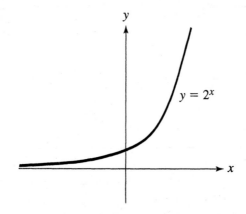

a. For what values of x is $f'(x)$ positive, negative, or zero?

b. What can you say about the values of $f'(x)$ as $x \to \infty$ and as $x \to -\infty$?

c. On another set of coordinates, sketch roughly the graph of $f'(x)$; use the geometric interpretation of the derivative at a point x as the slope of the tangent line to the graph of f at the point $(x, f(x))$.

d. Compare the graph of f with the graph of f'. What do you observe?

e. The two graphs look very similar. Is $f'(x)$ an exponential function? To answer this question, you must compute $f'(x)$ using the definition of the derivative. To do so, first write the average rate of change for the function $f(x) = 2^x$ over the interval with endpoints x and $x + h$.

f. Check that for each $h \neq 0$,

$$\frac{2^{x+h} - 2^x}{h} = 2^x \frac{2^h - 1}{h}. \tag{1}$$

g. Use your calculator to investigate $\lim\limits_{h \to 0} \frac{2^h - 1}{h}$. To do so, generate a table of values for $\frac{2^h - 1}{h}$ with $h \neq 0$ and varying from -0.01 to 0.01 with step size $\Delta h = 0.001$. What do you observe? Does $\lim\limits_{h \to 0} \frac{2^h - 1}{h}$ exist? If yes, give an estimate for its value.

h. Return to (1). What can you conclude about $f'(x)$?

2. Use the outline in a–h to determine the derivative of the function $g(x) = (\frac{1}{5})^x$.

3. Fix $b > 0$.

   a. Prove the following statement:

   $$\text{if } F(x) = b^x, \text{ then } F'(x) = b^x k, \text{ where } k = \lim_{h \to 0} \frac{b^h - 1}{h}.$$

   b. Use your calculator to estimate the values of k when $b = 0.2$, $b = 0.6$, $b = 1$, $b = 1.5$, $b = 2$, and $b = 5$. What do you observe? How do the values of k change as b increases? Compare k with 0 to fill in the blanks.

base b	constant k
$0 < b < 1$	k ___ 0
$b = 1$	k ___ 0
$b > 1$	k ___ 0

   c. Is the preceding table consistent with the geometric interpretation of the derivative? Think about how the shape of the graph of the exponential function b^x varies with the values of $b > 0$.

   It turns out that as b takes all the real positive values, k also takes all real values. Consequently, there is a value of b for which $k = 1$. This is precisely the case when b is the irrational number e. An approximate value for e is 2.71828. The number e arises naturally when modeling investments (see "Compound Interest" in Projects and Extensions 4.3). Using the formula you obtained in Problem 3a, you see that

   $$(e^x)' = e^x.$$

   d. Compute $(e^{-x})'$ using the formula for the derivative of reciprocals.

   e. Suppose that f is a differentiable function. Use the chain rule to compute $(e^{f(x)})'$. ◆

EXERCISES 4.3

In Exercises 1–14, compute the derivative of the function.

1. $f(x) = \frac{4}{5}e^x - 3x^5 + 6x - 4$

2. $f(x) = 4e^{-3x}$

3. $f(x) = e^{x^2 - x + 6}$

4. $f(x) = 2e^{x^3}$

5. $f(x) = \sqrt{e^{2x-1}}$

6. $f(x) = \sqrt{e^{x^2 - x + 4}}$

7. $f(x) = x\,e^x$

8. $f(x) = (x + 1)e^x$

9. $f(x) = \sqrt[3]{x\,e^{2x}}$

10. $f(x) = x\,e^{x^2 - x}$

11. $f(x) = \frac{e^{-5x+1}}{x^4 + x^2 + 4}$

12. $f(x) = \frac{x^1}{e^{1-x}}$

13. $f(x) = \frac{x}{1 + x\,e^x}$

14. $f(x) = \frac{xe^{-x}}{e^x + 2}$

In Exercises 15–22, sketch the graph of the function.

15. $f(x) = e^{-x}$

16. $f(x) = 2e^{-3x}$

17. $f(x) = -5 \cdot 3^{2x}$

18. $f(x) = 4 \cdot 5^{-\frac{1}{2}x}$

19. $f(x) = \left(\frac{3}{4}\right)^{-x}$

20. $f(x) = \left(\frac{1}{5}\right)^{x-3}$

21. $f(x) = 2^{2x+1}$

22. $f(x) = 3^{\frac{3}{2}x-1}$

In Exercises 23–26, determine the image of the function.

23. $f : [-1, 1] \to \mathbb{R}, f(x) = 2^{\sqrt{1-x^2}}$

24. $f : (-\infty, -1] \cup [1, \infty), f(x) = \left(\frac{1}{2}\right)^{\sqrt{x^2 - 1}}$

25. $f : [2, \infty) \to \mathbb{R}, f(x) = \sqrt{2^{2x} - 16}$

26. $f : \mathbb{R} \to \mathbb{R}, f(x) = \frac{1}{3^{|x|}}$.

In Exercises 27–28, determine the values of x for which the inequality holds.

27. $(2\sqrt{2})^{x+3} > \left(\frac{1}{8}\right)^{x-4}$

28. $3^{x^2 - 2x + 1} < 1$

PROJECTS AND EXTENSIONS 4.3

Compound Interest

1. Classroom Connection 4.3.2: Savings Accounts

The following exploration is from pages 502–503 in the eighth-grade textbook *Math Thematics*, *Book 3*. Work through Problems 12–16.

▶ Savings can grow as quickly as prices if you deposit money into a savings account that earns interest. The amount of money increases exponentially, even if you do not make any more deposits.

12 a. Suppose you deposit $1000 into an account that pays 8% annual interest. Find 8% of $1000 to determine the amount of interest you will earn in one year.

b. What is the new total in your account at the end of one year?

▶ How much money will you have in a savings account after 10, 20, or 30 years? To find out, look for a pattern.

EXAMPLE

Suppose you deposit $2000 into an account that earns 5% annual interest. After one year, you will have 100% of your deposit plus an additional 5% interest.

Initial deposit + Interest after one year

$(1.00)(2000) + (0.05)(2000)$

You can use the distributive property to rewrite this expression:

$(1.00)(2000) + (0.05)(2000) = (1.05)(2000)$

$$= 2100$$

After one year, there will be $2100 in the account.

13 Refer to the situation described in the Example.

a. Suppose you leave your money in the account for two years. How much money will you have at the end of the second year? Explain how you got your answer.

b. Copy and complete the table below, by continuing the pattern in the *Expression* column.

c. Discussion Describe the pattern in the *Expression* column.

Year	Amount in account at beginning of year	Expression	Amount in account at end of year
1	$2000	$1.05 \cdot 2000$	$2100
2	$2100	$1.05 \cdot 1.05 \cdot 2000$	?
3	?	?	?
4	?	?	?
5	?	?	?

14 a. At the end of two years, the amount of money in the account described in the Example is given by the expression 1.05 · 1.05 · 2000. Rewrite this expression using exponents.

b. 🖩 Calculator Use exponents to write an expression for the amount of money in your account at the end of 20 years. Use the 🔲 key to evaluate the expression.

▶ The growth of money in a savings account that earns annual interest is an example of exponential change. You can model exponential change with an **exponential equation** which has the form:

starting amount

amount after **x** years

$y = a \cdot b^x$

growth factor

15 Try This as a Class The equation $2923.08 = 1500 \cdot (1.1)^7$ gives the amount in an account after a certain number of years.

a. How much money was originally deposited into the account?

b. How many years was the money in the account? How much money is in the account after this amount of time?

c. What is the growth factor? What is the interest rate?

✔ **CHECKPOINT** Suppose you deposit $4000 into an account that earns 6% annual interest. Write an equation that models the amount *y* in the account after *x* years. How much will be in the account after 1 year? after 12 years? after 20 years?

✔ **QUESTION 16**

…checks that you can write and use an exponential equation to solve problems.

17 Discussion Look back at Question 11 on page 501.

a. Suppose the price of a candy bar starts at $.10 and increases by $.01 each year. Write an equation for the price after *x* years.

Price of candy bar	$.01 yearly increase	10% yearly increase
After 10 years	?	?
After 20 years	?	?
After 30 years	?	?
After 40 years	?	?

b. Suppose the price starts at $.10 and increases by 10% each year. Write an equation for the price after *x* years.

c. Use your equations to copy and complete the table. Compare the predicted price changes over time.

HOMEWORK EXERCISES ▶ See Exs. 6–17 on pp. 506–507.

Section 3   Modeling Exponential Change   **503**

◆

2. In all the problems in Classroom Connection 4.3.2, the annual interest was computed once at the end of the year. Suppose now that you deposit $1,000 in a bank at an annual interest rate of 8% and that the interest is compounded monthly. Each month, interest is computed at a rate of $\frac{0.08}{12}$ and is added to the account.

    a. How much money will be in the account at the year's end?

    b. How much money will be in the account at the end of 2 years?

    c. Compare your results in a and b with the case when the interest is computed once per year.

3. Suppose that your bank pays 8% annual interest compounded weekly and you deposit $1,000.

    a. How much money will you have in the account at the end of 1 year?

    b. How much money will you have in the account at the end of 2 years?

    c. Compare your results in a and b with the case when the interest is computed once per year or monthly.

    d. What if the interest is compounded daily?

4. If two banks offered the same interest rate but one compounded it yearly and the other daily, in which bank would you choose to open a savings account? Explain.

5. Now assume that the 8% annual interest is compounded x times per year. Write the formula that gives the amount of money in an account after 1 year if the original deposit is $1,000.

6. Write the formula that gives the amount of money $A(x)$ in an account after 1 year if the original deposit is P dollars, the annual interest rate is r (r is written in decimal representation), and the interest is compounded x times per year.

7. Write the formula for the amount of money $A(x)$ in an account after t years, if the original deposit is P, the annual interest rate is r (r is written in decimal representation), and the interest is compounded x times per year.

8. You have seen that it is to the investor's advantage to have the interest compounded as many times as possible per year. Suppose that the original deposit is $P = \$1$ and the interest is compounded x times per year. We want to look at what happens to the amount of money $A(x)$ in the account after 1 year as $x \to \infty$; in the limit, we say that the interest is being **compounded continuously**. This amounts to analyzing $\lim_{x \to \infty} A(x) = \lim_{x \to \infty} \left(1 + \frac{r}{x}\right)^x$. Using the rules for exponentials first, we have

$$A(x) = \left(1 + \frac{r}{x}\right)^x = \left[\left(1 + \frac{r}{x}\right)^{\frac{x}{r}}\right]^r.$$

9. Let $r > 0$ be fixed. What can you say about $\lim_{x \to \infty} \frac{x}{r}$?

10. Rewrite the expression for $A(x)$ in terms of $y = \frac{x}{r}$. Call this new function $B(y)$. From Problem 9, you expect that $\lim_{x \to \infty} A(x) = \lim_{y \to \infty} B(y)$.

11. Use your graphing calculator to graph the function $(1 + \frac{1}{y})^y$. Can you predict if the limit $\lim_{y \to \infty} (1 + \frac{1}{y})^y$ exists? Use your calculator's trace function to obtain estimates for the values of $(1 + \frac{1}{y})^y$ as y gets larger and larger.

12. Use this limit to give an alternative definition of the irrational number e:

$$e = \lim_{y \to \infty} \left(1 + \frac{1}{y}\right)^y \approx 2.71828.$$

Now, return to the limit of $B(y)$ and $A(x)$. What is the value of $\lim_{y \to \infty} B(y)$? How about $\lim_{x \to \infty} A(x)$? Interpret this result.

13. What is the amount of money in an account after 1 year if the original deposit is $1,000, the annual interest rate is r, and the interest is compounded continuously?

14. What is the amount of money in an account after 5 years if the original deposit is $1,000, the annual interest rate is r, and the interest is compounded continuously?

15. Conclusion

Fill in the blanks to make the following statement correct.

Let P be the amount deposited in a bank account paying an annual interest rate r (in decimal form). Then the balance B in the account after t years is

a. $B =$ _____ if the interest is compounded n times per year,

b. $B =$ _____ if the interest is compounded continuously.

16. Elizabeth's grandparents created a trust fund for her to help pay for her college education. They deposited $10,000 in an account on the day Elizabeth was born. Determine how much money Elizabeth will have from her grandparents when she turns 18 if the interest rate is

a. 0.05, and it is compounded monthly,

b. 0.05, and it is compounded continuously,

c. 8%, and it is compounded quarterly,

d. 8%, and it is compounded continuously.

17. Tim has received two job offers, one from NASA and one from Acme Airplanes. NASA has offered him a starting salary of $35,000 per year with a 7% raise each year. Acme will pay him $40,000 initially with a $3,000 raise each year.

a. If Tim chooses to work for Acme, what will be the amount of his raise after 1 year? What will his salary be during the second year? During the third year?

b. Write a function describing Tim's salary during his tth year at Acme. What kind of function is this? What will Tim's salary be in the tenth year at Acme? In the twentieth year?

c. If Tim chooses to work for NASA, what will be the amount of his raise after 1 year? What will his salary be during the second year? During the third year?

d. Write a function describing Tim's salary during his tth year at NASA. What kind of function is this? What will Tim's salary be in the tenth year at NASA? In the twentieth year?

CHAPTER 4 REVIEW

The relationship between the sign of the derivative of a function and the behavior of the function is very useful when solving optimization problems. These are problems in which we want to maximize or minimize certain quantities. The main steps to follow in such problems are:

1. Select the variables and write the expression for the function f to be optimized.
2. Write the constraint and use it to express the function f in terms of one variable.
3. Determine the domain of f.
4. Find all critical points of f, that is, all the points x_0 in the domain of f for which $f'(x_0) = 0$.
5. Determine the intervals where f' is positive and the intervals where f' is negative. Determine the intervals where f is increasing and the intervals where f is decreasing.
6. Determine maxima and/or minima for f.
7. Interpret and check your solution.

The second derivative of a function f is $f''(x) = (f'(x))'$, provided f' is differentiable at x. The sign of f'' also provides information regarding the shape of the graph of f. The graph of f is concave up on an interval if and only if $f'' \geq 0$ on that interval. Similarly, the graph of f is concave down on an interval if and only if $f'' \leq 0$ on that interval. The points x where the graph of f changes concavity are called *inflection points*. These are among the solutions of the equation $f''(x) = 0$. However, a solution of the equation $f''(x) = 0$ is not necessarily an inflection point.

Exponential functions are functions of the form $f(x) = b^x$, where the base b is positive, and x belongs to a subset of the real line. The shape of the graph of b^x depends on whether $0 < b < 1$ or $b > 1$ as seen here.

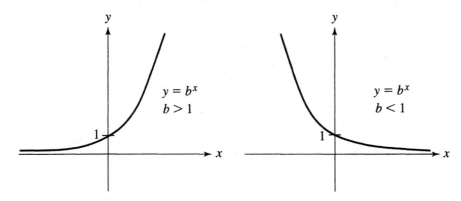

The derivative of the exponential function b^x is given by $(b^x)' = k \cdot b^x$, where k is a constant depending on b. If $0 < b < 1$, we have $k < 0$, if $b > 1$, then $k > 0$; and if $b = 1$, then $k = 0$. The constant k equals 1 precisely when $b = e$. A consequence of the chain rule and the formula for the derivative of an exponential function is the formula

$$(e^{f(x)})' = e^{f(x)} \cdot f'(x), \ x \in \mathbb{R},$$

defined for all differentiable functions f.

CHAPTER 4 REVIEW EXERCISES

1. **Chocolate Boxes:** A candy store is closed for remodeling. The owner has purchased square pieces of thin plastic of different colors in order to make new candy boxes with lids. The plastic squares have a 12-inch side length. To construct the boxes, he removes two equal squares from adjacent corners and two equal rectangles from the other two corners (see the following figure). What should be the side length of the squares he cuts so that the resulting boxes will have the largest volume?

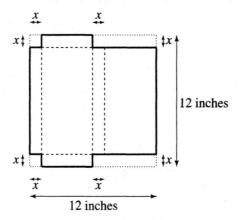

2. On Delta airlines, the sum of the length, height, and width of a checked item on a domestic flight can be at most 62 inches. Items over this size limit are charged an excess baggage fee. What dimensions should a box with a square base have in order to maximize its volume if we want to check it on a Delta domestic flight without having to pay an excess baggage fee?

In Exercises 3–6, sketch the graph of the function.

3. $f(x) = x^3 - 3x^2 + 3x$

4. $f(x) = x^3 - 6x^2 + 12x - 8$

5. $f(x) = x^5 - 5x$

6. $f(x) = x^4 - 2x^2$

In Exercises 7–18, compute the derivative of the function.

7. $f(x) = x^6 - \frac{1}{x} + 3e^x$

8. $f(x) = -e^{1-2x}$

9. $f(x) = e^{x^2 - 3x + 1}$

10. $f(x) = 5e^{5-x+x^3}$

11. $f(x) = e^{\sqrt{x}}$

12. $f(x) = e^{x - \sqrt{x}}$

13. $f(x) = (2x - 3)e^{-x}$

14. $f(x) = \sqrt{x}\, e^{x+2}$

15. $f(x) = \frac{e^x}{2 + e^x}$

16. $f(x) = \frac{e^{-x}}{x - e^{-x}}$

17. $f(x) = e^{\frac{x+1}{x^2+1}}$

18. $f(x) = e^{\frac{\sqrt{x}}{x+3}}$

In Exercises 19–22, sketch the graph of the function.

19. $f(x) = 2e^{3x}$

20. $f(x) = \frac{1}{2}e^{-x}$

21. $f(x) = \left(\frac{5}{6}\right)^x$

22. $f(x) = \left(\frac{6}{5}\right)^{-x}$

In Exercises 23–26, determine the image of the function.

23. $f : [0, 1] \to \mathbb{R}, f(x) = e^{\sqrt{1-x^2}}$

24. $f : [0, 1] \to \mathbb{R}, f(x) = e^{-\sqrt{1-x^2}}$

25. $f : (-\infty, 0] \to \mathbb{R}\, f(x) = \left(\frac{1}{3}\right)^{-x+1}$

26. $f : (-\infty, 0] \to \mathbb{R}\, f(x) = \left(\frac{1}{2}\right)^{x+1}$

In Exercises 27–28, determine the values of x for which the inequality holds.

27. $2^{x^2 - 3x + 2} < 1$

28. $\left(\frac{1}{2}\right)^{x^2 - 4} > 1$

CHAPTER 5

Integration

This chapter's goal is to establish a procedure that allows us to recover functions from their rates of change. This process is called *antidifferentiation*, since it reverses the process of differentiation. It is used to solve a variety of important real-life problems.

5.1 INDEFINITE INTEGRALS

Antiderivatives • Indefinite integrals • Antidifferentiation • Integration • Rules and basic techniques • Applications to the study of motion

You learned in Section 3.3 that if you know the position of an object moving along a straight line at any time, then you can determine its velocity simply by differentiating the position function with respect to time. Similarly, you can determine its acceleration at any time by differentiating the velocity function with respect to time. Quite often in practice, however, the acceleration is the known quantity, while the velocity and position might be unknown. This leads to the question: Can you find a moving object's position function $h(t)$ and its velocity function $v(t) = h'(t)$ from the information about its acceleration function $a(t) = h''(t)$?

To solve this problem, it is natural to first attempt to find $h'(t)$ from $h''(t)$, then to find $h(t)$ from $h'(t)$. To do this, we need to explore the process that undoes differentiation.

Classroom Discussion 5.1.1: Antiderivatives

1. Given the function $f(x) = 2x$ defined on the real line, can you find a function $F(x)$ satisfying $F'(x) = f(x)$ for all real numbers x?

2. Can you find other functions that satisfy the same property as $F(x)$?

3. Besides having the same derivative, what other relations are there between the functions you have found?

4. Fill in the blanks: The function f is _____ of the function F. The operation of finding the function f from the function F is called

   _____.

5. Try to find the appropriate words to fill in the blanks: The function F is _____ of the function f. The operation of finding the function F from the function f is called _____.

6. Recall that a function whose derivative is zero on a given interval must be constant on the interval. Now suppose an unknown function $G(x)$ satisfies $G'(x) = 2x$ for all real numbers. Can you determine the general expression of $G(x)$ in terms of $F(x)$? Justify your answer.

7. Determine all functions $G(x)$ satisfying $G(1) = 20$ and $G'(x) = 2x$ for all real numbers x. How many such functions are there?   ◆

Definition 5.1.1 A function F is called an **antiderivative** of a function f if $F'(x) = f(x)$ for all x in the domain of f. The process of finding a function from its derivative is called **antidifferentiation**.

Next, we check that our findings from Classroom Discussion 5.1.1 remain true for any function that has an antiderivative.

1. If a given function $f(x)$ defined on an interval I has an antiderivative $F(x)$, then it has infinitely many antiderivatives. Indeed, any function of type $G(x) = F(x) + C$, where C is a fixed constant, is also an antiderivative of $f(x)$ since it satisfies

$$G'(x) = [F(x) + C]' = F'(x) + C' = f(x) + 0 = f(x).$$

2. Conversely, every antiderivative of $f(x)$ is obtained in this manner. Indeed, for an arbitrary function $G(x)$ that is an antiderivative of $f(x)$, we have

$$G'(x) = f(x) = F'(x)$$

for all x in I. Thus, $G'(x) - F'(x) = 0$. Using the difference rule for differentiation, we then have

$$[G(x) - F(x)]' = 0.$$

This implies that the function $G(x) - F(x)$ is constant on the interval I; that is, $G(x) - F(x) = C$ for some numerical constant C. Therefore, $G(x) = F(x) + C$ on the interval I.

The following theorem states the results proved in 1 and 2.

Theorem 5.1.1. *Let f and F be two functions defined on an interval I with F being an antiderivative of f. If C is a constant, then the function $F(x) + C$ defined for all x in the interval I is also an antiderivative of f. Conversely, any antiderivative of f is of this form.*

3. Among all of the antiderivatives of $f(x)$, there is only one that takes a specific value y_0 at a given point x_0 in the interval I. To show this, let us look for all antiderivatives $G(x)$ that satisfy this additional requirement. By Theorem 5.1.1, $G(x) = F(x) + C$ for some arbitrary constant C. Moreover, we have $G(x_0) = y_0$. Since $G(x_0) = F(x_0) + C, F(x_0) + C = y_0$, which forces the constant C to be exactly $y_0 - F(x_0)$. Therefore, the desired function is uniquely determined by $G(x) = F(x) + [y_0 - F(x_0)]$, for all x in the interval I.

Notation and Terminology. Antidifferentiation is also referred to as **integration**. A function is said to be **integrable** if it has antiderivatives. The family of all antiderivatives of a given function f is called the **indefinite integral** of f; this is denoted by

$$\int f(x)\, dx,$$

which is read as "the integral of f with respect to x" or "the integral of $f(x)\, dx$." Theorem 5.1.1 says that if F is an antiderivative of f on the interval I, then

$$\int f(x)\, dx = \{F + C : C \text{ arbitrary real number}\}.$$

For simplicity's sake, we write $\int f(x)\, dx = F(x) + C$ instead. For example, $\int 2x\, dx = x^2 + C$. For additional terminology, see the following figure.

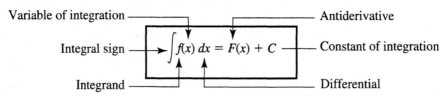

Classroom Discussion 5.1.2: Graphs of Antiderivatives

The function F is an antiderivative of the function f. Its graph is represented.

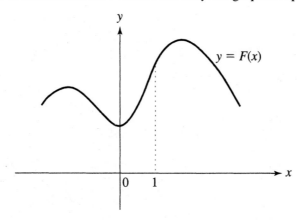

1. What information does the graph of F tell you about the sign of the function f?
2. What information does the graph of F tell you about the behavior (increasing or decreasing) of the function f?
3. How can you obtain the graph of an arbitrary antiderivative of f from the graph of F? Explain.
4. Sketch in the same system of coordinates, the graphs of the antiderivatives of f that take the value 0 at $x = 0$ and the value 0 at $x = 1$, respectively.
5. Now, starting from the graph of the function g shown, sketch roughly the graph of a couple of its antiderivatives.

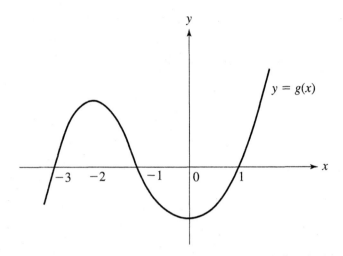

◆

Existence of Antiderivatives.   By now you might be wondering whether all functions have antiderivatives. The answer is no. There are functions that are not integrable just as there are functions that are not continuous or differentiable. However, most of the functions we consider in this book are continuous, and it turns out that all continuous functions are integrable. We will use this fact without proving it. On the other hand, it is good to keep in mind that continuity is not absolutely essential for integrability.

Exercise.   In light of the information you have gathered so far, draw a Venn diagram to show the relationships between the following families:

$\mathcal{F}(E)$: The family of all real-valued functions defined on a given subset E of $\mathbb{R}$.

$\mathcal{D}(E)$: The family of all real-valued differentiable functions defined on E.

$\mathcal{I}(E)$: The family of all real-valued integrable functions defined on E.

$\mathcal{C}(E)$: The family of all real-valued continuous functions defined on E.

Classroom Discussion 5.1.3: Integrating Nonnegative Powers of x

The goal here is to find the antiderivatives of functions of the form $f(x) = x^n$ where $n \geq 0$ is an arbitrary real number. We begin with an example. Consider the function

$f(x) = x^2$. Since differentiation reduces by 1 the exponents of functions that are powers of x, we naturally expect that integration increases their exponents by 1. Thus, a first guess for an antiderivative of x^2 might be x^3. Now, check this guess: $(x^3)' = 3x^2$. Apart from the extra factor of 3, the initial guess was correct. To eliminate the extra factor 3, just divide the initial guess by 3. That is, for the second guess, take $\frac{1}{3}x^3$ as a possible antiderivative of x^2. This guess is now correct since $\left(\frac{1}{3}x^3\right)' = \frac{1}{3} \cdot 3x^2 = x^2$. It follows that

$$\int x^2 \, dx = \frac{1}{3}x^3 + C.$$

The functions $\frac{1}{3}x^3$, $\frac{1}{3}x^3 - \sqrt{2}$, and $\frac{1}{3}x^3 + \pi$ are specific examples of antiderivatives of the function x^2 corresponding to the choices $C = 0$, $C = -\sqrt{2}$, and $C = \pi$, respectively.

Using the same strategy, evaluate the following indefinite integrals:

1. $\int 1 \, dx$
2. $\int 5x^7 \, dx$
3. $\int \sqrt{x} \, dx$. Hint: To evaluate this integral, first rewrite $\sqrt{x}$ as a power of x.
4. $\int x^n \, dx$ where $n \geq 0$ is an arbitrary given real number. ◆

Classroom Discussion 5.1.4: Integrating Negative Powers of x

1. Evaluate the indefinite integral $\int \frac{1}{x^2} \, dx$. What is the domain of the integrand?

2. Is the function $F(x)$ defined below an antiderivative of the function $\frac{1}{x^2}$?

$$F(x) = \begin{cases} -\dfrac{1}{x} + 1 & \text{if } x < 0, \\[2mm] -\dfrac{1}{x} + 3 & \text{if } x > 0. \end{cases}$$

3. Describe the function $F(x) - \left(-\frac{1}{x}\right)$.

4. In Problem 1, you showed that the function $-\frac{1}{x}$ is also an antiderivative of $\frac{1}{x^2}$; in Problem 3, you described the function $F(x) - \left(-\frac{1}{x}\right)$. Why doesn't this example provide a contradiction to Theorem 5.1.1?

5. How do you understand the answer to Problem 1 now?

6. Evaluate the indefinite integral $\int x^n \, dx$ where $n < 0$ is an arbitrary real number $\neq -1$.

7. Explain why the case $n = -1$ was excluded in Problem 6. ◆

Classroom Discussion 5.1.5: Integrating Exponential Functions

1. Find all antiderivatives of the function e^x. Then, select the one that equals 3 at $x = 0$.

2. Find all antiderivatives of the function e^{kx}. Then, select the one that equals 3 at $x = 0$. Here, $k \neq 0$ is an arbitrary real number.
3. Explain why the case $k = 0$ was excluded in Problem 2.   ◆

Classroom Discussion 5.1.6: Integrating Sums and Constant Multiples of Functions

Let F be an antiderivative of f, G an antiderivative of g, and k a given numerical constant.

1. Is it possible to express the antiderivatives of $kf(x)$ in terms of $F(x)$ and k? Explain.
2. Is it possible to express the antiderivatives of $f(x) + g(x)$ in terms of $F(x)$ and $G(x)$? Explain.
3. Fill in the blanks:

   a. $\int kf(x)\,dx = $ _____

   b. $\int [f(x) + g(x)]\,dx = $ _____

   c. $\int [f(x) - g(x)]\,dx = $ _____

4. Evaluate the indefinite integral $\int (3x^2 + 6x)\,dx$.   ◆

Classroom Discussion 5.1.7: Integrating Products of Functions

1. Evaluate the indefinite integral $\int x^2(x - 1)\,dx$.
2. Evaluate the indefinite integral $\int x^2\,dx$.
3. Evaluate the indefinite integral $\int (x - 1)\,dx$.
4. Compare $\int x^2(x - 1)\,dx$ with the product $\left(\int x^2\,dx\right)\left(\int(x - 1)\,dx\right)$. What can you conclude?
5. Fill in the blank:

   In general, $\int f(x)g(x)\,dx \neq$ _____.   ◆

We began Section 5.1 with the problem of determining a moving object's position function from its acceleration. Now we are ready to solve this problem.

Classroom Discussion 5.1.8: Motion and Antiderivatives

A penny is thrown upward from an initial height of 6 feet with an initial velocity of 10 feet per second.

1. Find the penny's velocity $v(t)$ at each instant t.
2. What is the penny's height $h(t)$ after t seconds?

3. What is the maximal height reached by the penny?
4. After how many seconds does the penny hit the ground?
5. What is the penny's velocity at impact?

To solve these problems, notice first that, in this case, the penny's acceleration can be found using Newton's second law of motion; that is,

$$\mathcal{F} = ma,$$

where $\mathcal{F}$ is the total force acting on the penny, m is its mass, and a is its acceleration. The main force acting on the penny here is due to gravity, so

$$\mathcal{F} = -mg$$

where $g = 32$ feet per second squared is the gravitational constant. It follows that $a = -g = -32$. This means that the position function $h(t)$ satisfies $h''(t) = -32$ at any time t. You are now ready to answer questions 1–5.

More generally, consider a coin that is thrown vertically from an initial height of h_0 feet with an initial velocity of v_0 feet per second.

6. What is the sign of v_0 if the penny is thrown upward? How about downward?
7. Find the coin's position function $h(t)$ at each instant t in terms of h_0, v_0, and t. ◆

Rules of Integration

1. $\displaystyle\int kf(x)\,dx = k\int f(x)\,dx$ (Constant multiple rule)

2. $\displaystyle\int [f(x) + g(x)]\,dx = \int f(x)\,dx + \int g(x)\,dx$ (Sum rule)

3. $\displaystyle\int [f(x) - g(x)]\,dx = \int f(x)\,dx - \int g(x)\,dx$ (Difference rule)

4. $\displaystyle\int x^n\,dx = \frac{x^{n+1}}{n+1} + C \ (n \neq -1)$ (Simple power rule)

5. $\displaystyle\int e^x\,dx = e^x + C$ (Simple exponential rule)

6. $\displaystyle\int e^{kx}\,dx = \frac{1}{k}e^{kx} + C \ (k \neq 0)$

EXERCISES 5.1

In Exercises 1–24, find the integrand's domain, then compute the indefinite integral. Use differentiation to justify your answers.

1. $\displaystyle\int x^3\,dx$ 2. $\displaystyle\int x^{100}\,dx$

3. $\displaystyle\int 2\,dx$ 4. $\displaystyle\int (1 + \sqrt{3})\,dx$

5. $\int 0\,dx$

6. $\int -dx$

7. $\int (t^2 + 1)\,dt$

8. $\int (-t^3 + 1)\,dt$

9. $\int t^{2/5}\,dt$

10. $\int t^{\sqrt{5}/3}\,dt$

11. $\int 3\sqrt{t}\,dt$

12. $\int 2\sqrt[3]{t}\,dt$

13. $\int x\sqrt{x}\,dx$

14. $\int x^3\sqrt[4]{x}\,dx$

15. $\int x^{-3}\,dx$

16. $\int \frac{1}{x^5}\,dx$

17. $\int (2 + x^{-2})\,dx$

18. $\int \left(1 + \frac{1}{x^3}\right)dx$

19. $\int e^t\,dt$

20. $\int (t + e^t)\,dt$

21. $\int (t\sqrt[3]{t} + e^t - \pi)\,dt$

22. $\int (t^2\sqrt[5]{t} + t - 2e^t)\,dt$

23. $\int \left(t^2 - \frac{1}{t\sqrt{t}}\right)(t + 1)\,dt$

24. $\int \left(2t + \frac{1}{t\sqrt{t}}\right)(t^2 + t)\,dt$

25. Find the antiderivative $F(x)$ of the function $f(x) = e^x + x - 1$ that satisfies $F(0) = 6$.

26. Find the antiderivative $F(x)$ of the function $f(x) = x^2 - e^x - 1$ that satisfies $F(1) = -e$.

27. Evaluate the indefinite integral $\int |x|\,dx$.

28. Evaluate the indefinite integral $\int |x + 1|\,dx$.

29. Find the antiderivative $F(x)$ of the function $f(x) = 1/x^2$ that satisfies $F(-1) = 1$ and $F(2) = 0$. Sketch the graph of F.

30. Find the antiderivative $F(x)$ of the function $f(x) = -2/x^3$ that satisfies $F(-2) = 0$ and $F(1) = 0$. Sketch the graph of F.

31. Keeping a constant acceleration, a motorcycle can go from 40 miles per hour to 60 miles per hour in 5 seconds.
  a. What is the motorcycle's acceleration in miles per second squared?
  b. What is the distance traveled by the motorcycle during these 5 seconds?

32. A diver jumps from a cliff with an upward initial velocity. The cliff is 60 feet above the sea.
  a. What is the diver's initial velocity if he reaches the maximal height after .25 second?
  b. What is the maximal height reached by the diver?
  c. After how many seconds does the diver hit the water?
  d. What is the diver's velocity at impact?

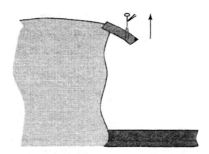

33. Samantha and Scott are standing on a diving platform 10 meters above the water's surface. At the same time that Samantha jumps straight up in the air with a speed of 1 meter per second, Scott dives into the water with a speed of 1 meter per second.
   a. Who hits the water first?
   b. Who has the fastest speed as they hit the water?

34. You proved earlier that for each real number $n \neq -1$, $\int x^n \, dx = \frac{1}{n+1} x^{n+1} + C$. The only remaining case is $\int 1/x \, dx$, corresponding to $n = -1$.
   a. Evaluate $\int 1/x \, dx$ using your calculator.
   b. Among the antiderivatives you obtained, use your calculator to graph the one that vanishes at $x = 1$.
   c. On the same system of coordinates, graph the function $y = e^x$.
   d. Is there any relation between these two graphs? If yes, describe the relation you see and explain.
   e. Use your calculator to evaluate the two compositions of the exponential function e^x and the function in Problem b. Do the results confirm your findings in Problem d?

   The function $\ln x$ you have discovered is called *the natural logarithmic function*. We investigate this function in full detail in the project at the end of Section 6.3.

35. A *differential equation* is an equation that is expressed in terms of x, y, y', y'', etc., where y is a function of x. For example, consider the equation $y' = 3x$. To solve for y, we integrate, obtaining the solution $y = \frac{3}{2}x^2 + C$. In fact, when evaluating integrals, we are actually solving differential equations of a particular type—namely, those involving only the first derivative.
   a. Solve the equation $3y^2 y' = 1$. Hint: Use the chain rule.
   b. Solve the equation $y' = 3y^6$.
   c. Solve the equation $4y^3 y' - 2x + 1 = 0$.

   Equations that involve a function's first derivative and possibly the function itself (such as those in Problems a, b, and c) are called *first-order differential equations*. Equations that involve the second derivative and possibly the function and its first derivative (such as the ones in Problems d and e below) are called *second-order differential equations*. And so on.
   d. Show that $y = e^{2x} + x^2$ satisfies the equation $y'' - 4y + 4x^2 - 2 = 0$.
   e. Show that $y = 3e^x + e^{2x}$ satisfies the equation $y'' - 3y' + 2y = 0$.

   We have not developed the mathematical machinery necessary to solve higher-order differential equations, such as those in Problems d and e; however, there do exist well-known strategies for solving a wide variety of differential equations. The theory of differential equations is rich with applications in many fields, such as business, economics, the social sciences, the physical sciences, and psychology.

5.2   DEFINITE INTEGRALS

Definite integrals • Functions defined by means of definite integrals • Properties of definite integrals

In the previous section, we studied the process of antidifferentiation, and we defined the indefinite integral of a function to be the family of all of its antiderivatives. In this section, we study the variations of a function's antiderivatives over a bounded interval; based on this, we introduce the *definite integral*. If f is a function defined on a bounded interval $[a,b]$, then **the variation** or **change of f over $[a,b]$** is, by definition, $f(b) - f(a)$. For convenience, this quantity is denoted by $f(x)\big|_a^b$. Thus,

$$f(x)\big|_a^b = f(b) - f(a).$$

For example, if $f(x) = x^3 + 1$, then $f(x)\big|_0^2 = f(2) - f(0) = (2^3 + 1) - (0^3 + 1) = 8$.

Classroom Discussion 5.2.1: Variations of Antiderivatives over Bounded Intervals

Consider the function $f(x) = 4x^3$ defined on the real line.

1. Choose an antiderivative $F(x)$ of $f(x)$, then evaluate the difference $F(2) - F(0)$.
2. Let $G(x)$ designate an arbitrary antiderivative of $f(x)$. Evaluate $G(2) - G(0)$.
3. What can you conclude from Problem 2?
4. Is your conclusion still true for arbitrary real numbers a and b instead of 0 and 2?
5. Consider now an arbitrary integrable function $f(x)$ defined over a bounded interval $[a, b]$. What can you say about the variations of the antiderivatives of $f(x)$ over $[a, b]$? Justify your answer. ◆

Definition 5.2.1 Assume that $F(x)$ is an antiderivative of $f(x)$ on the interval $[a, b]$. The number $F(b) - F(a)$ depends on f, a, and b, but not on the particular antiderivative F chosen to evaluate it. It is called the **definite integral** of f from a to b and is denoted by $\int_a^b f(x)\,dx$. Thus,

$$\int_a^b f(x)\,dx = F(x)\big|_a^b = F(b) - F(a).$$

The numbers a and b are called **lower** and **upper limits of integration**, respectively.

EXAMPLE Compute the definite integrals $\displaystyle\int_1^3 x\,dx$ and $\displaystyle\int_1^2 (3x^2 - 1)\,dx$.

Solution

$$\int_1^3 x\,dx = \frac{x^2}{2}\Big|_1^3 = \frac{3^2}{2} - \frac{1^2}{2} = \frac{9}{2} - \frac{1}{2} = 4.$$

$$\int_1^2 (3x^2 - 1)\,dx = (x^3 - x)\Big|_1^2 = (8 - 2) - (1 - 1) = 6. \qquad \blacksquare$$

Notice that the definite integral $\int_a^b f(x)\,dx$ is a number. It does not depend on the variable of integration since $\int_a^b f(x)\,dx = F(b) - F(a)$ where F is any antiderivative of f. Consequently, you may replace x with any other variable, and no difference will occur in the definite integral. In other words,

$$\int_a^b f(x)\,dx = \int_a^b f(y)\,dy = \int_a^b f(t)\,dt, \text{ etc.}$$

Since the variable of integration always disappears when you evaluate the definite integral, it is called a *dummy variable*.

Now, suppose that you allow the lower or the upper limit of a definite integral to vary. Then the variable appearing in the limit remains after you evaluate the definite integral. Thus, you will have defined a new function. The following Classroom Discussion's goal is to better illustrate this idea.

Classroom Discussion 5.2.2: Functions Defined by Means of Definite Integrals

1. Let $f(x) = x^2$ and consider the function $g(x) = \int_1^x f(t)\,dt$.

    a. Find a formula for $g(x)$.
    b. Evaluate $g'(x)$.
    c. Compute $g(1)$.
    d. Fill in the blanks:

        $g(x)$ is the _____ of $f(x) = x^2$ that takes the value ___ at $x = $ __.

2. Now, let $f(x)$ be an arbitrary function defined on the interval $[a, b]$, and assume $F(x)$ is an antiderivative of $f(x)$. Consider the function $g(x)$ defined as

$$g(x) = \int_a^x f(t)\,dt \quad \text{for all} \quad a \le x \le b.$$

    a. Express g in terms of F, then, using this expression, evaluate $g(a)$.
    b. How would you characterize g among all of the antiderivatives of f?
    c. Let $g_1(x) = \int_1^x \frac{1}{t}\,dt$ and $g_2(x) = \int_0^x \frac{1}{3t^2+e^t}\,dt$. Fill in the blanks:

        $g_1(x)$ is the _____ of $f_1(x) = $ ___ that takes the value __ at $x = $ __.

        $g_2(x)$ is the _____ of $f_2(x) = $ ___ that takes the value __ at $x = $ __.

                                                                                                                          ◆

Classroom Discussion 5.2.3: Properties of Definite Integrals

In the following properties, a, b, and c are fixed real numbers, k is a constant, and f and g are integrable functions defined on an interval containing a, b, and c. Use Definition 5.2.1 to prove that definite integrals satisfy these useful properties.

$$1. \quad \int_a^a f(x)\, dx = 0$$

$$2. \quad \int_a^b f(x)\, dx = -\int_b^a f(x)\, dx$$

$$3. \quad \int_a^c f(x)\, dx = \int_a^b f(x)\, dx + \int_b^c f(x)\, dx$$

$$4. \quad \int_a^b kf(x)\, dx = k \int_a^b f(x)\, dx$$

$$5. \quad \int_a^b [f(x) + g(x)]\, dx = \int_a^b f(x)\, dx + \int_a^b g(x)\, dx$$

$$6. \quad \int_a^b [f(x) - g(x)]\, dx = \int_a^b f(x)\, dx - \int_a^b g(x)\, dx \qquad \blacklozenge$$

EXAMPLE Compute the definite integral $\int_{-1}^2 |x|\, dx$. Recall that, by definition, $|x| = x$ for $x \geq 0$, and $|x| = -x$ for $x \leq 0$.

Solution   In order to integrate the function $|x|$ over the interval $[-1, 2]$, we must distinguish between negative x's and positive x's. Thus, we integrate this function over the subintervals $[-1, 0]$ and $[0, 2]$, after using Property 3 from Classroom Discussion 5.2.3.

$$\int_{-1}^2 |x|\, dx = \int_{-1}^0 |x|\, dx + \int_0^2 |x|\, dx = \int_{-1}^0 -x\, dx + \int_0^2 x\, dx.$$

Thus,

$$\int_{-1}^2 |x|\, dx = \frac{-x^2}{2}\Big|_{-1}^0 + \frac{x^2}{2}\Big|_0^2 = \left(0 - \frac{-1}{2}\right) + \left(\frac{4}{2} - 0\right) = \frac{5}{2}. \qquad \blacksquare$$

Practice Problem

Compute the definite integral $\int_0^2 f(x)\, dx$, where $f(x)$ is defined by

$$f(x) = \begin{cases} x^2 & \text{if } x \leq 1, \\ x^3 & \text{if } x \geq 1. \end{cases}$$

EXERCISES 5.2

In Exercises 1–17, compute the definite integral $\int_a^b f(x)\,dx$.

1. $f(x) = 0, a, b$ arbitrary

2. $f(x) = 1, a, b$ arbitrary

3. $f(x) = 7x^6, a = -1, b = 1$

4. $f(x) = x^3 + 1, a = -1, b = 1$

5. $f(x) = x^4(2x - 1), a = 0, b = -1$

6. $f(x) = (x + 1)(2 - x^2), a = 0, b = 1$

7. $f(x) = 1/x^4, a = 1, b = 3$

8. $f(x) = 1/x^3 - 1/x^2, a = 1, b = 2$

9. $f(x) = (x^4 + 1)/x^2, a = 1, b = 2$

10. $f(x) = (x^2 - 1)/(x^4 + x^3), a = 1, b = 2$

11. $f(x) = x\sqrt{x}, a = 1, b = 2$

12. $f(x) = \sqrt{x}(x^2 - 1), a = 0, b = 1$

13. $f(x) = |x|, a = -2, b = 5$

14. $f(x) = 2|x| + 1, a = -1, b = 4$

15. $f(x) = |2x + 1|, a = -1, b = 0$

16. $f(x) = e^x, a = 0, b = 2$

17. $f(x) = e^{x+2}, a = -2, b = 0$

18. Amy dives from a cliff with an upward initial velocity of 8 feet per second. Her acceleration (due to gravity) is -32 feet per second squared. Evaluate the definite integral corresponding to the change in Amy's velocity from the time she jumps off the cliff until she hits the water 1.95 seconds later. What is Amy's velocity at impact?

19. Without evaluating the indefinite integral $\int (3x - x^2)\,dx$, find the intervals on which this function is either increasing or decreasing. What values of a and b maximize the definite integral $\int_a^b (3x - x^2)\,dx$? Here, $a \le b$.

20. Without evaluating the indefinite integral $\int (x^6 - x^4)\,dx$, find the intervals on which this function is either increasing or decreasing. What values of a and b minimize the definite integral $\int_a^b (x^6 - x^4)\,dx$? Here, $a \le b$.

21. Let $F(x) = \int_0^x \frac{1}{t^2 + t + 1}\,dt$, then fill in the blanks:

$F(x)$ is the _____ of $f(t) =$ _____ that takes the value _____ at $x =$ _____.

22. Let $G(x) = \int_x^3 e^{t^2}\, dt$, then fill in the blanks:

    $G(x)$ is the _____ of $g(t) =$ _____ that takes the value _____ at $x =$ _____.

23. Let f be a function defined on the real line. Assume that f is integrable. We saw that the definite integral $F(x) = \int_a^x f(t)\, dt$ represents the function whose derivative is f and that vanishes at $x = a$. Here, we would like to understand expressions of the form $\int_{g(x)}^{h(x)} f(t)\, dt$, where $g(x)$ and $h(x)$ are given functions.

    a. Express the definite integral $H(x) = \int_a^{x^2} f(t)\, dt$ as the composition of F with another function.

    b. Determine $H'(x)$.

    c. Express the definite integral $G(x) = \int_{x^3}^a f(t)\, dt$ as a composite function.

    d. Determine $G'(x)$.

    e. Express $K(x) = \int_{x^3}^{x^2} f(t)\, dt$ in terms of $H(x)$ and $G(x)$. Then, evaluate $K'(x)$.

    f. Replace x^2 with $h(x)$ and replace x^3 with $g(x)$, then solve Problems a–e. In Problems b, d, and e, assume that the functions g and h are differentiable.

    g. Find the derivative of the function $\int_x^{x^4} te^t\, dt$ using your findings in Problem f.

    h. Find the derivative of the function $\int_{e^x}^{2x} \frac{1}{t^2+1}\, dt$ using your findings in Problem f.

5.3 SOME TECHNIQUES OF INTEGRATION

General power rule • General exponential rule • Substitution or change of variable

You have seen that finding the antiderivatives of polynomials and some exponential functions is fairly easy. However, these are very special cases. Computing the antiderivatives of a function is, in general, a difficult task. In this section, we discuss techniques of integration that are suitable for certain types of functions. Namely, we discuss *the general power* and *exponential rules*, and, more generally, we discuss *the method of substitution*. These rules are based on the chain rule for differentiation.

Classroom Discussion 5.3.1: General Power Rule for Integration

Let $n \neq -1$ be an arbitrary real number, and consider the function $f(x) = 3x^2(x^3 + 1)^n$. Our goal is to find the antiderivatives of the function $f(x)$.

    1. Evaluate $\int f(x)\, dx$ in the case $n = 2$.

    2. Based on the techniques of integration you have learned in Section 5.1, describe a strategy for evaluating $\int f(x)\, dx$ in the case where n is an arbitrary whole number.

    Now our goal is to find an alternative method for integrating the function $f(x)$ without expanding the binomial $(x^3 + 1)^n$ in a polynomial form. To do so, observe first that if you let $u(x) = x^3 + 1$, then $f(x) = u'(x)u^n(x)$ since $u'(x) = 3x^2$. This expression is reminiscent of the general power rule of differentiation discussed in Section 3.5, which says

$$[u^{n+1}(x)]' = (n + 1) u^n(x) u'(x),$$

and which holds for all real numbers n (that is, n need not be a whole number). When $n \neq -1$, you can divide both sides by $n + 1$ and use the constant multiple rule of differentiation to obtain

$$\left[\frac{1}{n + 1} u^{n+1}(x) \right]' = u^n(x) u'(x).$$

Consequently,

$$\int u^n(x) u'(x) \, dx = \frac{1}{n + 1} u^{n+1}(x) + C \text{ for } n \neq -1.$$

As you can see, this reasoning is valid for any function $u(x)$ that has a derivative. The last formula is referred to as the **general power rule for integration** or simply the **general power rule**.

3. Use the general power rule to evaluate the indefinite integral in Problem 1. Compare your solution with the former one.

4. Use the general power rule to evaluate the indefinite integral in Problem 2.

5. Is your answer to Problem 4 still valid if you allow n to be an arbitrary real number? Explain.

6. Use the general power rule to evaluate the indefinite integral $\int 3x^4 \sqrt{x^5 + 6} \, dx$. To do so, first express $\sqrt{x^5 + 6}$ as a power of $x^5 + 6$.

7. Use the general power rule to evaluate the definite integral $\int_0^1 \frac{5x}{(x^2+1)^2} \, dx$. Start first by evaluating the corresponding indefinite integral.   ◆

Classroom Discussion 5.3.2: General Exponential Rule for Integration

Consider the function $f(x) = 2x \, e^{x^2}$ defined on the entire real line. Our goal is to find the antiderivatives of $f(x)$.

1. Determine $u(x)$ such that $f(x) = u'(x) e^{u(x)}$.

The expression $e^{u(x)} u'(x)$ is reminiscent of the derivative of the composite function $e^{u(x)}$. Indeed, using the chain rule discussed in Section 4.3, you see that

$$[e^{u(x)}]' = e^{u(x)} u'(x).$$

Consequently,

$$\int e^{u(x)} u'(x) \, dx = e^{u(x)} + C.$$

These two formulas are valid for any function $u(x)$ that has a derivative; the first one is referred to as the *general exponential rule for differentiation* and the second one as the **general exponential rule for integration** or simply the **general exponential rule**.

2. Use Problem 1 and the general exponential rule to evaluate the indefinite integral $\int 2x\, e^{x^2}\, dx$.

3. Using the general exponential rule, evaluate the indefinite integral $\int \frac{e^{\sqrt{x}}}{\sqrt{x}}\, dx$. Start by making an appropriate choice for the function $u(x)$.   ◆

Classroom Discussion 5.3.3: The Method of Substitution

Quite often, introducing a new variable, say u, that depends on the original one, say x, simplifies the process of integration. This procedure is called **substitution** or **change of variables**. The general power and exponential rules are actually special cases of this method. Substitution allows us to integrate a variety of other types of functions.

1. Let $u = u(x)$ be a function with derivative $u'(x)$. Since you will be dealing with two variables u and x, you must find the relation between the differentials du and dx.

   a. Evaluate $\int u'(x)\, dx$.
   b. Evaluate $\int 1\, du$.
   c. Compare the expressions obtained in Problems 1a and 1b.
   d. What is the relation between du and dx?

2. The goal is to evaluate the integral $I = \int 2x\sqrt{x^2 - 3}\, dx$ using the substitution $u = x^2 - 3$.

   a. Compute $u'(x)$.
   b. Express du in terms of x and dx using the relation found in Problem 1d.
   c. Express the integral in terms of u only.
   d. Evaluate the integral using the new expression obtained in Problem 2c.
   e. Rewrite your previous answer in terms of x only.
   f. Compute directly the integral using the general power rule for antidifferentiation.
   g. Compare your findings from Problems 2e and 2f.

3. The goal is to evaluate the integral $J = \int (3x^2 + 1)\, e^{x^3 + x + 2}\, dx$ using the method of substitution.

   a. Which change of variable $u = u(x)$ do you suggest to simplify the integral?
   b. Compute $u'(x)$, then express du in terms of x and dx only.
   c. Express the integral in terms of u only.
   d. Evaluate the integral using the new expression obtained in Problem 3c.
   e. Rewrite your previous answer in terms of x only.

  f. Compute directly the integral using the general exponential rule for antidifferentiation.

  g. Compare your findings from Problems 3e and 3f.

4. Evaluate the integral $K = \int \frac{3x^2 + 2x + 1}{(x^3 + x^2 + x + 1)^4} dx$ using an appropriate change of variable.

Summary

To evaluate an indefinite integral $\int f(x)\, dx$ using the method of substitution, follow these steps:

  1. Choose a new variable $u = u(x)$ that would simplify integration.
  2. Compute $u'(x)$.
  3. Express the integral in terms of u by using the fact that $du = u'(x)\, dx$.
  4. Evaluate the integral using this new expression.
  5. Rewrite your final answer in terms of x. ◆

Additional Rules of Integration

1. $\displaystyle\int u^n(x)\, u'(x)\, dx = \int u^n\, du = \frac{1}{n+1} u^{n+1} + C, \quad n \neq -1$

                 (General power rule)

2. $\displaystyle\int e^{u(x)}\, u'(x)\, dx = \int e^u\, du = e^u + C$ (General exponential rule)

EXERCISES 5.3

In Exercises 1–16, compute the given integral. Use differentiation to justify your answers.

1. $\displaystyle\int \sqrt{x + 2}\, dx$

2. $\displaystyle\int \sqrt{3 - x}\, dx$

3. $\displaystyle\int 2x(x^2 + 1)^2\, dx$

4. $\displaystyle\int x^4(1 - x^5)^3\, dx$

5. $\displaystyle\int x^2\sqrt{x^3 + 9}\, dx$

6. $\displaystyle\int x(x^2 + 5)\sqrt{x^2 + 5}\, dx$

7. $\displaystyle\int_{-1}^{1} \frac{x + 1/2}{(x^2 + x + 1)^5}\, dx$

8. $\displaystyle\int \frac{x + 1}{(x^2 + 2x - 2)^3}\, dx$

9. $\displaystyle\int \frac{x^2 + 7}{\sqrt{x^3 + 21x + 8}}\, dx$

10. $\displaystyle\int_{0}^{1} \frac{x^3}{\sqrt{(x^4 + 1)^3}}\, dx$

11. $\displaystyle\int_{1}^{4} \frac{1}{\sqrt{x}(1 + \sqrt{x})^2}\, dx$

12. $\displaystyle\int_{0}^{1} \frac{\sqrt{x}}{(2 + \sqrt{x^3})^2}\, dx$

13. $\displaystyle\int \frac{1}{x^2} e^{1/x}\, dx$

14. $\displaystyle\int (x + 1) e^{x^2 + 2x - 2}\, dx$

15. $\displaystyle \int \frac{e^x}{(e^x - 2)^3}\, dx$

16. $\displaystyle \int \frac{e^x + x}{\sqrt{2e^x + x^2 + 1}}\, dx$

17. Integration by Parts

Using the sum, constant multiple, and chain rules for differentiation, we derived rules and techniques of integration. Here, we investigate a technique of integration that is derived from the product rule. Let u and v be two differentiable functions on the real line.

a. Write the formula for $(uv)'$ in terms of u, v, u', and v'.

b. Using Problem a, show that $\int uv'\, dx = uv - \int u'v\, dx$.

Integration by parts is a technique of integration based on the formula in Problem b. This is a very useful technique when dealing with integrands involving products of polynomial and exponential functions.

c. Can you find $\int x e^x\, dx$ using the techniques of integration you learned in Sections 5.1 or 5.3?

d. Use integration by parts to evaluate $\int x e^x\, dx$ with $u = x$ and v such that $v' = e^x$.

e. Evaluate $\int x e^{3x}\, dx$ using integration by parts.

f. Evaluate $\int x^2 e^x\, dx$ using integration by parts twice.

g. Evaluate $\int (x^2 + x + 1) e^x\, dx$ using integration by parts twice.

h. Evaluate $\int x^3 e^x\, dx$ using integration by parts repeatedly.

i. Given a polynomial function $P(x)$, explain how you can find the antiderivatives of the function $P(x) e^x$. How many times would you need to use integration by parts in this case?

j. Given a polynomial function $P(x)$ and a numerical constant k, explain how you can find the antiderivatives of the function $P(x) e^{kx}$. How many times would you need to use integration by parts in this case?

Integration by parts can be used to integrate a variety of functions other than those of the form $P(x)e^{kx}$.

18. The Method of Partial Fractions

This is a technique for integrating rational functions that is based on the fact that any rational function can be decomposed into partial fractions; these are simple rational functions with simple antiderivatives. To give you an idea of how the method works, we discuss one example.

a. Consider the function $f(x) = \dfrac{2x - 1}{x^2(x - 1)^2}$. Can you use the techniques of integration learned so far to find the antiderivatives of $f(x)$?

b. A theorem from advanced algebra says that there are numerical constants A, B, C, and D such that $f(x) = \dfrac{A}{x} + \dfrac{B}{x^2} + \dfrac{C}{x - 1} + \dfrac{D}{(x - 1)^2}$. Find A, B, C, and D. Hint: rewrite the sum on the right as a single quotient, identify its numerator with $2x - 1$, then solve for A, B, C, and D.

c. Use the decomposition obtained in Problem b to integrate $f(x)$.

CHAPTER 5 REVIEW

A function F is an antiderivative of the function f if $F' = f$; a function that has an antiderivative is antidifferentiable or integrable.

If F is an antiderivative of f, then any function of the form $F + C$, where C is a numerical constant, is also an antiderivative of f. Conversely, any antiderivative of f is of this form.

The set of all antiderivatives of f is denoted by $\int f(x)\, dx$; for the sake of simplicity, we write $\int f(x)\, dx = F(x) + C$, where F is an arbitrary antiderivative of f.

Any formula obtained earlier when discussing differentiation leads now to a formula for antidifferentiation:

1. $\int x^n\, dx = \frac{1}{n+1}x^{n+1} + C, \; n \neq -1$ (Power rule)

2. $\int \frac{1}{x}\, dx = \ln|x| + C$ (Logarithmic rule)

3. $\int e^x\, dx = e^x + C$ (Exponential rule)

4. $\int kf(x)\, dx = k \int f(x)\, dx$ (Scalar multiplication rule)

5. $\int (f + g)(x)\, dx = \int f(x)\, dx + \int g(x)\, dx$ (Addition rule)

6. $\int (f - g)(x)\, dx = \int f(x)\, dx - \int g(x)\, dx$ (Difference rule)

7. $\int u^n(x)u'(x)\, dx = \frac{1}{n+1}u^{n+1}(x) + C, \; n \neq -1$ (General power rule)

8. $\int e^{u(x)}u'(x)\, dx = e^{u(x)} + C$ (General exponential rule)

9. $\int f[g(x)]g'(x)\, dx = f[g(x)] + C$ (Chain rule)

The definite integral of f from a to b, denoted by $\int_a^b f(x)\, dx$, is by definition the variation of any antiderivative F of f over the interval $[a, b]$; that is, $F(b) - F(a)$. The definite integral does not depend on the particular antiderivative chosen to evaluate it. Thus, to evaluate $\int_a^b f(x)\, dx$, first find an antiderivative F for f, then evaluate the difference $F(b) - F(a)$.

Applying the definition of definite integrals, we get Properties 1–3; applying further the properties of indefinite integrals, we get Properties 4–6.

1. $\int_a^a f(x)\, dx = 0$

2. $\int_b^a f(x)\, dx = -\int_a^b f(x)\, dx$

3. $\int_a^c f(x)\, dx = \int_a^b f(x)\, dx + \int_b^c f(x)\, dx$

4. $\int_a^b kf(x)\, dx = k \int_a^b f(x)\, dx$

5. $\int_a^b (f + g)(x)\, dx = \int_a^b f(x)\, dx + \int_a^b g(x)\, dx$

6. $\int_a^b (f - g)(x)\, dx = \int_a^b f(x)\, dx - \int_a^b g(x)\, dx$

CHAPTER 5 REVIEW EXERCISES

In Exercises 1–16, find the domain of the integrand, then evaluate the indefinite integral.

1. $\int 9x^4\, dx$

2. $\int 2x^{1/3}\, dx$

3. $\int \frac{3}{10}x^{9/2}\, dx$

4. $\int 6\sqrt{x}\,dx$

5. $\int -\frac{1}{x^2}\,dx$

6. $\int \frac{32}{7}x^{-4}\,dx$

7. $\int 5\sqrt[3]{x}\,dx$

8. $\int (x^9 + \frac{7}{6}x^3 + 4)\,dx$

9. $\int \left(e^x - \frac{1}{\sqrt[3]{x}}\right)\,dx$

10. $\int 5x\sqrt{x}\,dx$

11. $\int (x + 1)(x^4 + 9)\,dx$

12. $\int (\sqrt{x} - 1)\left(2 + \frac{1}{\sqrt{x}}\right)\,dx$

13. $\int (-x^{2/3} + x)\left(\frac{1}{x} - 2\right)\,dx$

14. $\int \frac{1}{\sqrt[5]{x}}(2x^3 + x - 3)\,dx$

15. $\int \frac{x^2 + 2x - 1}{x^4}\,dx$

16. $\int \frac{3x^{3/4} + 2}{\sqrt{x}}\,dx$

17. Use differentiation to show that
  a. $\int xe^x\,dx = (x - 1)e^x + C$
  b. $\int x^2 e^x\,dx = (x^2 - 2x + 2)e^x + C$
  c. $\int x^3 e^x\,dx = (x^3 - 3x^2 + 6x - 6)e^x + C$
  d. More generally, for any positive integer n, we have

$$\int x^n e^x\,dx = \Big[x^n - nx^{n-1} + n(n - 1)x^{n-2} - \cdots + (-1)^{n-1}n!\,x$$
$$+ (-1)^n n!\Big]e^x + C$$

  where $n! = n(n - 1) \cdots 3 \cdot 2 \cdot 1$.

18. Find the antiderivatives of $f(x) = |x - 1|$.

19. Find the antiderivatives of $f(x) = |2x - 1|$.

20. Find the antiderivative of $f(x) = x^3\sqrt{x}$ that takes the value 1 at $x = 1$.

21. Find the antiderivative of $f(x) = |2x|$ that takes the value 0 at $x = 3$.

22. Find the antiderivative of $f(x) = \frac{1}{x\sqrt{x}}$ that takes the value -2 at $x = 1$.

23. Find the antiderivative of $f(x) = x + \frac{1}{x^2}$ that takes the values $1/2$ at $x = 1$ and $3/2$ at $x = -1$.

24. Find the position function of a projectile fired straight up from a height of 10 meters above the ground and with an initial velocity of 40 meters per second. What is the position function if this were to happen on the moon? The acceleration of gravity on Earth is -9.8 meters per second squared. The acceleration of gravity on the moon is -1.6 meters per second squared.

25. A ball is dropped with a downward initial velocity from a roof 30 feet high. The ball hits the ground after 1 second. What is the initial velocity of the ball? What is the velocity of the ball at impact?

26. An object moves along a straight line with acceleration $a(t) = 20\sqrt{t} - 4/\sqrt{t}$. At $t = 9$, the object's position is 36 feet, and its velocity is 300 feet per second. Find the object's position function.

In Exercises 27–35, evaluate the definite integral.

27. $\int_{-1}^{1} x^{100} \, dx$

28. $\int_{9}^{16} (\sqrt{x} - \frac{1}{\sqrt{x}}) \, dx$

29. $\int_{1}^{-3} dx$

30. $\int_{-1}^{0} (x^2 - 1)(x^2 + 1) \, dx$

31. $\int_{1}^{2} x^3 (\frac{1}{x} + 2\sqrt{x} - 1) \, dx$

32. $\int_{1}^{2} \frac{2x^2 + 9x + 8}{x^5} \, dx$

33. $\int_{0}^{1} 3(e^x - 1) \, dx$

34. $\int_{3}^{5} \frac{x^2 - 1}{x + 1} \, dx$

35. $\int_{-1}^{1} |x| \, dx$

36. Consider the functions $F(x) = \int_{1}^{x} \frac{y}{y^3 + 2} \, dy$, $G(x) = \int_{x}^{1} t\sqrt{t + 1} \, dt$, and $H(x) = \int_{x}^{5} \frac{s + 2}{s - 1} \, ds$. Fill in the blanks:
   a. $F(x)$ is the _____ of $f(y) =$ _____ that takes the value ___ at $x =$ ___.
   b. $G(x)$ is the _____ of $g(t) =$ _____ that takes the value ___ at $x =$ ___.
   c. $H(x)$ is the _____ of $h(s) =$ _____ that takes the value ___ at $x =$ ___.

37. Given $\int_{1}^{2} f(x) \, dx = 4$, $\int_{2}^{3} f(x) \, dx = 2$, $\int_{1}^{2} g(x) \, dx = -1$, and $\int_{2}^{3} g(x) \, dx = 1$, evaluate
   a. $\int_{1}^{3} f(x) \, dx$;
   b. $\int_{3}^{1} g(x) \, dx$;
   c. $\int_{1}^{3} [2f(x) - 5g(x)] \, dx$.

In Exercises 38–50, evaluate the integral.

38. $\int \sqrt{x + 4} \, dx$

39. $\int_{0}^{1} \sqrt{x + 2} \, dx$

40. $\int \sqrt[4]{x - 5} \, dx$

41. $\int \sqrt{2x + 3} \, dx$

42. $\int (x + 3)(x^2 + 6x - 1)^{5/7} \, dx$

43. $\int x^2(x^3 + 5)^{98}\, dx$

44. $\int xe^{x^2}\, dx$

45. $\int (2x - 1)e^{x^2 - x + 2}\, dx$

46. $\int_{-1}^{0} 3x^2(x^3 + 1)^9\, dx$

47. $\int_0^1 \frac{x}{(x^2+1)^2}\, dx$

48. $\int_0^1 e^{3x}\, dx$

49. $\int_{-1}^{1} x\sqrt{x^2 - 1}(x^2 - 1)\, dx$

50. $\int_{-1}^{0} xe^{1-x^2}\, dx$

CHAPTER
6

Applications of Integration to Area

The word *area* is frequently used in real life. For example, the word might be used to describe the size of a given piece of property. How would you informally define the meaning of *area*?

Since ancient times, people have had to devise methods for measuring the areas of plane regions in order to solve real-life problems, such as dividing land among heirs. This task is considerably harder than that of measuring the lengths of plane curves. Why do you think this is so?

Our objective in this chapter is to develop, step by step and starting from scratch, a procedure for measuring the area of a plane region that has a continuous boundary and to produce a formula for the area of such a region. So, forget all the formulas related to area that you have learned in the past, and let us go on a discovery adventure!

6.1 AREAS OF POLYGONAL REGIONS

Areas of squares • Areas of rectangles, triangles, and trapezoids • Areas of polygonal regions

In order to develop an approach for measuring the areas of regions in the plane, let us start with the simplest geometric figure: a unit square (a square whose sides each have a length of 1 unit). By definition, we take its area to be 1 *square unit*. This definition is reasonable, and it is unambiguous since all unit squares are congruent.

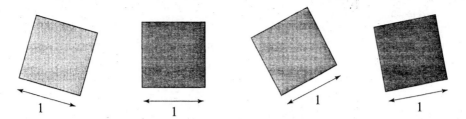

So, from now on, 1 square unit is the area occupied by a unit square. Based on this definition, let us compute the area of a square whose sides each have length a, where a is a positive real number.

Case 1: a is a whole number (in the following figure, $a = 3$). The a-by-a square can be divided into a^2 unit squares, and thus its area is a^2 times the area of a unit square, or a^2.

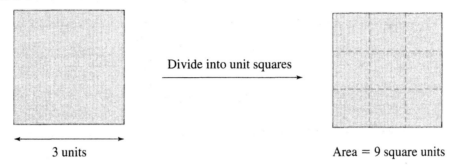

Divide into unit squares

3 units   Area = 9 square units

Case 2: a is a rational number that is not a whole number, so $a = p/q$, where p and q are two relatively prime whole numbers with $q \neq 0, 1$ (in the following figure, $a = 2/3$). At this point, you possess only a formula for the areas of squares whose side lengths are whole numbers. Therefore, you must think of a way to reduce this Case 2 to Case 1. If you duplicate the sides q times, as in the following figure, you get q^2 squares, each of which is congruent to the original square. Since each side of the resulting large square has length p (a whole number), it follows from Case 1 that the large square's area is p^2. Therefore, the original square's area is p^2/q^2, or a^2.

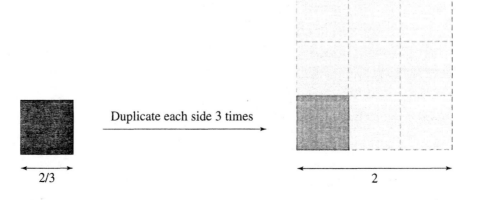

Duplicate each side 3 times

2/3   2

Case 3: a is an irrational number (for example, $a = \sqrt{2}$). The only formula you have so far for computing areas allows you to deal only with squares whose side lengths are rational numbers. Here, again, you must find a way to reduce the current case to the former one. Recall that, since a is a positive irrational number, there exists an increasing sequence $\{a_n\}_n$ of positive rational numbers satisfying $a = \lim_{n \to \infty} a_n$ (for $a = \sqrt{2}$, you may choose the sequence $a_1 = 1.4, a_2 = 1.41, a_3 = 1.414, \ldots$). Consider the squares whose sides are $a_1, a_2, \ldots a_n, \ldots$ (all rational numbers). Their areas are equal to $a_1^2, a_2^2, \ldots a_n^2, \ldots$, respectively, by Case 2. Since these squares approach the original square more and more as n increases, the original square's area must be $\lim_{n \to \infty} a_n^2 = a^2$.

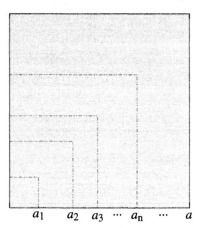

$$a_1 \quad\quad a_2 \quad a_3 \;\cdots\; a_n \;\cdots\; a$$

Are you convinced that the area of an a-by-a square is a^2 for any real number $a > 0$? From now on, you can use this result whenever necessary. To evaluate the areas of arbitrary polygonal regions, answer the questions in the following Classroom Discussion.

Classroom Discussion 6.1.1: Area of an Arbitrary Polygonal Region

1. What is the area of a rectangle whose length is a and whose width is b? To answer this question, you may proceed either by arguing as in the case of squares or by considering the square whose sides each have length $a + b$.

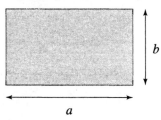

2. What is the area of a triangle whose height is h and whose base has length b? To answer this question, determine first a right triangle's area using your finding from Problem 1.

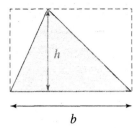

3. What is the area of a trapezoid whose height is h and whose bases have lengths b_1 and b_2? To answer this question, you may decompose the trapezoid into two triangles and use your findings from Problem 2.

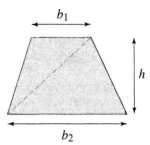

4. Explain how to compute the area of any region that has a polygonal boundary.

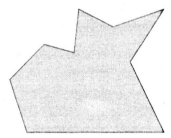

◆

EXERCISES 6.1

1. A middle-school student is confused when he realizes that the area of a square whose sides each have length 0.1 unit is 0.01 square unit; it is smaller than the common length of the sides! Another student doesn't see why the area of a square, whose sides each have length π units, is π^2 square units. Write to each student a paragraph explaining why the area is 0.01 square unit in the former case and why it is π^2 square units in the latter.

2. The following parallelogram has height h, and its base has length b. Based only on the area formulas proved so far, find the area of the parallelogram in terms of h and b using each of these ideas:

  a. Divide the parallelogram into two triangles along one of its diagonals.

  b. Divide the parallelogram into two triangles and a rectangle.

  c. Imagine, as Cavalieri did, that the parallelogram is made up of a stack of line segments of the same length.

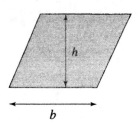

Historical Note: Bonaventura Cavalieri (1598–1647; from Habsburg Empire [now Italy])

Cavalieri was born and raised in Milan, where he entered a monastery at the age of 18. His interest in mathematics was inspired by Euclid's writings, and after a meeting with Galileo, he regarded himself a disciple of the astronomer. Cavalieri's theory of indivisibles, presented in his 1635 book *Geometria Indivisibilibus Continuorum Nova Quadam Ratione Promota* ("Geometry, Advanced in a New Way by the Indivisibles of the Continua"), was based on Archimedes's method of exhaustion, and it incorporated Kepler's theory of infinitesimally small geometric quantities. Cavalieri is best known for the slicing method, used for computing areas and volumes of geometric figures.

3. **Areas of Regular Polygons**

   a. A regular polygon has perimeter p and apothem a. Based only on the area formulas proved so far, determine its area in terms of p and a. Recall that a regular polygon's apothem is the distance between its center and one of its sides.

   b. Let n denote the number of sides of the regular polygon in problem a. Using trigonometry, express p in terms of n and a. Then, express the area of the regular n-gon in terms of n and a only.

   c. A regular hexagon has apothem a. Express its area in terms of a only.

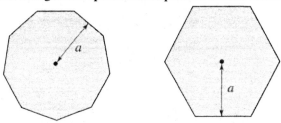

4. Find an approximate value for the area of the following polygonal region; use ruler measurements and any of the area formulas proved so far.

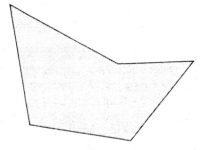

6.2   APPROXIMATE VALUES FOR THE AREA OF AN IRREGULAR SHAPE

Approximating areas • Riemann sums • Trapezoidal method • Left, right, and midpoint rectangular methods

Can you find the areas of the following figures by applying the techniques used in Section 6.1? Explain why or why not.

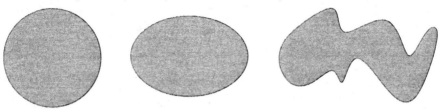

As you can see from these figures, the boundary of a plane region is *not* necessarily polygonal. Throughout this book, we refer to a region like the ones here as an *irregular* shape. Our goal in this section is to discuss ways to measure the area of irregular shapes.

Classroom Connection 6.2.1: Area of Madagascar

The following exploration is taken from page 464 in the sixth-grade textbook *Math Thematics*, *Book 1*. An approximate value for the area of the island Madagascar is to be found therein. Discuss this exploration in small groups.   ◆

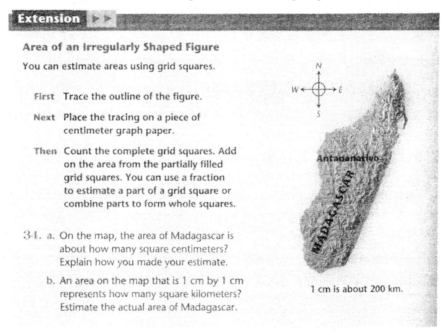

Extension ▶ ▶

Area of an Irregularly Shaped Figure

You can estimate areas using grid squares.

First   Trace the outline of the figure.

Next   Place the tracing on a piece of centimeter graph paper.

Then   Count the complete grid squares. Add on the area from the partially filled grid squares. You can use a fraction to estimate a part of a grid square or combine parts to form whole squares.

34. a. On the map, the area of Madagascar is about how many square centimeters? Explain how you made your estimate.

b. An area on the map that is 1 cm by 1 cm represents how many square kilometers? Estimate the actual area of Madagascar.

1 cm is about 200 km.

464   **Module 7**   Wonders of the World

Classroom Connection 6.2.2: Area of a Foot

In the following problem, taken from page 20 in the sixth-grade textbook *Connected Mathematics, Covering and Surrounding*, approximate values for the area of a foot are to be found by counting unit squares and/or fractions of unit squares. Work in small groups and solve Problems 1–5 to approximate the area of the printed foot, using its magnified image.

1. Find a *lower bound* for the area of the printed foot. A **lower bound** simply means a specific value that is less than or equal to the exact value of the printed foot's area.

2. Find an *upper bound* for the area of the printed foot. An **upper bound** simply means a value that is greater than or equal to the exact value of the printed foot's area.

3. Alice approximates the area of the printed foot by 131 square units. An approximate value differs from the true value by a certain amount, called the **error**. What is the *magnitude* of the error generated by Alice's approximation using your answers to Problems 1 and 2? The **magnitude** of an error is an upper bound for the error; it gives an idea about the accuracy of the approximation to the area's exact value.

4. Bob approximates the area of the printed foot by the average value between the lower and upper bounds found in Problems 1 and 2. What is the magnitude of the error in this case?

5. If you were to choose between using Alice's approximation or Bob's approximation, which one would you pick? Explain the reasons behind your choice.

6. What do you think will happen to the error if you use a finer grid?

7. How about if you continue using finer and finer grids? Explain.   ◆

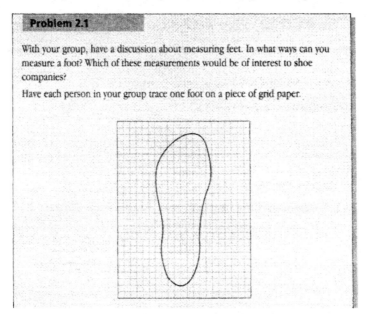

Problem 2.1

With your group, have a discussion about measuring feet. In what ways can you measure a foot? Which of these measurements would be of interest to shoe companies?

Have each person in your group trace one foot on a piece of grid paper.

For each person's foot, estimate the length, width, area, perimeter, and any other measures your group thinks should be included. Record your data in a table with these column headings:

Student	Shoe size	Foot length	Foot width	Foot area	Foot perimeter

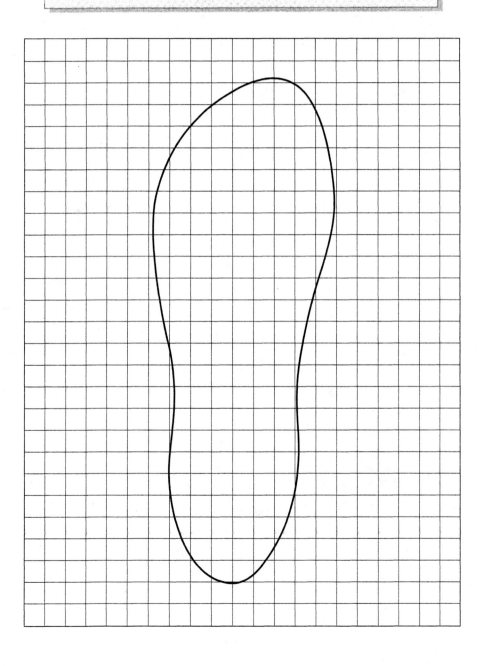

By examining the problems from the preceding Classroom Connections more closely, you can see that the underlying idea in finding approximate values for an irregular shape's area is to approximate the shape itself by a region whose boundary is polygonal, compute the area of the approximating polygonal region using the results of Section 6.1, and then use the obtained value to approximate the irregular shape's area. The smaller the "gaps" between the given irregular shape and the approximating polygonal region, the better your estimate will be for the area. Thus, if the approximating polygonal regions are chosen to approach the given irregular shape more closely, their areas will in turn approach more and more closely the area of the region of interest. The exact value of the area can thus be obtained by passing to the limit:

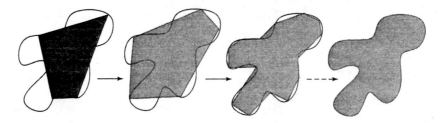

To illustrate this process more precisely, consider the region of the plane bounded by the graph of a nonnegative continuous function $y = f(x)$, the x-axis, and the vertical lines $x = a$ and $x = b$.

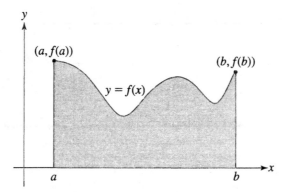

Draw examples of plane regions to see that you can decompose the vast majority of irregular shapes into subregions, each of which is a region bounded by the graph of some function.

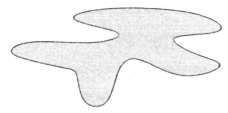

Return to the region below the graph of f and above the x-axis. The only part of the boundary creating a difficulty here is the curved path between points $(a, f(a))$ and $(b, f(b))$. Before looking at Classroom Discussion 6.2.1, think of a natural way to approximate an irregular shape of this type using polygonal regions.

Classroom Discussion 6.2.1: The Trapezoidal Method

Pick $n - 1$ arbitrary points $P_1, P_2, \dots, P_{n-1}$ on the curved path between the endpoints $(a, f(a))$ and $(b, f(b))$ (in the figure, $n = 5$). Then, construct the line segments connecting any two consecutive points lying on this path. The polygonal region underneath these line segments and above the x-axis can serve as an approximating region for the irregular one. Think about how you can compute the area of this polygonal region.

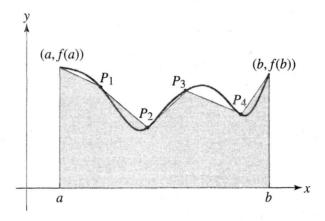

As you may have already noticed, the preceding polygonal region can be decomposed into n trapezoidal tiles (in the figure, there are five such tiles since $n = 5$).

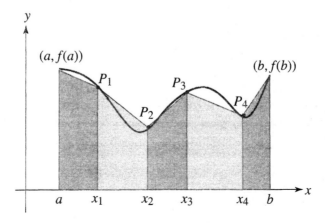

The area of each of these tiles can be computed easily using the formula obtained in Section 6.1. The result is summarized in this table:

$$\text{Area of each tile} \ = \ \frac{1}{2} \ \text{width (base 1 + base 2)}$$

$$\text{Area of Tile 1} \ = \ \frac{1}{2}(x_1 - a)\big[f(a) + f(x_1)\big]$$

$$\text{Area of Tile 2} \ = \ \frac{1}{2}(x_2 - x_1)\big[f(x_1) + f(x_2)\big]$$

$$\vdots \qquad\qquad \vdots$$

$$\text{Area of Tile } i \ = \ \frac{1}{2}(x_i - x_{i-1})\big[f(x_{i-1}) + f(x_i)\big]$$

$$\vdots \qquad\qquad \vdots$$

$$\text{Area of Tile } n \ = \ \frac{1}{2}(b - x_{n-1})\big[f(x_{n-1}) + f(b)\big]$$

The area of the polygonal region denoted by T_n is the sum of the areas of all these trapezoidal tiles. It is an approximate value for the area $\mathcal{A}$ of the irregular shape.

1. How can you reduce the error that is generated when approximating the area of the irregular shape with T_n? Explain.
2. Louise claims that the exact value of the irregular shape's area can be obtained by taking the limit of T_n as the number of points you select *throughout* the curved path goes to ∞. Do you agree with Louise? Explain.
3. For which functions are the areas T_n necessarily lower bounds for the areas of the resulting irregular shapes? (Hint: Think about the shape of their graphs.)
4. For which functions are the areas T_n necessarily upper bounds for the areas of the resulting irregular shapes?

Convenient Choice.   Originally, you randomly picked points $P_1, P_2, \dots, P_{n-1}$ along the curved path. However, if you are seeking a simple expression for the approximating areas T_n, it is helpful to select these points so that their x-coordinates $x_1, x_2, \dots, x_{n-1}$ are evenly spaced along the interval $[a, b]$; that is,

$$x_1 - a = x_2 - x_1 = \dots = b - x_{n-1}.$$

Denote this common value by δ.

5. Show that in this case,

$$T_n = \frac{\delta}{2}\left[f(a) + 2\sum_{i=1}^{n-1} f(x_i) + f(b)\right].\tag{1}$$

6. Find the value of δ in terms of a, b, and n only. Then replace this value in the preceding expression of T_n.

7. Express each coordinate x_i in terms of a, b, and i.

8. Write a calculator program that will compute T_n for arbitrary n. Your program should be set up to provide the value of T_n once you enter information about f, a, b, and n.

This method of finding approximate values for the area of the region below the graph of a function and above the x-axis is called **the trapezoidal method**. Formulas such as (1) are attributed to the mathematician Bernhard Riemann and are usually referred to as **Riemann sums for the function f on the interval $[a, b]$.** ◆

Historical Note: Bernhard Riemann (1826–1866; from Hanover [now Germany])

Riemann received his Ph.D. under the direction of Gauss at the University of Göttingen. Although his work often lacked in rigor, Riemann made major contributions to the theory of functions of a complex variable, mathematical physics, number theory, and the foundations of geometry. In calculus, *Riemann sums* provide a mechanism for computing areas below functions that are not necessarily continuous.

EXAMPLE Let $\mathcal{A}$ be the area of the plane region bounded by the graph of $f(x) = 1/x^2$, the x-axis, and the vertical lines $x = 1$ and $x = 5$, let T_n be the area of the polygonal region with n trapezoids of equal width, as described in the preceding section "Convenient Choice".

1. Compute by hand T_4 and T_8. Are these approximations upper or lower bounds for $\mathcal{A}$?

2. Compute $T_{10}, T_{20}, \dots, T_{200}$ using your calculator program from Classroom Discussion 6.2.1. Tabulate your results.

3. Based on your table and on your analysis of what happens as n increases, conjecture an approximate value for $\mathcal{A}$ with an error of magnitude less than 10^{-3}.

Solution

1. The area T_4 is obtained by applying the formula (1) or by directly computing the area of each of the trapezoidal tiles involved.

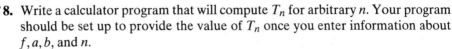

$$T_4 = \frac{5-1}{8}\left(1 + 2\cdot\frac{1}{4} + 2\cdot\frac{1}{9} + 2\cdot\frac{1}{16} + \frac{1}{25}\right) \approx 0.94361.$$

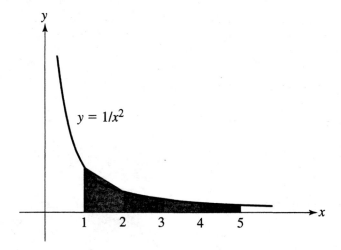

Since the graph of the function is concave upward, T_4 is an upper bound for $\mathcal{A}$; that is, $\mathcal{A} \leq T_4 \approx 0.94361$.

Similarly, the area T_8 may be obtained by applying directly the formula (1) as follows:

$$T_8 = \frac{5 - 1}{16}\left(1 + 2 \cdot \frac{4}{9} + 2 \cdot \frac{1}{4} + 2 \cdot \frac{4}{25}\right.$$

$$\left. + 2 \cdot \frac{1}{9} + 2 \cdot \frac{4}{49} + 2 \cdot \frac{1}{16} + 2 \cdot \frac{4}{81} + \frac{1}{25}\right) \approx 0.83953.$$

For the same reason, $T_8 \approx 0.83953$ is also an upper bound for $\mathcal{A}$.

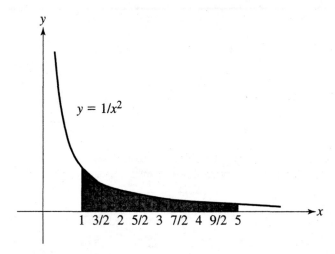

2.

n	T_n
10	0.82568
20	0.80656
30	0.80293
40	0.80165
50	0.80106
60	0.80073
70	0.80054
80	0.80041
90	0.80033
100	0.80026

n	T_n
110	0.80023
120	0.80018
130	0.80016
140	0.80013
150	0.80012
160	0.80010
170	0.80009
180	0.80008
190	0.80007
200	0.80006

3. By analyzing the table in the preceding problem, it seems reasonable to conjecture that the approximation $A \approx 0.800$ generates an error of magnitude less than 10^{-3}; that is, $0.799 < A < 0.801$.

In Section 6.3, you will learn how to compute the exact value of the area A, and thus you will have the opportunity to check whether your conjecture is true or false. ■

Classroom Discussion 6.2.2. The Rectangular Methods

This Classroom Discussion's goal is to approximate the region in the plane that is under the graph of a continuous nonnegative function with appropriate *rectangular* tiles instead of trapezoidal tiles in order to find approximate values for its area. Pick $n - 1$ arbitrary points $P_1, P_2, \ldots, P_{n-1}$ on the curved path between the endpoints $(a, f(a))$ and $(b, f(b))$ (in the figure $n = 5$). Then, think about how you can use the points that are lying on the curved path to form rectangular tiles to approximate the irregular shape.

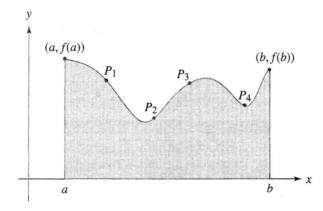

Let $x_1, x_2, \ldots, x_{n-1}$ be the x-coordinates of points $P_1, P_2, \ldots, P_{n-1}$. For convenience, set $a = x_0$ and $b = x_n$.

1. **The Left Rectangular Method:** Guided by the following problems, discuss this first variant of the rectangular method in small groups.

   a. For each $1 \leq i \leq n$, the ith rectangle is defined by three vertices $(x_{i-1}, 0)$, $(x_i, 0)$, and $(x_{i-1}, f(x_{i-1}))$. Draw the first, the ith and the nth rectangles.

   b. Find the area of each rectangular tile.

   c. Write down the Riemann sums for f on the interval $[a, b]$ corresponding to your rectangular tiles.

   d. Assume $P_1, P_2, \ldots, P_{n-1}$ are selected such that the coordinates $x_0, x_1, \ldots, x_n$ are equally spaced. Show that in this case, the Riemann sums are given by the formula

   $$L_n = \frac{b-a}{n} \sum_{i=1}^{n} f(x_{i-1}). \tag{2}$$

   e. Analyze what happens as $n \longrightarrow \infty$. Then, write down a formula for the area $\mathcal{A}$ of the irregular shape.

   f. For which functions are the areas L_n necessarily lower bounds for the areas of the resulting irregular shapes? (Hint: Think about the shape of their graphs.)

   g. For which functions are the areas L_n necessarily upper bounds for the areas of the resulting irregular shapes?

   h. Write a calculator program for computing the areas L_n. Set up your program to provide the value of L_n once you enter the given information on $f, a, b,$ and n.

2. **The Right Rectangular Method**

   a. Describe the right rectangular method by analogy with the left one, following similar steps. (Denote by R_n the Riemann sum corresponding to the right rectangular method with n rectangular tiles.)

   b. Is there a link between the trapezoidal method and the left and right rectangular methods? Explain.

   c. As n becomes very large, which approximate values for the area do you think become more accurate, T_n, L_n, or R_n?

3. **The Midpoint Rectangular Method:** Describe this method by analogy with the left and right rectangular methods, following similar steps. (Denote by M_n the Riemann sum corresponding to the midpoint rectangular method with n rectangular tiles.)

4. Generalization

   a. How can you describe *simultaneously* the left, the right, and the midpoint methods? Your description should include all three methods as particular cases.

   b. Based on this generalization, describe what needs to be done in order for the Riemann sums to be lower/upper bounds for the area of the irregular shape. For instance, how should the heights of the rectangular tiles be chosen?   ◆

EXAMPLE Let A be the area of the plane region bounded by the graph of $f(x) = 1/x^2$, the x-axis, and the vertical lines $x = 1$ and $x = 5$; let L_n and R_n be the areas of the polygonal regions with n rectangular tiles as described in the left and right rectangular methods, respectively.

   1. Compute by hand L_4 and L_8. How do these values compare to the (unknown) value of the area A?

   2. Compute by hand R_4 and R_8. How do these values compare to the area A?

   3. Using your calculator program from Classroom Discussion 6.2.2, compute L_n and R_n for $n = 10, 20, \ldots, 200$. Tabulate your results.

   4. Among the integers n in your table, determine, if possible, the smallest one that allows you to find an approximate value for A with an error of magnitude less than 10^{-1}, 10^{-2}, and 10^{-3}, respectively.

Solution

   1. The area L_4 is obtained by applying the formula (2) or by directly computing the area of each of the rectangular tiles involved.

$$L_4 = \frac{5-1}{4}\left(1 + \frac{1}{4} + \frac{1}{9} + \frac{1}{16}\right) \approx 1.42361.$$

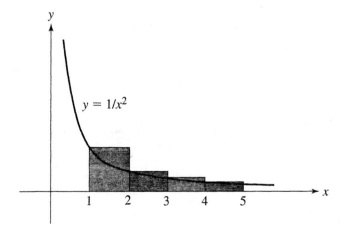

Since the function is decreasing, L_4 is an upper bound for $\mathcal{A}$; that is, $\mathcal{A} \leq L_4 \approx 1.42361$.

Similarly, the area L_8 may be obtained by directly applying the formula (2).

$$L_8 = \frac{5 - 1}{8}\left(1 + \frac{4}{9} + \frac{1}{4} + \frac{4}{25} + \frac{1}{9} + \frac{4}{49} + \frac{1}{16} + \frac{4}{81}\right) \approx 1.07954$$

For the same reason, the area $L_8 \approx 1.07954$ is also an upper bound for $\mathcal{A}$.

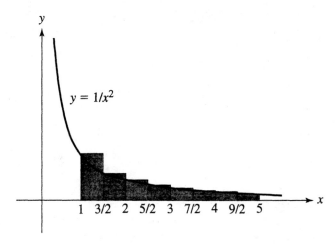

2. The area R_4 is obtained by directly applying the formula for the Riemann sums obtained in part 2 of Classroom Discussion 6.2.2 or by directly computing the area of each of the rectangular tiles involved.

$$R_4 = \frac{5 - 1}{4}\left(\frac{1}{4} + \frac{1}{9} + \frac{1}{16} + \frac{1}{25}\right) \approx 0.46361.$$

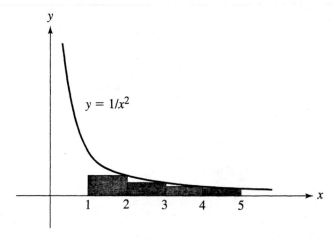

Since the function is decreasing, R_4 is a lower bound for $\mathcal{A}$; that is, $\mathcal{A} \geq R_4 \approx 0.46361$.

Similarly, the area R_8 may be obtained by directly applying the formula for the Riemann sums obtained in part 2 of Classroom Discussion 6.2.2.

$$R_8 = \frac{5-1}{8}\left(\frac{4}{9} + \frac{1}{4} + \frac{4}{25} + \frac{1}{9} + \frac{4}{49} + \frac{1}{16} + \frac{4}{81} + \frac{1}{25}\right) \approx 0.59953$$

For the same reason, the area $R_8 \approx 0.59953$ is also a lower bound for $\mathcal{A}$.

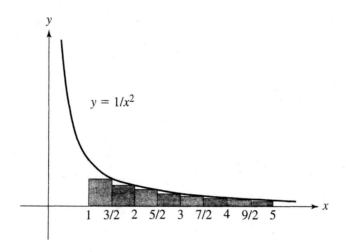

3.

n	L_n	R_n
10	1.01768	0.63368
20	0.90256	0.71056
30	0.86693	0.73893
40	0.84965	0.75365
50	0.83946	0.76266
60	0.83273	0.76873
70	0.82797	0.77311
80	0.82441	0.77641
90	0.82166	0.77899
100	0.81946	0.78106

n	L_n	R_n
110	0.81767	0.78276
120	0.81618	0.78418
130	0.81493	0.78539
140	0.81385	0.78642
150	0.81292	0.78732
160	0.81210	0.78810
170	0.81139	0.78880
180	0.81075	0.78941
190	0.81018	0.78997
200	0.80967	0.79047

4. For each value of n, we have $R_n \leq \mathcal{A} \leq L_n$. The error generated when estimating $\mathcal{A}$ with any value in the range $[R_n, L_n]$ has magnitude less than or equal to $L_n - R_n$. The error incurred in approximating $\mathcal{A}$ with the average value $(R_n + L_n)/2$ has magnitude less than or equal to $(L_n - R_n)/2$.

By analyzing the table in the preceding problem, you can see that $n = 20$ is the smallest integer for which $L_n - R_n < 0.2$. Therefore, the average value $(R_{20} + L_{20})/2 = 0.80656$ must approximate A with an error of magnitude less than 0.1.

Similarly, by analyzing the table, you can see that $n = 200$ is the smallest integer for which $L_n - R_n < 0.02$. Therefore, the average value $(R_{200} + L_{200})/2 = 0.80007$ must approximate A with an error of magnitude less than 0.01.

To find an approximate value for A with an error of magnitude less than 10^{-3}, you would need to compute R_n and L_n for larger values of n. The best conclusion you can draw from the current table is that $0.79047 \le A \le 0.80967$. ∎

Classroom Discussion 6.2.3: Riemann Sums for Functions Taking Negative Values

So far, we have considered Riemann sums only for nonnegative functions. However, the expressions for Riemann sums make complete sense for functions that also take negative values. For instance, the expression obtained for the Riemann sum corresponding to the function f over the interval $[a, b]$ using n trapezoidal tiles having the same width is

$$T_n = \frac{b - a}{2n} \left[f(a) + 2 \sum_{i=1}^{n-1} f(x_i) + f(b) \right],$$

where $x_i = a + \frac{b-a}{n} i$. This algebraic expression is well defined even if f takes negative values. Recall that, in the case of continuous functions taking only nonnegative values, the Riemann sums with a large number of tiles provide approximate values for the areas of the regions in the plane that are below their graphs and above the x-axis.

1. Let $f(x)$ be a continuous function defined on the interval $[a, b]$ taking only nonpositive values. Give a geometric interpretation for the corresponding Riemann sums. Explain your reasoning.

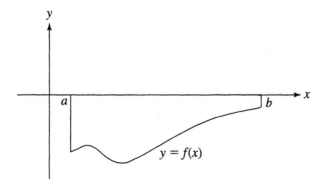

2. Denote by A_1, A_2, and A_3 the areas of the shaded regions in the following figure. Find a geometric interpretation for the corresponding Riemann sums in terms of the areas A_1, A_2, and A_3. Explain your reasoning.

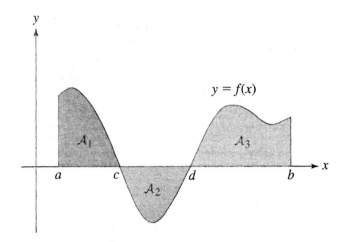

Hint: One approach is to consider the functions f^+ and f^- where $f^+(x) = f(x)$ if $f(x) > 0$ and 0 otherwise, and where $f^-(x) = f(x)$ if $f(x) < 0$ and 0 otherwise. Express f in terms of f^+ and f^-, then find the relationship between the Riemann sums of f, f^+, and f^-. ◆

EXERCISES 6.2

Classroom Connection 6.2.3: Area of an Arbitrary Shape

In the following problem, taken from page 86 in the sixth-grade textbook *Connected Mathematics, Covering and Surrounding*, approximate values for the area of an arbitrary shape are to be found by counting squares. Work Problems a–f below.

1. **a.** Find a lower bound α and an upper bound β for the area A of the shape under consideration.

   b. What is the magnitude γ of the error generated when approximating the area A with some value in the range $[\alpha, \beta]$?

   c. Which approximate value for the area A generates an error of at most $\gamma/2$?

   d. Use a finer grid and then answer Questions a–c.

   e. Which of the two grids leads to better approximations for the area A?

   f. What do you think will happen if you continue using finer and finer grids? Explain. ◆

3. How would you estimate the area and perimeter of an irregular figure such as the one drawn on the grid below? What is a reasonable estimate for the area of this particular figure?

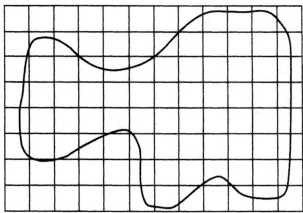

Area and perimeter are among the most useful concepts for measuring the size of geometric figures. You will use strategies for estimating and calculating the size of geometric figures in many future units of *Connected Mathematics*, especially those that deal with surface area and volume of solid figures, similarity, and the Pythagorean Theorem. You will also find that area and volume estimates and calculations are required in a variety of practical and technical problems.

2. Let $\mathcal{A}$ be the area of the plane region bounded by the graph $y = \sqrt{x}$, the x-axis, and the vertical lines $x = 1$ and $x = 4$.

Part 1. Let T_n be the area of the polygonal region with n trapezoidal tiles.

a. Compute by hand T_3. How does this value compare to the (unknown) value of $\mathcal{A}$?

b. Compute $T_{30}, T_{60}, \ldots, T_{300}$ using your calculator program from Classroom Discussion 6.2.1. Tabulate your results.

c. Based on your table and on your analysis of what happens as n increases, conjecture a value for $\mathcal{A}$ that is correct to three decimal places.

Part 2. Now, let L_n and R_n be the areas of the polygonal regions with n rectangular tiles corresponding to the left and right rectangular methods, respectively.

a. Compute by hand L_3. How does this value compare to the area $\mathcal{A}$?

b. Compute by hand R_3. How does this value compare to the area $\mathcal{A}$?

c. Using your calculator program from Classroom Discussion 6.2.2, compute $L_{30}, L_{60}, \ldots, L_{300}$ and $R_{30}, R_{60}, \ldots, R_{300}$. Tabulate your results.

d. Among the integers $30, 60, \ldots, 300$, determine, if possible, the smallest one that allows you to find an approximate value for the area $\mathcal{A}$ with an error of magnitude less than 10^{-1}, 10^{-2}, and 10^{-3}, respectively.

3. Let $\mathcal{A}$ be the area of the plane region bounded by the graph of the function $f(x)$, the x-axis, and the vertical lines $x = -1$ and $x = 1$, where

$$f(x) = \begin{cases} 1 + x^3 & \text{if } -1 \le x \le 0 \\ 1 - x^2 & \text{if } 0 \le x \le 1. \end{cases}$$

a. Sketch the graph of the function f.

b. Using the rectangular method with four rectangular tiles, find lower and upper bounds for the area $\mathcal{A}_1$ of the subregion for which $x \le 0$.

c. Using the rectangular method with four rectangular tiles, find lower and upper bounds for the area $\mathcal{A}_2$ of the subregion for which $x \ge 0$.

d. Use your findings in b and c to derive lower and upper bounds for the area $\mathcal{A}$.

4. a. Using ruler measurements and nine trapezoidal tiles, find an approximate value for the area of the shaded region in the following figure.

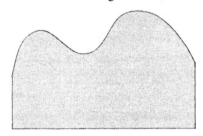

b. Using ruler measurements and nine rectangular tiles, find a lower and an upper bound for the area of the figure's shaded region.

5. Let A be the area of the plane region bounded by the graph $y = e^{-x}$, the x-axis, and the vertical lines $x = 0$ and $x = 5$; let M_n be the area of the polygonal region obtained using the midpoint rectangular method with n tiles. Compute M_5.

6. Let A be the area of the plane region bounded by the graph $y = 1/x^2$, the x-axis, and the vertical lines $x = 1$ and $x = 4$; let M_n be the area of the polygonal region obtained by using the midpoint rectangular method with n tiles. Compute M_3 by hand.

7. Consider the region bounded by the graphs of the functions $f(x) = x^2$ and $g(x) = \dfrac{1}{x}$, the x-axis, and the vertical lines $x = 0$ and $x = 2$. Denote by A its area.

   a. Sketch the region previously described.
   b. Write down a calculator program that enables you to compute the approximating values for the area A when using an even number of trapezoidal tiles; that is, T_{2n}.
   c. Fill in the following table.

n	T_{2n}
20	
40	
60	
80	
100	
120	
140	
160	
180	
200	
220	
240	
260	

   d. Based on your table and on your analysis of what happens as n increases, conjecture a value for A that is correct to three decimal places.

8. Consider the region bounded by the graphs of the functions $f(x) = 2x^3$ and $g(x) = 1 + \sqrt{x}$, the x-axis, and the vertical lines $x = 0$ and $x = 2$. Denote by A its area, then answer Questions a–d in Problem 7.

PROJECTS AND EXTENSIONS 6.2

I. Riemann Sums for the Function $f(x) = x^2$

This project's goal is to find the area A of the region above the x-axis and below the parabola $y = x^2$, where $0 \le x \le 1$.

1. Approximate values for A: Denote by L_n and R_n the areas of the polygonal regions with n rectangular tiles as described in the left and right rectangular methods, respectively.

   a. Compute by hand L_5 and L_{10}. How do these values compare to the area A?
   b. Compute by hand R_5 and R_{10}. How do these values compare to the area A?
   c. Using your calculator program from Classroom Discussion 6.2.2, compute L_n and R_n for $n = 40, 80, \ldots , 400$. Tabulate your results.
   d. Among the integers in your table, determine, if possible, the smallest one that allows you to find an approximate value for A correct to two decimal places.

2. Exact value of the area A: Here you will simplify the expression for the Riemann sums R_n, then compute the actual value of A by taking the limit as n goes to ∞.

   a. Show that for each integer $n \ge 1$,

   $$R_n = \frac{1}{n^3}\left(1^2 + 2^2 + \cdots + n^2\right).$$

   b. To find a simple expression for the sum $1^2 + 2^2 + \cdots + n^2$, consider the identity $(k + 1)^3 = k^3 + 3k^2 + 3k + 1$ and follow these steps:
   - Sum each side of the identity from $k = 1$ up to $k = n$.
   - Simplify the resulting equality by canceling the terms that appear on both sides.
   - Evaluate the sum $\displaystyle\sum_{k=1}^{n} k$. See Section 1.1 for adding consecutive terms of an arithmetic sequence.
   - Use the preceding steps to deduce the value of the sum $\displaystyle\sum_{k=1}^{n} k^2$.
   - Simplify the expression of R_n.

   c. Find the exact value of the area A.

II. Riemann Sums for the Function $f(x) = x^3$

The goal of this project is to find the area A of the region above the x-axis and below the graph $y = x^3$ where $0 \le x \le 1$. Answer the same questions as in Project I with the following changes:

In Problem 2a, show that

$$R_n = \frac{1}{n^4}\left(1^3 + 2^3 + \cdots + n^3\right).$$

In Problem 2b, consider the identity $(k + 1)^4 = k^4 + 4k^3 + 6k^2 + 4k + 1$ and use similar ideas to derive a simple expression for R_n.

3. For which other functions do you think a reasoning along the same lines would work? Explain.

III. Riemann Sums for Increasing and Decreasing Functions

Consider the plane region bounded by the graph $y = f(x)$, the x-axis, and the vertical lines $x = a$ and $x = b$, where f is an increasing, continuous, and nonnegative function defined on the interval $[a, b]$.

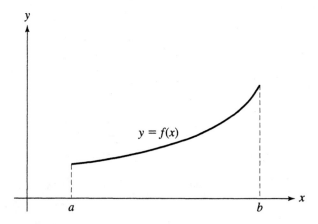

1. Sketch the rectangular tiles approximating the given region, as described in the left rectangular method, with $n = 5$.

2. Sketch the rectangular tiles approximating the given region, as described in the right rectangular method, with $n = 5$.

3. Shade the gap that is between the region formed of all the rectangles obtained in 1 and the one formed of all those obtained in 2. What does the area of this gap represent?

4. Stack up all of the rectangles forming the shaded region. What are the height and width of the resulting rectangle?

5. What would be the height and width of the resulting rectangle if $n = 6, 7, \ldots$?

6. What happens to the area of the rectangle in 4 as $n \to \infty$?

7. Use your findings to convince your classmates that the limit of the Riemann sums is indeed the area of the region below the graph $y = f(x)$ and above the x-axis.

8. Does this reasoning still work if the function $f(x)$ is decreasing instead of increasing?

6.3   EXACT VALUE OF THE AREA OF AN IRREGULAR SHAPE

Fundamental Theorem of Calculus • Areas of regions bounded by graphs

Consider the graph $y = f(x)$ where $f(x)$ is a continuous function defined on the interval $[a, b]$. Our task in this section is to find the exact value of the area $\mathcal{A}$ of the region of the plane bounded by the graph of f, the x-axis, and the vertical lines $x = a$ and $x = b$.

Classroom Discussion 6.3.1: The Fundamental Theorem of Calculus

Throughout this Classroom Discussion, the function f is assumed to be nonnegative. Fix $a \leq x \leq b$ and denote by $\mathcal{A}(x)$ the area of the region in the plane bounded by the graph of f, the x-axis, and the vertical lines passing through points $(a, 0)$ and $(x, 0)$. The function $\mathcal{A}(x)$ is sometimes called the *area-so-far* function.

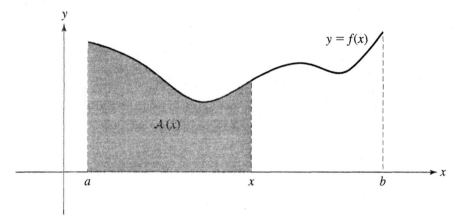

In order to familiarize yourself with the function $\mathcal{A}(x)$, answer the following two questions.

1. What are the values of $\mathcal{A}(a)$, $\mathcal{A}(\frac{a+b}{2})$, and $\mathcal{A}(b)$?

2. Is the function $\mathcal{A}(x)$ increasing, decreasing, or neither? Explain.

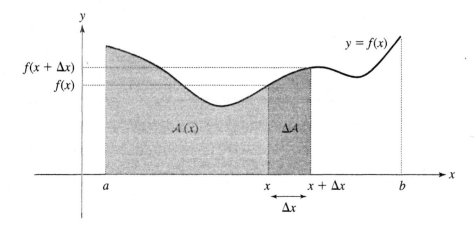

Let x increase by a small amount Δx; then the area $A(x)$ increases by an amount ΔA corresponding to the additional region.

3. If Δx is very small, which polygonal region would you suggest to approximate this additional region?

4. Express the corresponding approximation for ΔA in terms of Δx, $f(x)$, and $f(x + \Delta x)$.

5. Use your findings from Problem 4 to approximate the average rate of change $\dfrac{\Delta A}{\Delta x}$.

6. Evaluate $\lim\limits_{\Delta x \to 0} f(x + \Delta x)$. Recall that $f(x)$ is continuous over the interval $[a, b]$.

7. Show that $\lim\limits_{\Delta x \to 0} \dfrac{\Delta A}{\Delta x} = f(x)$.

8. What can you say about the derivative $A'(x)$?

9. Express $A(x)$ as a definite integral.

10. Evaluate the area A of the plane region below the graph of f and above the x-axis. ◆

Theorem 6.3.1 (The Fundamental Theorem of Calculus). *Let $f(x)$ be a nonnegative and continuous function defined on the interval $[a, b]$. Denote by $A(x)$ the area of the plane region bounded by the graph of f, the x-axis, and the vertical lines through points $(a, f(a))$ and $(x, f(x))$. Then,*

$$A'(x) = f(x), \quad \text{for all } x \text{ in } [a, b]. \tag{3}$$

Therefore,

$$A(x) = \int_a^x f(t)\, dt, \quad \text{for all } x \text{ in } [a, b]. \tag{4}$$

In particular, the area of the plane region bounded by the graph of f, the x-axis, and the vertical lines $x = a$ and $x = b$ is given by

$$A(b) = \int_a^b f(t)\, dt. \tag{5}$$

Recall that the Riemann sums for a nonnegative and continuous function f defined on an interval $[a, b]$ (when using a large number of tiles) provide approximate values for the area A of the region that is below the graph of f and above the x-axis. Their limit, as the number of tiles goes to ∞, coincides with the area A. Using the Fundamental Theorem of Calculus, you now see that the Riemann sums for f, when using a large number of tiles, provide approximate values for the definite integral $\int_a^b f(x)\,dx$. Their limit, as the number of tiles goes to ∞, coincides with $\int_a^b f(x)\,dx$.

Historical Note: Sir Isaac Newton (1643–1727; from England)

While still under 25 years old, Newton made revolutionary advances in mathematics, physics, optics, and astronomy. He laid the foundations for differential and integral calculus several years before its independent discovery by Leibniz. Newton's *De Methodis Serierum et Fluxionum* ("On the Methods of Series and Fluxions"), the first book on calculus, was written in 1671, but it did not appear in print until 1736. His later work *Principia* is considered to be the greatest scientific book ever written. In 1705, Newton was knighted by Queen Anne; he was the first scientist ever to receive such an honor for his work.

EXAMPLE Compute the area of the plane region bounded by the graph of $f(x) = 3$, the x-axis, and the vertical lines $x = 1$ and $x = 4$, first using the Fundamental Theorem of Calculus, and then using the area formulas obtained in Section 6.1.

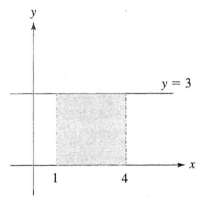

Solution Using the Fundamental Theorem of Calculus, the area of the shaded region is

$$\int_1^4 3\,dx = 3x\big|_1^4 = (12 - 3) = 9.$$

On the other hand, the shaded region is a square whose sides each have length 3. Using the formula for the area of a square, its area is $3^2 = 9$. ∎

Practice Problems

1. Compute the area of the plane region bounded by the graph of $f(x) = 2x$, the x-axis, and the vertical line $x = 2$, first using the Fundamental Theorem of Calculus, and then using the area formulas obtained in Section 6.1.

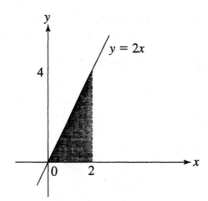

2. Compute the area of the plane region bounded by the graph of $f(x) = x + 1$, the x-axis, and the vertical lines $x = 2$ and $x = 4$, first using the Fundamental Theorem of Calculus, and then using the area formulas obtained in Section 6.1.

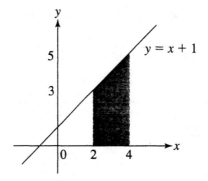

3. Compute the area of the plane region bounded by the graph of $f(x) = x^2$, the x-axis, and the vertical lines $x = 0$ and $x = 2$ using the Fundamental Theorem of Calculus.

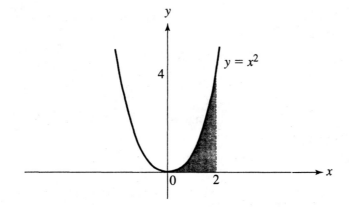

4. Compute the area of the plane region bounded by the graph of $f(x) = 1/x^2$, the x-axis, and the vertical lines $x = 1$ and $x = 5$ using the Fundamental Theorem of Calculus. Compare this exact value with the approximate values you obtained using Riemann sums in Section 6.2.

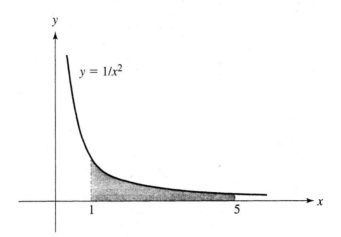

EXAMPLE **A Region to Be Divided into Suitable Subregions** Compute the area of the region bounded by the graph of the parabola $y = x^2 + 1$, the line $y = 3 - x$, the x-axis, and the y-axis.

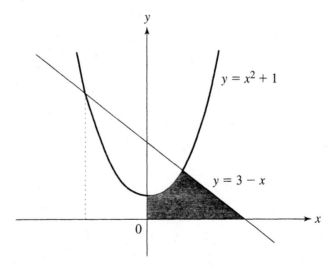

Solution First, determine the point in the half-plane $x \geq 0$, where the parabola $y = x^2 + 1$ and the line $y = 3 - x$ intersect. To do this, you need to solve the equation

$$x^2 + 1 = 3 - x$$

$$x^2 + x - 2 = 0$$

$$(x + 2)(x - 1) = 0.$$

Thus, $x = 1$ (the solution $x = -2$ corresponds to the intersection point lying in the left half-plane). Now, partition the shaded region into two subregions as follows:

1. The subregion bounded by the parabola, the vertical lines $x = 0$ and $x = 1$, and the x-axis; its area $\mathcal{A}_1$ is given by

$$\mathcal{A}_1 = \int_0^1 (x^2 + 1)\, dx = \left(\frac{x^3}{3} + x\right)\Big|_0^1 = \frac{1}{3} + 1 = \frac{4}{3}.$$

2. The subregion bounded by the graph $y = 3 - x$, the vertical line $x = 1$, and the x-axis; since the graph $y = 3 - x$ intersects the x-axis at $x = 3$, the area $\mathcal{A}_2$ of this subregion is given by

$$\mathcal{A}_2 = \int_1^3 (3 - x)\, dx = \left(3x - \frac{x^2}{2}\right)\Big|_1^3 = \left(9 - \frac{9}{2}\right) - \left(3 - \frac{1}{2}\right)$$

$$= 9 - \frac{9}{2} - 3 + \frac{1}{2} = 2.$$

Therefore, the area of the entire shaded region is $\mathcal{A}_1 + \mathcal{A}_2 = \frac{4}{3} + 2 = \frac{10}{3}.$

■

Classroom Discussion 6.3.2: Area of a Region above a Curve and below the x-axis

1. Consider the function $f(x) = -1$ defined on the interval $[1, 4]$.

   a. Graph the function $f(x)$.
   b. Compute the definite integral $\int_1^4 f(x)\, dx$.
   c. What is the geometric interpretation of $\int_1^4 f(x)\, dx$ in this case? Explain.

2. Now let $f(x)$ be an arbitrary continuous function defined on the interval $[a, b]$ satisfying $f(x) \leq 0$ for all $a \leq x \leq b$. What is the geometric interpretation of $\int_a^b f(x)\, dx$? Explain your reasoning.

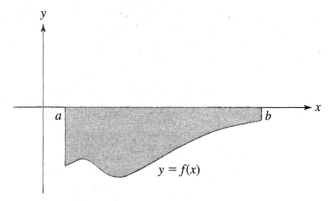

3. Consider the function $f(x) = x^3$ defined on the interval $[-1, 1]$.
   a. Graph the function $f(x)$.
   b. Compute the definite integrals $\int_{-1}^{0} f(x)\, dx$ and $\int_{0}^{1} f(x)\, dx$.
   c. Deduce the value of $\int_{-1}^{1} f(x)\, dx$ from your findings in 3b.
   d. What is the area of the region that is between the graph of f and the x-axis?
   e. What is the geometric interpretation of $\int_{-1}^{1} f(x)\, dx$? Justify your answer.

4. Let A_1, A_2, and A_3 be the areas of the shaded regions in the following figure.

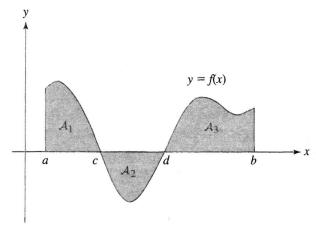

   a. Express the area A of the entire shaded region in terms of definite integrals.
   b. Express $\int_{a}^{b} f(x)\, dx$ in terms of the areas A_1, A_2, and A_3.

    c. Describe the relationship between $\int_a^b f(x)\,dx$ and the Riemann sums for f over the interval $[a, b]$. ◆

Numerical Integration. As we have seen, the area of the plane region bounded by the x-axis and the graph of a function $f(x)$, over an interval $[a, b]$, can be found either by determining the limit of Riemann sums or by using the Fundamental Theorem of Calculus. In cases where $f(x)$ has an elementary antiderivative, it is often easier and faster to apply the Fundamental Theorem of Calculus, since it is quite difficult to evaluate the limit of Riemann sums directly for most functions. On the other hand, there are many functions $f(x)$ whose antiderivatives cannot be expressed in a simple way (for example, $f(x) = \sqrt{x^4 + 1}$ or $f(x) = e^{x^2}$); for such functions, the Fundamental Theorem of Calculus cannot be used to evaluate the areas of the regions below their graphs and above the x-axis. In such cases, one can instead use the method of Riemann sums to compute a good approximation of the area: one splits the interval into a large number of subintervals, and the corresponding Riemann sum is then calculated numerically with a computer. If the number of subintervals is sufficiently large, one can obtain an approximation of the area to any desired degree of accuracy. This technique, known as numerical integration, is available with many existing software packages, such as Mathematica.

EXERCISES 6.3

In Exercises 1–12, sketch the graph $y = f(x)$, where $a \leq x \leq b$. Then, using the Fundamental Theorem of Calculus, compute by hand the area of the plane region below this graph and above the x-axis.

    1. $f(x) = 4$, $a = 0$, $b = 10$

    2. $f(x) = \sqrt{2}$, $a = 1$, $b = 5$

    3. $f(x) = x$, $a = 0$, $b = 3$

    4. $f(x) = -x$, $a = -1$, $b = 0$

    5. $f(x) = 3x + 2$, $a = 0$, $b = 10$

    6. $f(x) = -\frac{1}{5}x + 1$, $a = -5$, $b = 5$

    7. $f(x) = (x - 1)^2$, $a = 1$, $b = 4$

    8. $f(x) = -(x + 1)^3$, $a = -2$, $b = -1$

    9. $f(x) = e^x$, $a = 0$, $b = 1$

    10. $f(x) = -e^x + 1$, $a = -1$, $b = 0$

    11. $f(x) = x^3 + 2x$, $a = 1$, $b = 2$

    12. $f(x) = -x^4 + 4x$, $a = 0$, $b = \sqrt[3]{4}$

In Exercises 13–22, sketch the graph $y = f(x)$ where $a \leq x \leq b$. Then, using the Fundamental Theorem of Calculus, write down a formula for the area of the plane region that is bounded by the graph and the x-axis. Use your calculator to find the actual value of the area.

    13. $f(x) = \sqrt{1 - x^2}$, $a = 0$, $b = 1$

14. $f(x) = -\sqrt{4 - x^2},\ a = -2,\ b = 0$

15. $f(x) = \sqrt{2 + x^2},\ a = 0,\ b = 2$

16. $f(x) = -\sqrt{1 + 8x^2},\ a = -1,\ b = 1$

17. $f(x) = x\,e^x,\ a = 0,\ b = 1$

18. $f(x) = xe^{-x},\ a = -1,\ b = 2$

19. $f(x) = x/(x^2 + 1),\ a = -1/2,\ b = 3/2$

20. $f(x) = (1 - x^2)e^x,\ a = -3,\ b = 3$

21. $f(x) = x^3 e^x,\ a = 0,\ b = 1$

22. $f(x) = (x^3 - x^2)e^x,\ a = -2,\ b = 3$

23. Using the Fundamental Theorem of Calculus, compute the exact value of the area of the regions described in Exercises 6.2, Problems 2, 3, 5, 6, and 8.

PROJECTS AND EXTENSIONS 6.3

The Natural Logarithmic Function

Here, we introduce the natural logarithmic function, study its algebraic and analytic properties, investigate its link with the exponential function e^x, and derive a new technique of integration suitable for a certain type of functions.

I. The Function ln x

1. Graph the function $f(x) = 1/x$ defined on the interval $(0, \infty)$.

2. For a given $x > 1$, what is the geometric interpretation of the definite integral $\int_1^x 1/t\,dt$?

3. For a given $0 < x < 1$, what is the geometric interpretation of the definite integral $\int_1^x 1/t\,dt$?

The function that assigns the value $\int_1^x 1/t\,dt$ to each real number $x > 0$ is called the **natural logarithmic function**. It is denoted by $\ln x$ and is read as "the natural logarithm of x" or simply "ln of x."

4. Fill in the blanks using one of the symbols $=, <,$ or $>$.

$$\ln x \ \rule{1cm}{0.4pt}\ 0 \text{ for } 0 < x < 1.$$
$$\ln 1 \ \rule{1cm}{0.4pt}\ 0.$$
$$\ln x \ \rule{1cm}{0.4pt}\ 0 \text{ for } x > 1.$$

5. What is the derivative of the function $\ln(x)$?

6. Is the function $\ln(x)$ continuous? Explain.

7. On which intervals is the function $\ln x$ increasing? On which intervals is it decreasing?

8. Evaluate $(x \ln x)'$, then use the result to find the antiderivatives of the function $\ln x$.

9. In order to compare $\ln x$ with x, consider the function $g(x) = x - \ln x$ defined for all $x > 0$.

    a. Using the techniques from Section 4.1, determine the minimum of $g(x)$ on the interval $(0, \infty)$.

b. Use your findings in 9a to determine the sign of $g(x)$ on the interval $(0, \infty)$.

c. Can you determine the sign of $g(x)$ on the interval $(0, 1)$ directly without using 9a?

d. Fill in the blank using one of the symbols $=$, $<$, or $>$: $\ln x \underline{\quad} x$ for all $x > 0$.

II. Algebraic Properties of ln x

1. Let a, b be two positive real numbers. The goal is to compare $\ln(ab)$ with $\ln a$ and $\ln b$. To do this, consider the function $g(x) = \ln(xb)$.

a. Using the chain rule for differentiation, evaluate the derivative $g'(x)$.

b. Show that $\ln(xb) = \ln x + C$ for some constant C.

c. Make an appropriate choice for x to determine the exact value of the constant C.

d. Fill in the blanks: $\ln(ab) \underline{\quad} \ln a \underline{\quad} \ln b$.

2. The goal is to now find the relation between $\ln a$ and $\ln(1/a)$ for a given $a > 0$.

a. Evaluate $\ln a + \ln(1/a)$ using Project II, Problem 1d.

b. Fill in the blank: $\ln(1/a) \underline{\quad} \ln a$.

c. Using Project II, Problems 1d and 2b, find the relation between $\ln(a/b)$, $\ln a$, and $\ln b$, then fill in the blanks: $\ln(a/b) \underline{\quad} \ln a \underline{\quad} \ln b$.

3. The goal here is to find a relation between $\ln(a^n)$, n, and $\ln a$, where $a > 0$ and n is an arbitrary real number.

a. Assume n is a whole number. Using Project II, Problem 1d, compare $\ln(a^n)$ with $n \ln a$.

b. Assume n is a negative integer. Using Project II, Problems 2b and 3a, compare $\ln(a^n)$ with $n \ln a$.

c. Assume $n = p/q$ is a rational number. Compare $\ln(a^{qn})$ with $q \ln(a^n)$ and with $p \ln a$. Then, derive the relation between $\ln(a^n)$ and $n \ln a$.

d. Assume n is an irrational number. Using Project II, Problem 3c together with the continuity of the natural logarithmic function, find the relation between $\ln(a^n)$ and $n \ln a$.

e. Fill in the blank: Given $a > 0$, $\ln(a^n) \underline{\quad} n \ln a$, for arbitrary real numbers n.

III. The Graph $y = \ln x$

1. In order to graph the function $\ln x$, let us analyze first $\lim\limits_{x \to \infty} \ln x$ and $\lim\limits_{x \to 0^+} \ln x$.

a. Consider the sequence $\{2^n\}_n$. What is $\lim\limits_{n \to \infty} 2^n$?

b. What is the sign of $\ln 2$?

c. Use Project II, Problem 3e to express $\ln(2^n)$ in terms of n and $\ln 2$. Then, evaluate $\lim\limits_{n \to \infty} \ln(2^n)$.

d. Using Project I, Problem 7 and Project III, Problem 1c, evaluate $\lim\limits_{x \to \infty} \ln x$.

e. Using Project II, Problem 2b and Project III, Problem 1d, evaluate $\lim\limits_{x \to 0^+} \ln x$.

2. Study the concavity of the function $\ln x$.

3. Using your findings from Project I and Project III, Problems 1–2, sketch the graph of the function $y = \ln x$.

IV. Relation between ln x and e^x

1. Let $f(x)$ be a positive function. Consider the composite function $g(x) = \ln[f(x)]$.

   a. Why does the composite function $g(x)$ make sense?

   b. Using the chain rule for differentiation, express the derivative $g'(x)$ in terms of $f(x)$ and its derivative $f'(x)$.

2. Consider the special case $g(x) = \ln(e^x)$.

   a. For which values of x does $g(x)$ make sense?

   b. Using Project IV, Problem 1b, evaluate $g'(x)$.

   c. Using Project IV, Problem 2b, show that $g(x) = x + C$ for some constant C.

   d. To determine the constant C, evaluate $g(0)$.

3. To summarize your findings from Project IV, answer the following questions.

   a. Is $\ln x$ the inverse function of e^x? For the notion of inverse functions, see Project V in Section 2.1.

   b. Fill in the blanks:
   $\ln(e^x)$ ___ x for all real numbers x. $e^{\ln x}$ ___ x for all real numbers x ___ 0.

   c. Based on your previous answer, how can the graph of $\ln x$ be obtained from the graph of e^x?

   d. Sketch the graph $y = e^x$ on the same coordinate system as the earlier graph $y = \ln x$.

   e. Are the two graphs as you described them in Project IV, Problem 3c?

V. The Logarithmic Rule for Antidifferentiation

Consider the function $k(x) = \dfrac{2x+1}{x^2+x+1}$. The goal here is to evaluate the antiderivatives of k.

1. Determine, if possible, a function $f(x)$ satisfying $k(x) = \dfrac{f'(x)}{f(x)}$.

2. What differentiation formula can you derive from Project IV, Problem 1b?

3. Using your answers to Problems 1 and 2, evaluate the indefinite integral $\int k(x)\,dx$.

4. In Exercises a–f, find the integrand's domain. Then, using the techniques of integration you have learned so far, evaluate the given integral.

a. $\displaystyle\int \frac{1}{x+5}\,dx$

b. $\displaystyle\int \frac{2x+1}{x^2+x-1}\,dx$

c. $\displaystyle\int \frac{3x^2+x+1}{2x^3+x^2+2x-1}\,dx$

d. $\displaystyle\int \frac{e^x}{e^x-1}\,dx$. Then, determine $\displaystyle\int \frac{1}{1-e^{-x}}\,dx$

e. $\displaystyle\int \frac{1}{x(\ln x)^2}\,dx$

f. $\displaystyle\int \frac{1}{x\ln x}\,dx$

VI. Antiderivatives of the Function $f(x) = \dfrac{1}{x(x+1)}$ and the Area below Its Graph

1. Sketch the graph of the function $f(x) = \frac{1}{x(x+1)}$ defined for $x > 0$.

2. Check that $\frac{1}{x(x+1)} = \frac{1}{x} - \frac{1}{x+1}$ for all $x > 0$.

3. Use the identity in Problem 2 to evaluate $\int \frac{1}{x(x+1)}\,dx$.

4. Simplify the expression obtained in Problem 3 using Project II, Problem 2c.

Consider the region above the x-axis and below the graph of f for $x \geq 1$; denote its area by A_1. Since the region under consideration is unbounded, it is not clear at this point whether A_1 is finite or infinite. The goal is to find the value of A_1. To do so, let $b > 1$ and denote by $A_1(b)$ the area of the subregion for which $1 \leq x \leq b$.

5. Use the Fundamental Theorem of Calculus and your answer to Problem 4 to evaluate $A_1(b)$.

6. Compute $\displaystyle\lim_{b\to\infty} \frac{b}{b+1}$, then $\displaystyle\lim_{b\to\infty} A_1(b)$.

7. What is the area A_1? Explain your answer.

Now consider the region above the x-axis and below the graph of f for $0 < x \leq 1$; denote its area by A_2. Since the region under consideration is again unbounded, it is not clear whether A_2 is finite or infinite. The goal is to find the value of A_2. To do so, let $0 < b < 1$, and denote by $A_2(b)$ the area of the subregion for which $b \leq x \leq 1$.

8. Use the Fundamental Theorem of Calculus and your answer to Problem 4 to evaluate $A_2(b)$.

9. Compute $\displaystyle\lim_{b\to 0^+} A_2(b)$.

10. What is the area A_2? Explain your reasoning.

6.4   AREA OF A DISC FROM DIFFERENT POINTS OF VIEW

Computing the area of a disc in middle-school, high school, and college

In your own words, how would you define the terms *circle* and *disc*?

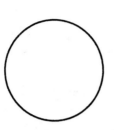 

Our goal in this section is to find a formula for the area of a disc of radius $r > 0$. We present four different approaches: one for sixth-grade middle-school students based on elementary estimates, one for seventh-grade middle-school students based on dividing the disc into sectors, one for high school and college students based on approximating the circle by regular n-gons, and one for college students based on the Fundamental Theorem of Calculus.

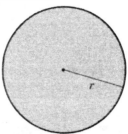

Classroom Connection 6.4.1: An Elementary Estimate for the Area of a Disc

This exploration is taken from pages 499–500 and page 7–55 in the sixth-grade textbook *Math Thematics, Book 1*. Discuss it in small groups. ◆

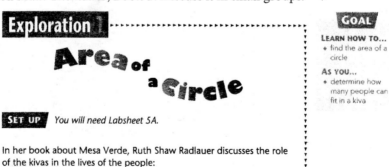

Exploration

Area of a Circle

SET UP   *You will need Labsheet 5A.*

In her book about Mesa Verde, Ruth Shaw Radlauer discusses the role of the kivas in the lives of the people:

> "When children were old enough, they were initiated, or proclaimed adults in a ceremony. Then they could spend some of the winter in a warm kiva. The kiva was a sort of clubhouse for adults and a place for ceremonies."

GOAL

LEARN HOW TO...
◆ find the area of a circle

AS YOU...
◆ determine how many people can fit in a kiva

▶ To find out how many people fit in a kiva, you need to find the area of the floor. The floor of a kiva is shaped like a circle.

3 Use Labsheet 5A. Follow the directions for *Estimating the Area of a Circle* by finding the areas of the inner and outer squares.

▶ You can use the method in **Question 3** to estimate the area of any circle with radius *r*.

4 Try This as a Class The diagram below can help you see the relationship between the area of the circle and *r*.

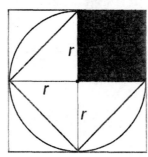

a. Use the variable *r* to write an expression for the area of a small red square.

b. How many small red squares fit in the green outer square? Use your answer to write an expression for the area of the green outer square.

c. How many small red squares fit in the blue inner square? (*Hint:* Each small red square is made up of two triangles.) Use your answer to write an expression for the area of the blue inner square.

d. Use your answers to parts (b) and (c) to write an expression that can be used to estimate the area of the circle.

5 Use your answer to Question 4(d) to estimate the area of a circle with a radius of 4 cm. How does your estimate compare with the estimate you found in Question 3?

▶ **Formula for the Area of a Circle** In Question 4, you found an expression that can be used to estimate the area of a circle. To find the actual area of a circle with radius *r*, multiply pi by *r* to the second power.

$$A = \pi r^2$$

You can read r^2 as "*r* squared."

6 How does the formula above compare with the expression you found in Question 4(d)?

Name _____   **Date** _____

Estimating the Area of a Circle

(Use with Question 3 on page 499.)

Directions Complete parts (a)–(d) to estimate the area of the circle in the figure. Each grid square is 1 cm by 1 cm, or 1 cm².

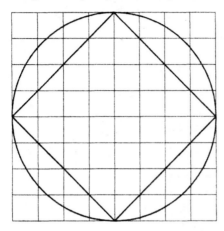

a. Is the area of the circle *greater than*, *less than*, or *equal* to the area of the outer square? the area of the inner square?

b. Find the area of the outer square. Describe the method you used.

c. Find the area of the inner square. Describe the method you used.

d. Use your results from parts (b) and (c) to estimate the area of the circle. Explain how you made your estimate.

Classroom Connection 6.4.2: Cutting the Disc into Sectors

This exploration is taken from pages 401–403 and page 6-51 in the seventh-grade textbook *Math Thematics, Book 2*. Discuss it in small groups. ◆

Exploration ▶

AREA OF A CIRCLE

SET UP   *You will need: • Labsheet 2A • scissors • tape • ruler*

GOAL

LEARN HOW TO...
• find the area of
  a circle

AS YOU...
• investigate kite
  designs

23 **Discussion** A centipede kite is made from circular pieces of silk. How can you estimate the amount of silk in a centipede kite?

In Section 1 you used rectangles to develop a formula for the area of a parallelogram. You can use the same idea to find the area of one of the circles in a centipede kite.

Section 2 Square Roots, Surface Area, and Area of a Circle   **401**

24 **Use Labsheet 2A.** Follow the directions below.

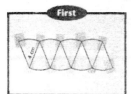

First

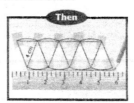

Then

Cut out the *Circle*. Then cut apart the eight sectors and arrange them to form the figure shown. Tape the figure to a sheet of paper.

Use a ruler to draw segments across the top and bottom of your figure. Extend the sides of the figure to meet the bottom segment.

25 What kind of polygon is the new figure you drew in Question 24?

26 **Try This as a Class** Use the figure you made in Question 24.

a. Estimation Explain how you could use the new figure you drew to estimate the area of the circle. Then estimate the area. Do you think this is a good estimate? Why or why not?

b. How is the length of the base of the figure related to the circumference of the circle?

c. How is the height of the figure related to the radius of the circle?

27 **Discussion** Examine the drawings shown.

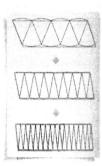

    a. As a circle is cut into more and more sectors and put back together as shown, what begins to happen to its shape?

    b. The area A of a parallelogram is found by multiplying the length of its base b by the height h, or $A = bh$. Use your figure to explain why this formula can be written as $A = \frac{1}{2}Cr$ to find the area of a circle, where C is the circumference and r is the length of the radius of the circle.

28 The circumference C of a circle is equal to $2\pi r$, where r is the length of the radius. Rewrite the formula $A = \frac{1}{2}Cr$ by substituting $2\pi r$ for C.

29 The area of a circle with radius r is $A = \pi r^2$. Compare πr^2 with the expression you wrote for Question 28. Do they have the same value? Explain.

▶ **You can use the equation $A = \pi r^2$ to find the area of a circle when you know the length of its radius.**

EXAMPLE

Find the area of a circle with radius 3.2 cm. Use 3.14 for π to find the approximate area.

3.2 cm

SAMPLE RESPONSE

Exact Area

$A = \pi r^2$

$\quad = \pi(3.2)^2$

$\quad = \pi(10.24)$

$\quad = 10.24\pi$

The exact area is 10.24π cm^2.

Approximate Area

$A = \pi r^2$

$\quad = \pi(3.2)^2$

$\quad \approx 3.14(10.24)$

$\quad \approx 32.15$

This is an approximation because 3.14 is an approximation for π.

The approximate area is 32.2 cm^2.

For Questions 30–31, use 3.14 or the key on a calculator.

30 Try This as a Class Use your figure from Question 24.

    a. Use $A = \pi r^2$ to write an expression that represents the exact area of the *Circle*.

    b. Find the approximate area of the *Circle*.

    c. If you were planning to make a circle kite with a 4 cm radius, would you use your answer from part (a) or from part (b) to order the material? Why?

    d. How does the area of the circle from part (b) compare with the estimated area of the figure in Question 26(a)?

31   ✔ **CHECKPOINT** A centipede kite has 10 in. diameter circles.

    a. Find the exact area of one circle.

    b. Find the approximate area of one circle.

    c. About how many square inches of silk were used to make all 11 circles of the kite?

✔ **QUESTION 31**
...checks that you can find the area of a circle.

HOMEWORK EXERCISES ▶ See Exs. 16–23 on p. 407.

Name _____   **Date** _____

| **MODULE 6** | **LABSHEET 2A** |

Circle (Use with Question 24 on page 402.)

Directions

• Cut out the circle.

• Cut apart the eight sectors and arrange them to form the figure shown below.

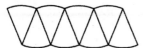

• Tape the figure to a sheet of paper.

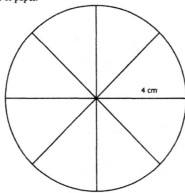

4 cm

    Math Thematics, Book 2 **6-51**

Classroom Discussion 6.4.1: Approximating the Circle by Inscribed Regular *n*-gons

To find the area of a disc, we use here the *method of exhaustion*, which was invented by Eudoxus (similar ideas may be found in the project "Archimedes's Computation of π" in Section 1.1). To see how the method works, use the following outline.

1. For each $n \geq 3$, let a_n and p_n denote the apothem and the perimeter of an inscribed regular *n*-gon, respectively. How can you express the area A_n inside the *n*-gon in terms of a_n and p_n?

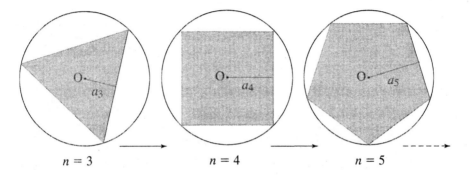

$$n = 3 \qquad\qquad n = 4 \qquad\qquad n = 5$$

2. What happens to the *n*-gons as *n* increases? What happens to the values of a_n, p_n, and A_n as *n* increases? What can you say about their limits as $n \longrightarrow \infty$?
3. Using your findings in Problems 1 and 2, find the area A of the disc.
4. Why do you think this method is called the *method of exhaustion*? ◆

Historical Note: Eudoxus of Cnidus (408–355 BC; from Asia Minor [now Turkey])

Eudoxus, a contemporary of Plato, had a rich and varied academic background in mathematics, music, medicine, astronomy, theology, and meteorology. Early in his career, he developed a theory of *proportion*, which appears in Euclid's *Elements* and facilitated his early work on finding areas. Eudoxus introduced the *method of exhaustion*, which led to important developments in calculus by Archimedes and others; Eudoxus himself was the first to prove that a cone's volume is one-third the volume of a cylinder having the same base and equal height.

Classroom Discussion 6.4.2: Area of a Disc Using the Fundamental Theorem of Calculus

The Fundamental Theorem of Calculus allows us to find the area of a region bounded by the graph of a function and the *x*-axis. The disc is not such a region, but since it has a reflectional symmetry about the *x*-axis, its area is twice that of the upper semidisc. The upper semidisc is a region for which the Fundamental Theorem of Calculus applies.

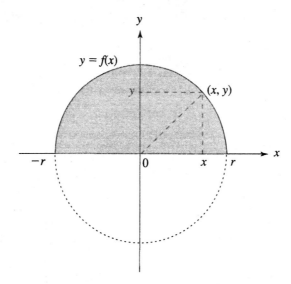

1. Use the Pythagorean Theorem to find a relation between x and y for the point (x, y) to lie on the circle centered at the origin with radius r.
2. What additional constraint must be placed on y for the point (x, y) to lie on the upper semicircle?
3. Solve for y in your equation from Problem 1 to obtain the equation of the upper semicircle.
4. Use the Fundamental Theorem of Calculus to express the disc's area as a definite integral.
〈T〉 5. Compute the area $\mathcal{A}$ by using your calculator to evaluate the definite integral.
◆

EXERCISES 6.4

1. Write a paragraph about the Fundamental Theorem of Calculus. In particular, write about your understanding of this theorem, and why you think it is (or is not) useful.
2. In Classroom Connection 6.4.1, we used the formula $A_1 = 3r^2$ to estimate the area A of a disc of radius r. In the old Babylonian civilization, the formula $A_2 = \dfrac{C^2}{12}$ was used for the area A inside a circle of circumference C.
   a. Express A_2 in terms of r using the classical formula for the circumference of a circle.
   b. Express the errors $A_1 - A$ and $A_2 - A$ in terms of r using the formula obtained for A in Section 6.4. Interpret the results.
   c. Evaluate the relative error in each estimate. The *relative error*, when estimating an exact value v with an approximate value v_0, is the quantity $\dfrac{v_0 - v}{v} = \dfrac{v_0}{v} - 1$.
   d. Evaluate the percentage error in each estimate. The *percentage error* in an estimate is 100% times the relative error in the estimate.

The goal of the following problems is to use what you have already done in the case of a disc to find the area of the region inside an ellipse.

3. What is an ellipse? Research its definition and its Cartesian equation.

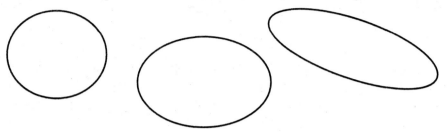

4. Can the methods in Classroom Connection 6.4.2 and in Classroom Discussion 6.4.1 be adapted to ellipses? Why or why not?

5. Adapt the method used in Classroom Connection 6.4.1 to the case of a region inside an ellipse.

6. Use the Fundamental Theorem of Calculus to compute the area of the region inside an ellipse. To do so, follow the same steps as in the case of a circle.

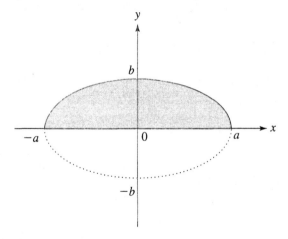

PROJECTS AND EXTENSIONS 6.4

I. History of Finding Areas

The problem of finding areas of plane regions has a rich and fascinating history, which started with the early Greek philosophers and continued for thousands of years afterward. Research the history of this problem and write a detailed report describing, in chronological order, the main advances that were made on the problem. Be sure to include biographical information about the mathematicians responsible for these advances.

II. Leibniz vs. Newton

The controversy over who discovered calculus, Leibniz or Newton, caused bitter disputes among their followers for many years. Eventually, both mathematicians were credited with its discovery. Research the history of this controversy and write a detailed report of the relevant events. Be sure to include a prehistory of the controversy, its effect on Leibniz and on Newton, and the way in which it was eventually resolved.

6.5 AREA OF A REGION BOUNDED BY TWO GRAPHS

Our goal in this section is to derive from the Fundamental Theorem of Calculus a formula for computing the area of an irregular shape that is bounded by the graphs of two functions. Let $f(x)$ and $g(x)$ be two given continuous functions defined on the interval $[a, b]$. Denote by A the area of the region bounded by the graphs $y = f(x)$ and $y = g(x)$ and the vertical lines $x = a$ and $x = b$.

Classroom Discussion 6.5.1: Area of a Region Bounded by Graphs

1. Assume that the graph of f lies above the graph of g and that both graphs lie in the plane's upper half; that is,

$$0 \le g(x) \le f(x) \text{ for all } a \le x \le b.$$

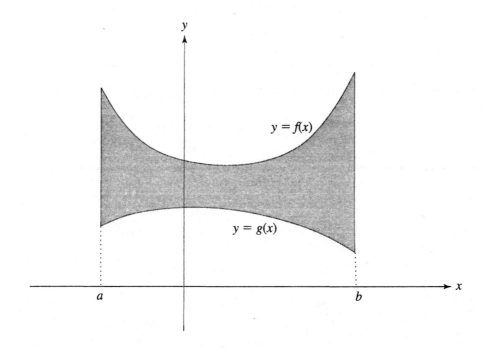

a. Denote by A_f the area of the region below the graph $y = f(x)$ and above the x-axis. Express A_f as a definite integral.

b. Similarly, denote by A_g the area of the region below the graph $y = g(x)$ and above the x-axis. Express A_g as a definite integral.

c. How can you express $\mathcal{A}$ in terms of $\mathcal{A}_f$ and $\mathcal{A}_g$?

d. Use the properties of definite integrals to express the area $\mathcal{A}$ as one definite integral.

e. Apply what you have just discovered to find the area of the plane region bounded by the parabola $y = x^2$ and the line $y = 4$.

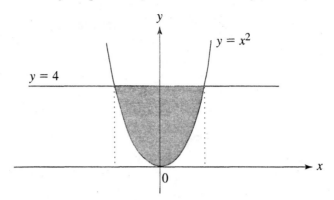

2. Assume now that the graph of f lies above the graph of g but that the latter does not lie above the x-axis; that is, $g(x) \le f(x)$ for all $a \le x \le b$, but the condition $g(x) \ge 0$ is no longer satisfied for all $a \le x \le b$.

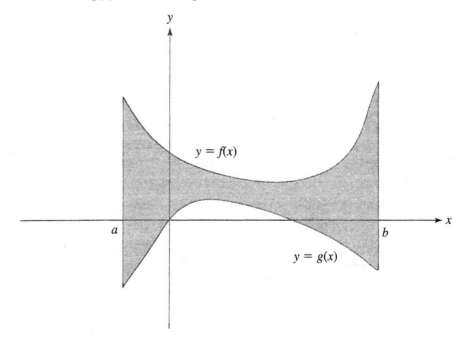

a. Find a transformation that moves the shaded region above the x-axis while preserving its shape.

b. The new region located above the x-axis lies below the graph of some function $F(x)$ and above the graph of some function $G(x)$ where now $F(x)$ and $G(x)$ satisfy the conditions in Problem 1. Express the functions $F(x)$ and $G(x)$ in terms of $f(x)$ and $g(x)$, respectively.

c. Using Problem 1d, express the area of the new region as a definite integral involving the new functions F and G.

d. Express the area $\mathcal{A}$ as a definite integral in terms of the original functions f and g.

e. Find the area of the region bounded by the parabola $y = x^2$ and the line $y = 1 - x$ for $1 \leq x \leq 3$.

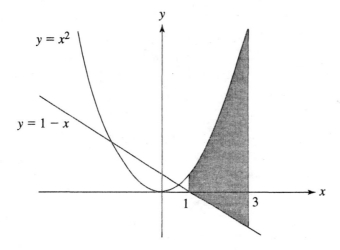

3. Assume now that the graphs of f and g intersect as in the following figure:

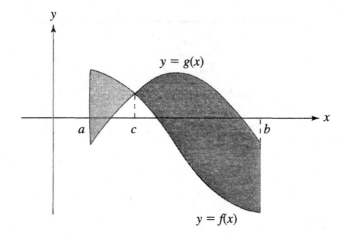

a. Explain how the area $\mathcal{A}$ of the shaded region can be computed in this case.

b. Express the area A as a sum of definite integrals.

c. Show that the formula obtained in Problem 3b can be rewritten as

$$A = \int_a^b |f(x) - g(x)| \, dx.$$

d. In the following figure, compute the area of the shaded region that is enclosed by the graph $y = x^3 - 1$ and the line $y = x - 1$. (Hint: Start by finding the points of intersection between the two graphs.)

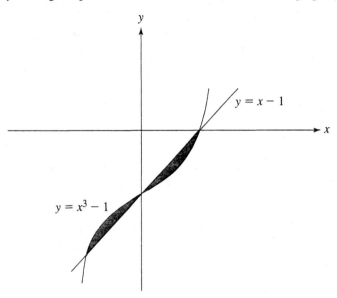

EXERCISES 6.5

1. Find the area of the region enclosed by the curve $y = \sqrt{x}$ and by the lines $x = 9$, $y = 0$.
2. Find the area of the region for which $x \geq 0$ and that is enclosed by the curve $y = x^5$ and the line $y = x$.
3. Find the area of the region enclosed by the parabolas $y = x^2 + 2x + 1$ and $-x^2 - 2x + 1$.
4. Find the area of the region enclosed by the parabolas $y = x^2 - 3x + 1$ and $-x^2 + 3x + 1$.
5. Find the area of the region enclosed by the parabola $y = 1 - x^2$ and the line $y = x - 1$.
6. Find the area of the region enclosed by the parabola $y = x^2 - 3x + 2$ and the line $y = 2x - \frac{1}{4}$.
7. Find the area of the region enclosed by the graphs $y = 6x^2 - 6x^3$ and $y = x^2 - x$. (Hint: This region is composed of two subregions.)
8. Find the area of the region enclosed by the graphs $y = x^2$ and $y = x^4 - x^2$.
9. Consider the triangle with vertices $A(0,0)$, $B(2,1)$, and $C(3,3)$. Find the area of the region inside triangle ABC using the results from Section 6.5.

10. Consider the quadrilateral with vertices $A(0,0)$, $B(2,1)$, $C(2,5)$, and $D(4,4)$. Find the area of the region inside triangle ABC using the results from Section 6.5.

11. In two different ways, find the area of the region inside the circle of radius 1 centered at the origin and below the line $y = x - 1$:
    a. Use directly the formulas derived from Sections 6.1 and 6.4
    b. Use the Fundamental Theorem of Calculus and the ideas discussed in Section 6.5. (You may use your calculator to integrate.)

12. Find the area of the region inside the ellipse $x^2/4 + y^2/9 = 1$ and above the line $y = -3x/2 + 3$.

13. While holding a piece of chalk of length b vertically, Anna draws two ribbons on the board: one by sliding the chalk horizontally and the other by allowing her hand to go up and down when sliding the chalk. See the following figure.

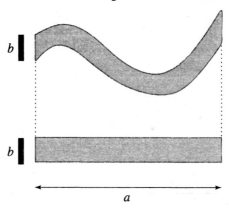

Which ribbon has the largest area? Think of some different arguments to answer this question.

PROJECTS AND EXTENSIONS 6.5

Areas of Infinite Regions in the Plane and Improper Integrals

In Chapter 5, we mainly considered integrals of functions that are continuous and defined over finite intervals. In this chapter, we discussed the problem of finding areas of finite plane regions, and we established the relationship between integrals and areas. The goal of this project is to find the areas of plane regions that are infinite and to extend the definition of an integral to functions that have *infinite discontinuities* or those defined over intervals that are infinite; such integrals are said to be *improper integrals*. We say that a function f has an infinite discontinuity at a point c if $\lim_{x \to c^-} f(x) = \pm\infty$, or $\lim_{x \to c^+} f(x) = \pm\infty$. To simplify the discussion, we consider only nonnegative functions, but improper integrals can be defined for functions that also take negative values.

I. Areas of Infinite Regions in the Horizontal Direction

1. Let f be an arbitrary continuous nonnegative function defined on the interval $[a, \infty)$. The region below the graph of f and above the x-axis extends indefinitely in the horizontal direction to the right. Denote by $\mathcal{A}$ the area of the region.

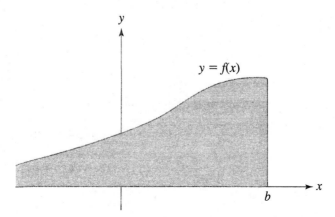

a. Can you think of a condition on the function f that is necessary for the area $\mathcal{A}$ to be finite?

b. What happens to the value of $\int_a^t f(x)\,dx$ as $t \to \infty$?

c. What would you suggest as a definition for $\int_a^\infty f(x)\,dx$? Explain.

d. Use your strategy to evaluate the area of the region above the x-axis and below the graph $y = \dfrac{1}{x^4}$, where $x \geq 1$.

e. Use your strategy to evaluate the area of the region above the x-axis and below the graph $y = \dfrac{3}{\sqrt{x}}$, where $x \geq 2$.

f. Is the necessary condition found in 1a also sufficient? Explain.

g. Evaluate $\int_1^\infty \frac{1}{x^4}\,dx$ and $\int_2^\infty \frac{3}{\sqrt{x}}\,dx$ based on your definition in 1c.

2. Now let f be an arbitrary continuous nonnegative function defined on the interval $(-\infty, b]$. The region below the graph of f and above the x-axis extends indefinitely in the horizontal direction to the left. Denote by $\mathcal{A}$ the area of the region.

    a. Can you think of a condition on the function f that is necessary for the area $\mathcal{A}$ to be finite?

    b. Can you come up with a natural strategy for finding the area $\mathcal{A}$ using the Fundamental Theorem of Calculus?

    c. What would you suggest as a definition for $\int_{-\infty}^{b} f(x)\, dx$? Explain.

    d. Use your strategy to evaluate the area of the region above the x-axis and below the graph $y = \dfrac{-1}{\sqrt[3]{x}}$, where $x \leq -1$.

    e. Use your strategy to evaluate the area of the region above the x-axis and below the graph $y = e^x$, where $x \leq 0$.

    f. Is the necessary condition found in 2a also sufficient? Explain.

    g. What are the values of $\int_{-\infty}^{-1} \dfrac{-1}{\sqrt[3]{x}}\, dx$ and $\int_{-\infty}^{0} e^x\, dx$ based on your definition in 2c?

3. Let f be an arbitrary continuous nonnegative function defined on the interval $(-\infty, \infty)$. The region below the graph of f and above the x-axis extends indefinitely in the horizontal direction to the right and to the left. Denote by $\mathcal{A}$ the region's area.

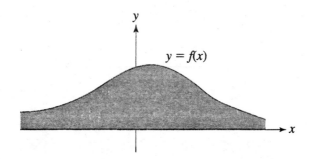

    a. Give necessary conditions on the function f for the area $\mathcal{A}$ to be finite.

    b. Divide the region into appropriate subregions and use your strategies from Problems 1b and 2b to come up with a formula for the area $\mathcal{A}$.

    c. Suggest a definition for $\int_{-\infty}^{\infty} f(x)\, dx$ in line with your definitions in Problems 1c, 2c, and 3b.

    d. Use your strategy from Problem 3b and your findings from Problems 1d and 2e to evaluate the area of the region above the x-axis and below the graph of the function f defined here. Then, evaluate $\int_{-\infty}^{\infty} f(x)\, dx$.

$$f(x) = \begin{cases} e^x & \text{if } x \leq 0, \\ 1 & \text{if } 0 \leq x \leq 1, \\ \dfrac{1}{x^4} & \text{if } x \geq 1. \end{cases}$$

e. Use your strategy from Problem 3b and your results from Problems 1e and 2e to determine the area of the region above the x-axis and below the graph of the function f defined here. Then, evaluate $\int_{-\infty}^{\infty} f(x)\,dx$.

$$f(x) = \begin{cases} e^x & \text{if } x \le 2, \\ \dfrac{3}{\sqrt{x}} & \text{if } x \ge 2. \end{cases}$$

f. Use your strategy from Problem 3b and your results from Problems 1d and 2d to determine the area of the region above the x-axis and below the graph of the function f defined here. Then, evaluate $\int_{-\infty}^{\infty} f(x)\,dx$.

$$f(x) = \begin{cases} \dfrac{-1}{\sqrt[3]{x-2}} & \text{if } x \le 1, \\ \dfrac{1}{x^4} & \text{if } x \ge 1. \end{cases}$$

II. Areas of Infinite Regions in the Vertical Direction

1. Let f be an arbitrary continuous nonnegative function defined on the interval $(a, b]$ and that satisfies $\lim\limits_{x \to a^+} f(x) = \infty$. The region below the graph of f and above the x-axis extends indefinitely in the vertical direction. Denote by $\mathcal{A}$ the area of the region.

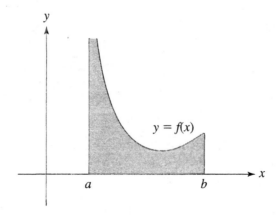

a. Describe a natural strategy for finding the area $\mathcal{A}$ using the Fundamental Theorem of Calculus.

b. What would you suggest as a definition for $\int_a^b f(x)\,dx$? Explain.

c. Use your strategy to evaluate the area of the region above the x-axis and below the graph $y = \dfrac{1}{\sqrt{x}}$, where $0 < x \le 1$.

d. Use your strategy to evaluate the area of the region above the x-axis and below the graph $y = \dfrac{1}{x^2}$, where $0 < x < 2$.

e. Evaluate $\int_0^1 \frac{1}{\sqrt{x}}\, dx$ and $\int_0^2 \frac{1}{x^2}\, dx$ based on your definition in 1c.

2. Let f be an arbitrary continuous nonnegative function defined on the interval $[a, b)$ and that satisfies $\lim_{x \to b^-} f(x) = \infty$. The region below the graph of f and above the x-axis again extends indefinitely in the vertical direction. Denote by $\mathcal{A}$ the area of the region.

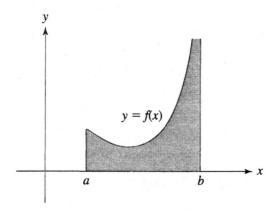

a. Describe a natural strategy for finding the area $\mathcal{A}$ using the Fundamental Theorem of Calculus.

b. Suggest a definition for $\int_a^b f(x)\, dx$ in this case. Explain.

c. Use your strategy to evaluate the area of the region above the x-axis and below the graph $y = \frac{-1}{\sqrt[3]{x}}$, where $-1 \leq x < 0$.

d. Use your strategy to evaluate the area of the region above the x-axis and below the graph $y = \frac{1}{(2-x)^5}$, where $1 \leq x < 2$.

e. Evaluate $\int_{-1}^0 \frac{-1}{\sqrt[3]{x}}\, dx$ and $\int_1^2 \frac{1}{(2-x)^5}\, dx$ based on your definition in 2c.

3. Let f be an arbitrary continuous nonnegative function defined on the interval (a, b) and satisfying $\lim_{x \to a^+} f(x) = \lim_{x \to b^-} f(x) = \infty$.

a. Sketch a graph for one such function and explain your strategy for evaluating the area of the region above the x-axis and below the graph of f.

b. Suggest a definition for $\int_a^b f(x)\, dx$ in this case.

4. Let f be an arbitrary continuous nonnegative function defined on the interval $[a, c) \cup (c, b]$ and that satisfies $\lim_{x \to c} f(x) = \infty$.

a. Answer Problems 3a and 3b in this case.

b. Evaluate $\int_{-1}^2 \frac{1}{\sqrt{|x|}}\, dx$.

c. Evaluate $\int_0^3 \frac{1}{(x-1)^2}\, dx$.

III. Areas of Infinite Regions in Both Directions

Consider a nonnegative function f that is continuous on an infinite interval except at a few points where it has infinite discontinuities. Explain how you compute the area of the region above the x-axis and below the graph of such a function.

IV. Improper Integrals

The integrals in Problems I, II, and III are called *improper integrals*. We say that an improper integral *converges* if it is equal to a real number; otherwise it *diverges*.

Let p be an arbitrary real number (assume $p \neq 1$ if you did not work on the project "The Natural Logarithmic Function" from Section 6.3).

1. For which values of the exponent p does the improper integral $\int_1^\infty \frac{1}{x^p}$ converge or diverge?

2. For which values of the exponent p does the improper integral $\int_0^1 \frac{1}{x^p}$ converge or diverge?

3. For which values of the exponent p does the improper integral $\int_0^\infty \frac{1}{x^p}$ converge or diverge?

CHAPTER 6 REVIEW

The area of a region in the plane measures the amount of the plane the region occupies.

To develop a strategy for measuring the areas of regions in the plane, we start by defining a unit measure: 1 square unit is the area of any square whose sides each have length 1 unit.

Based on this definition, we can derive formulas for the areas of arbitrary squares, rectangles, triangles, parallelograms, trapezoids, and regular polygons using basic ideas from geometry. Using only the formula for the areas of triangles, we can compute the areas of arbitrary polygonal regions, as the latter can be always divided into triangles.

For the areas of irregular shapes, Euclidean geometry can no longer lead to formulas but calculus can. To describe how calculus might be used to compute the areas of irregular shapes, consider, for simplicity, the region in the plane that is below the graph of a nonnegative function f defined on an interval $[a, b]$ and that is above the x-axis. For approximate values of its area $\mathcal{A}$, use the Riemann sums. These are the areas of some polygonal regions that approximate the region of interest. The polygonal regions are formed by either trapezoids or rectangles.

The Riemann sum T_n, which is the area of the polygonal region with n trapezoidal tiles with equal widths, is given by the formula

$$T_n = \frac{b-a}{2n}\left[f(a) + 2\sum_{i=1}^{n-1} f(x_i) + f(b)\right],$$

where the x_i's are

$$x_i = a + i\frac{b-a}{n}, 0 \le i \le n.$$

The Riemann sum L_n, which is the area of the polygonal region with n rectangular tiles with the same width corresponding to the left rectangular method, is given by the formula

$$L_n = \frac{b-a}{n} \sum_{i=0}^{n-1} f(x_i).$$

The Riemann sum R_n, which is the area of the polygonal region with n rectangular tiles with the same width corresponding to the right rectangular method, is given by the formula

$$R_n = \frac{b-a}{n} \sum_{i=1}^{n} f(x_i).$$

As n becomes large, the values T_n, L_n, and R_n yield good approximations for the area A of the irregular shape. The exact value of A is obtained by taking the limit as n goes to ∞ in each case. There are several results that give the magnitudes of the errors when estimating A with T_n, L_n, or R_n, but they are beyond the scope of this book.

Another way to compute the area of a region bounded by the x-axis and the graph of a nonnegative function f defined over an interval $[a, b]$ is given by the Fundamental Theorem of Calculus, which says

$$A = \int_a^b f(x)\, dx.$$

It follows that for a function f whose sign changes on the interval $[a, b]$, the area A of the region between the graph of f, the x-axis, and the vertical lines $x = a$ and $x = b$ is

$$A = \int_a^b \left| f(x) \right| dx.$$

More generally, the area of the region lying between the graphs of two functions f and g both defined on the interval $[a, b]$ is

$$A = \int_a^b \left| f(x) - g(x) \right| dx.$$

When we need to find the area of a region in the plane, we can sometimes determine its exact value using the Fundamental Theorem of Calculus, but in some instances, we can only find approximations for the area using the method of Riemann sums; this primarily depends on the nature of the region's boundary (see the note on numerical integration at the end of Section 6.3).

CHAPTER 6 REVIEW EXERCISES

1. Write a paragraph in which you explain to an elementary-school student that the area of a rectangle whose length is 1/2 and whose width is 1/3 is equal to 1/6.

2. Write a paragraph in which you explain to a middle-school student that the area of a square whose sides each have length $\sqrt{2}$ is equal to 2. Use two different arguments, one based on geometry and one on the idea of approximating the number $\sqrt{2}$ by rational numbers.

3. Use measurements and the formulas obtained in Section 6.1 to find an approximate value for the area of the following geometric figure.

4. Using the given grids, find a lower and an upper bound for the area of the following design.

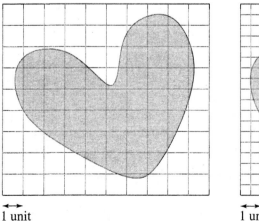

1 unit

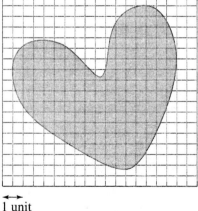

1 unit

5. In order to estimate the area of a piece of land he just inherited, a farmer took the following measurements (in meters). What approximate value for the area does the farmer obtain using the trapezoidal method?

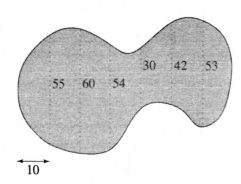

$\xleftrightarrow{\hspace{0.8cm}}$
10

6. Using the trapezoidal method with four tiles, approximate the area of the plane region that is below the graph $y = 1/x$ and above the x-axis for $1 \le x \le 5$.

7. Using the left rectangular method with six tiles, approximate the area of the plane region that is below the graph $y = e^{-x}$ and above the x-axis for $0 \le x \le 3$.

8. Using the right rectangular method with five tiles, approximate the area of the plane region that is below the graph $y = \dfrac{1}{x^2+1}$ and above the x-axis for $0 \le x \le 5$.

9. Using the midpoint rectangular method with four tiles, approximate the area of the plane region that is below the graph $y = 1/x$ and above the x-axis for $1 \le x \le 5$.

10. Denote by A the area of the plane region below the graph $y = \sqrt{\dfrac{x}{x+1}}$ and above the x-axis, where $0 \le x \le 2$. Denote by T_n, L_n, and R_n the areas of the polygonal regions with n tiles that approximate the plane region using the trapezoidal, left rectangular, and right rectangular methods, respectively.

   a. Compute by hand T_4. How does this value compare to the area A?
   b. Compute $T_{40}, T_{80}, \ldots, T_{280}$ using your calculator program from Classroom Discussion 6.2.1. Tabulate your results.
   c. Based on your table, conjecture a value for A that is correct to two decimal places.
   d. Compute by hand L_4. How does this value compare to the area A?
   e. Compute by hand R_4. How does this value compare to the area A?

   f. Using your calculator program from Classroom Discussion 6.2.2, compute $L_{40}, L_{80}, \ldots, L_{280}$ and $R_{40}, R_{80}, \ldots, R_{280}$. Tabulate your results.
   g. Find an approximate value for the area A correct to two decimal places.

11. Using the rectangular methods with a total of six tiles, find a lower and an upper bound for the area of the region that is above the x-axis and below the graph $y = f(x)$, where

$$f(x) = \begin{cases} x^4 & \text{if } -3 \le x \le 0, \\ \sqrt{x} & \text{if } 0 \le x \le 3. \end{cases}$$

12. Using the rectangular methods with a total of five tiles, find a lower and an upper bound for the area of the region that is above the x-axis and below the graph $y = f(x)$, where

$$f(x) = \begin{cases} 2x - x^2 & \text{if } 0 \le x \le 1, \\ \dfrac{1}{x} & \text{if } 1 \le x \le 2. \end{cases}$$

In Exercises 13–15, determine a function $f(x)$, two real numbers a and b, and a whole number n such that the given sum is the Riemann sum of f over the interval $[a, b]$; use either the trapezoidal, the left rectangular, or the right rectangular methods with n tiles.

13. $\displaystyle\sum_{i=1}^{9} 1/i^2$

14. $\displaystyle\frac{1}{2}\sum_{i=1}^{20}\left(\frac{i}{2}\right)^3$

15. $\displaystyle\frac{1}{20}\left(\frac{3}{2} + 2\sum_{i=1}^{9}\frac{1}{1 + \frac{i}{10}}\right)$

In Exercises 16–26, sketch the graph $y = f(x)$. Then, using the Fundamental Theorem of Calculus, compute by hand the area of the plane region below this graph and above the x-axis for $a \le x \le b$.

16. $f(x) = \sqrt{2}, a = -5, b = 4$

17. $f(x) = 2x - 1, a = 1, b = 3$

18. $f(x) = |x|, a = -10, b = 10$

19. $f(x) = x^4, a = -1, b = 1$

20. $f(x) = x\sqrt{x}, a = 4, b = 9$

21. $f(x) = 1/x^3, a = 1, b = 2$

22. $f(x) = 3x^2 - 2x, a = 2/3, b = 1$

23. $f(x) = 5x - x^2, a = 0, b = 5$

24. $f(x) = x^3 - x, a = -1, b = 0$

25. $f(x) = x^2(1 - x^2), a = -1, b = 1$

26. $f(x) = e^x - x, a = 0, b = 3$

In Exercises 27–33, compute the area of the plane region bounded by the graph $y = f(x)$ and by the x-axis for $a \le x \le b$. You may use your calculator to sketch the graph of f and to evaluate the definite integral corresponding to the area of the region of interest.

27. $f(x) = \sqrt{x + 1}, a = 2, b = 3$

28. $f(x) = x(x^2 + 1)^{10}, a = 0, b = 1$

29. $f(x) = \frac{2x}{(x^2+1)^2}, a = 1, b = 2$

30. $f(x) = \frac{x^2}{x^3+2}, a = -1, b = 1$

31. $f(x) = x\sqrt{x^2 + 2}, a = -2, b = -1$

32. $f(x) = e^{-x}, a = -3, b = 3$

33. $f(x) = xe^{x^2}, a = 0, b = 1$

In Exercises 34–38, describe the region whose area corresponds to the given definite integral.

34. $\int_2^5 (x^3 - x^2)\, dx$

35. $\int_{-1}^1 (7 - x^2)\, dx$

36. $\int_0^1 (x - x^4)\, dx$

37. $\int_3^{10} \left(\frac{1}{x+1} - \frac{1}{x^2+1} \right) dx$

38. $\int_{-2}^0 [(x^2 + 3) + x] \, dx$

In Exercises 39–50, sketch the graphs $y = f(x)$ and $y = g(x)$ for $a \leq x \leq b$. Then, using the Fundamental Theorem of Calculus, compute by hand the area of the plane region lying between the two graphs.

39. $f(x) = 3x + 1, g(x) = -x + 2, a = -2, b = 3$

40. $f(x) = |x|, g(x) = 4, a = -4, b = 3$

41. $f(x) = \sqrt{x}, g(x) = x^2, a = 0, b = 1$

42. $f(x) = x^2 + x, g(x) = x + 9, a = -3, b = 2$

43. $f(x) = x\sqrt{x}, g(x) = x^2, a = 0, b = 1$

44. $f(x) = x^2 + x + 1, g(x) = 2, a = -1 - \sqrt{5}, b = -1 + \sqrt{5}$

45. $f(x) = x^3 + x^2 + x, g(x) = -1, a = -1, b = 0$

46. $f(x) = |x|, g(x) = 4, a = -5, b = 10$

47. $f(x) = \sqrt{x}, g(x) = x^2, a = 0, b = 2$

48. $f(x) = x^2 + x, g(x) = x + 9, a = -3, b = 6$

49. $f(x) = x\sqrt{x}, g(x) = x^2, a = 0, b = 3$

50. $f(x) = x^2 + x + 1, g(x) = 2, a = -3, b = 2$

Further Applications of Integration

CHAPTER

7

7.1 **LENGTHS OF CURVES IN THE PLANE**
7.2 **AREAS OF SURFACES OF REVOLUTION**
7.3 **VOLUMES OF SOLIDS**

In this chapter, we continue exploring applications of integration to various topics from the middle-school mathematics curricula: finding lengths, surface areas, and volumes. In Sections 7.2 and 7.3, we begin by using Euclidean geometry to compute the surface area and volume of simple figures. Then, building on the results obtained from geometry, we use calculus to produce formulas that hold for more general shapes. Our presentation again stresses the interplay between geometry and calculus, and it shows that calculus is a powerful and essential tool for computing areas, lengths, surface areas, and volumes.

7.1 LENGTHS OF CURVES IN THE PLANE

Approximating lengths • Exact values of lengths of graphs • Computing the circumference of a circle in middle-school and college

When a curve is straight, its *length* can be easily measured using a ruler. When a curve is not straight, this is no longer possible. Our objective in this section is to demonstrate how to measure the lengths of plane curves and how to find a corresponding computational formula.

Classroom Discussion 7.1.1: Setting the Stage

Jason asked Katie, "How can the *length* of an arbitrary curve be measured?" Katie answered, "By measuring the amount of string it takes to exactly cover the curve."

1. Do you agree with Katie?
2. How practical is Katie's solution?
3. Can Katie's answer serve as a definition for the length of an arbitrary curve?
4. Can you answer Jason's question using a ruler if a string is not available? ◆

Classroom Discussion 7.1.2: Approximate Values for the Length of an Arbitrary Plane Curve

To find the length of the following curve, we use polygonal paths that increasingly better fit the curve, an idea we repeatedly used in Chapter 6 and in the projects "The Number Called π" and "Archimedes's Computation of π" in Section 1.1.

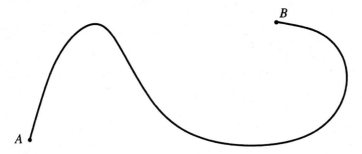

1. Pick a point P on the curve between points A and B, then, using your ruler, measure the length of line segments $\overline{AP}$ and $\overline{PB}$. Compute $\mathcal{L}_1$, the total length of these two line segments.
2. Pick a point P_0 on the curve between points A and P and pick a point P_1 on the curve between points P and B, then measure the length of line segments $\overline{AP_0}, \overline{P_0P}, \overline{PP_1}$, and $\overline{P_1B}$. Compute $\mathcal{L}_2$, the total length of these four line segments.
3. Let $\mathcal{L}_3$ and $\mathcal{L}_4$ be defined in a similar manner. Fill in Table 1.

n	$\mathcal{L}_n$
1	
2	
3	
4	

Table 1

4. Let $\mathcal{L}_n$ be defined in a similar way for $n \geq 4$. Describe what happens to $\mathcal{L}_n$ as n becomes larger and larger. ◆

Classroom Discussion 7.1.3: Approximate Values for the Length of a Graph in the Plane

Your task here is to find approximate values for the length $\mathcal{L}$ of a curve that represents the graph $y = f(x)$, where $a \le x \le b$; this time, don't use a ruler. Before looking at Problems 1–10, think of an organized way to achieve this task.

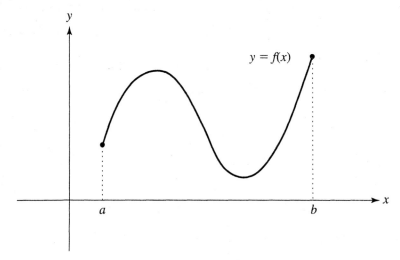

1. Pick $n - 1$ arbitrary points $P_1, P_2, \ldots, P_{n-1}$ on the graph of f, and let $P_0 = (a, f(a))$ and let $P_n = (b, f(b))$, then construct the line segments joining each two consecutive points on the graph.

2. For each fixed index $0 \le i \le n$, denote by x_i the x-coordinate of the point P_i. Express the length of the line segment $\overline{P_i P_{i+1}}$ in terms of the first coordinates of points P_i and P_{i+1} and the function f.

3. Express the length $\mathcal{L}_n$ of the path composed of the preceding n line segments.

4. How does the length $\mathcal{L}$ of the given graph compare to each $\mathcal{L}_n$?

5. Describe the behavior of the sequence $\{\mathcal{L}_n\}_n$ as the number of points you select *throughout* the graph goes to ∞.

6. How can you select points $P_1, P_2, \ldots, P_{n-1}$ in order to obtain simpler expressions for the lengths $\mathcal{L}_n$?

7. In the case of your choice of points in Problem 6, express each length $\mathcal{L}_n$ in terms of a, b, n, and f only.

8. Write a calculator program that will compute $\mathcal{L}_n$ for arbitrary n. Set up your program to provide the value of $\mathcal{L}_n$ once you enter the information about f, a, b, and n.

⟨T⟩ 9. Compute the values of $\mathcal{L}_1, \mathcal{L}_2, \ldots, \mathcal{L}_{10}$ in the case of the portion of the parabola $y = x^2$, where $0 \le x \le 1$. Fill in Table 2.

n	$\mathcal{L}_n$
1	
2	
3	
4	
5	
6	
7	
8	
9	
10	

Table 2

10. Conjecture an approximate value for the length $\mathcal{L}$ of this curve that is correct to one decimal place. (You will have the opportunity to check your conjecture later in this section.)   ◆

Classroom Discussion 7.1.4: Exact Value of the Length of a Graph in the Plane

Consider the curve $y = f(x)$ where $f(x)$ is a continuous function defined on the interval $[a, b]$. Your task here is to find, if possible, the exact value of the length $\mathcal{L}$ of this curve. Fix $a < x < b$ and denote by $\mathcal{L}(x)$ the length-so-far function, that is, $\mathcal{L}(x)$ is the length of the portion of the graph of f between the points $(a, f(a))$ and $(x, f(x))$.

1. What are the values of $\mathcal{L}(a)$, $\mathcal{L}(\frac{a+b}{2})$, and $\mathcal{L}(b)$?

2. Is the function $\mathcal{L}(x)$ increasing, decreasing, or neither?

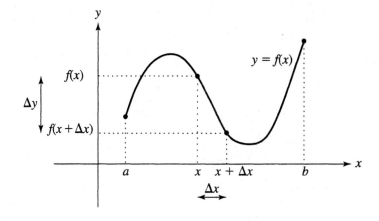

Let x increase by a small amount $\triangle x$; then $\mathcal{L}(x)$ increases by an amount $\triangle \mathcal{L}$ that corresponds to the additional portion of the graph between the points $(x, f(x))$ and $(x + \triangle x, f(x + \triangle x))$.

3. If $\triangle x$ is very small, how might you view this additional portion?

4. Express the corresponding approximation for $\triangle \mathcal{L}$ in terms of $\triangle x$ and $\triangle y$, where $\triangle y = f(x + \triangle x) - f(x)$.

5. Using Problem 4, find an approximation for the rate of change $\dfrac{\triangle \mathcal{L}}{\triangle x}$.

6. Assume that the function $f(x)$ is differentiable on the interval $[a, b]$. Show that in this case

$$\mathcal{L}'(x) = \sqrt{1 + [f'(x)]^2}.$$

7. Assume in addition that the function $f'(x)$ is continuous on the interval $[a, b]$. This implies that the function $\sqrt{1 + [f'(x)]^2}$ is also continuous and thus has an antiderivative.

Explain why the function $\sqrt{1 + [f'(x)]^2}$ is continuous, then, in this case, express $\mathcal{L}(x)$ as a definite integral.

8. Deduce a formula for computing $\mathcal{L}$. ◆

Theorem 7.1.1. *Let f be a function defined on the interval $[a, b]$ whose derivative f' is a continuous function. Then, the length $\mathcal{L}$ of the graph $y = f(x)$ is given by the formula*

$$\mathcal{L} = \int_a^b \sqrt{1 + [f'(x)]^2}\, dx. \tag{7.1}$$

Historical Note: Pierre de Fermat (1601–1665; from France)

Fermat was a lawyer, a skilled linguist, and one of his century's greatest mathematicians. In addition to his work in physics and number theory, Fermat made numerous contributions to the development of calculus, including a dynamic approach to tangency, a method for finding centroids, the standard formula for the first derivative of a polynomial function, and formulas for finding the arc length and for finding the area of a surface of revolution.

EXAMPLE 1 Find the length $\mathcal{L}$ of the line segment $y = x/2 + 1$ between points $(0, 1)$ and $(4, 3)$ using Theorem 7.1.1; then, check your result using the Pythagorean Theorem.

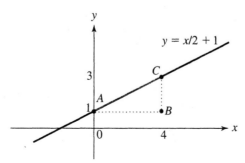

Solution In order to use Theorem 7.1.1, you must first compute the derivative $y'(x)$, then evaluate and simplify the function $\sqrt{1 + [y'(x)]^2}$ as much as possible.

$$\sqrt{1 + [y'(x)]^2} = \sqrt{1 + \left(\frac{1}{2}\right)^2} = \frac{\sqrt{5}}{2}.$$

Therefore,

$$\mathcal{L} = \int_0^4 \sqrt{1 + [y'(x)]^2}\, dx = \int_0^4 \frac{\sqrt{5}}{2}\, dx = \frac{\sqrt{5}}{2} x \Big|_0^4 = 2\sqrt{5}.$$

On the other hand, since $\triangle ABC$ has a right angle at vertex B, it satisfies $\mathcal{L}^2 = AB^2 + BC^2$ (Pythagorean Theorem). Since $AB = 4$ and $BC = 2$, it follows that

$$\mathcal{L} = \sqrt{4^2 + 2^2} = \sqrt{20} = 2\sqrt{5}. \qquad \blacksquare$$

EXAMPLE 2 Find the length $\mathcal{L}$ of the graph $y = \frac{x^3}{6} + \frac{1}{2x}$ between points $(1, \frac{2}{3})$ and $(2, \frac{19}{12})$ using Theorem 7.1.1.

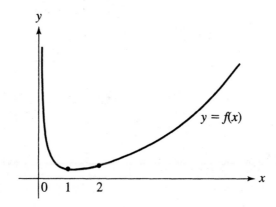

Solution In order to use Theorem 7.1.1, you must first compute the derivative $y'(x)$, then evaluate and simplify the function $\sqrt{1 + [y'(x)]^2}$ as much as possible:

$$y'(x) = \left(\frac{x^3}{6} + \frac{1}{2x}\right)' = \frac{3x^2}{6} - \frac{1}{2x^2} = \frac{x^2}{2} - \frac{1}{2x^2} = \frac{x^4 - 1}{2x^2}.$$

$$1 + [y'(x)]^2 = 1 + \frac{(x^4 - 1)^2}{4x^4} = \frac{4x^4 + x^8 - 2x^4 + 1}{4x^4} = \frac{x^8 + 2x^4 + 1}{4x^4}$$

$$= \frac{(x^4 + 1)^2}{4x^4}.$$

Thus,

$$\sqrt{1 + [y'(x)]^2} = \frac{x^4 + 1}{2x^2} = \frac{x^2}{2} + \frac{1}{2x^2}.$$

Therefore,

$$\mathcal{L} = \int_1^2 \sqrt{1 + [y'(x)]^2}\, dx = \int_1^2 \left(\frac{x^2}{2} + \frac{1}{2}x^{-2}\right) dx = \left(\frac{x^3}{6} - \frac{1}{2x}\right)\Big|_1^2.$$

Hence,

$$\mathcal{L} = \left(\frac{8}{6} - \frac{1}{4}\right) - \left(\frac{1}{6} - \frac{1}{2}\right) = \frac{17}{12}. \qquad \blacksquare$$

Practice Problems

1. Find the length $\mathcal{L}$ of the portion of the parabola $y = x^2$ between points $(0,0)$ and $(1,1)$. You must compute by hand the integrand in Theorem 7.1.1 and simplify it as much as possible, but you may use your calculator to evaluate the resulting definite integral. Compare your result with the approximate values in Table 2 from Classroom Discussion 7.1.3.

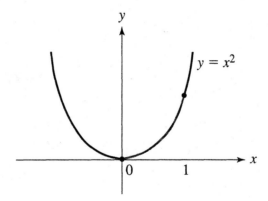

2. A cable hanging between two poles of the same height usually takes the shape of a catenary. What is the length $\mathcal{L}$ of a cable that is hanging between two poles 40 meters apart if this cable takes the shape of the catenary $y = 10\left(e^{x/20} + e^{-x/20}\right)$? To answer this question, follow these steps:

    a. Evaluate the derivative $y'(x)$.

    b. Expand and simplify the expression of the function $1 + [y'(x)]^2$.

    c. Use the simplified expression obtained in Problem b to rewrite $1 + [y'(x)]^2$ as a perfect square.

    d. Evaluate $\mathcal{L}$.

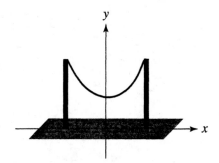

Classroom Discussion 7.1.5: Circumference of a Circle from Different Points of View

The number π is defined as $\pi = \dfrac{C}{2r}$, where C is the circumference of any circle and r is its radius. For more on this, see the project "The Number Called π" in Section 1.1. Suppose, however, that you do not know this relation and you want to find the circumference of a circle of radius $r > 0$. We present two different approaches: one for middle-school students based on measurements, and one for college students based on calculus.

1. Classroom Connection 7.1.1: Circumference of a Circle Using Measurements

The following exploration is taken from pages 39–41 in the eighth-grade textbook *Math Thematics*, *Book 3*. Discuss it in small groups.   ◆

Section 4   **Circumference and Volume**

Setting the Stage ·····························

Amazing fiction has a way of wandering into tall tales. The legend of Paul Bunyan, a giant lumberjack, came out of the logging camps of the northern United States in the 1800s. The story below tells of some amazing appetites.

> The lumber crews liked pancakes best, but they would gobble up and slurp down the pancakes so fast that the camp cooks couldn't keep up with them, even when the cooks got up twenty-six hours before daylight. The main problem was that the griddles the cooks used for frying the pancakes were too small. . . .
>
> [Paul Bunyan] went down to the plow works at Moline, Illinois, and said, "I want you fellows here to make me a griddle so big I won't be able to see across it on a foggy day."
>
> The men set to work. When they were finished, they had built a griddle so huge there was no train or wagon large enough to carry it.
>
> "Let me think what to do," said Paul. "We'll have to turn the griddle up on end, like a silver dollar, and roll it up to Michigan." He hitched [Babe] the Blue Ox to the upturned griddle, and away they went. . . . A few miles from the Big Onion lumber camp, Paul unhitched Babe and let the griddle roll on by itself.
>
> *American Tall Tales*, Adrien Stoutenberg

Section 4   Circumference and Volume   **39**

Think About It

1   What amazing "facts" in the story let you know that it is a tall tale?

2   Paul Bunyan's pancake griddle was circular. According to one version of the story, its diameter was 236 ft.

    a. Sketch a circle and draw a diameter. Is this the only diameter the circle has? Explain.

    b. Would Paul Bunyan's griddle fit in any room in your school?

GOAL

LEARN HOW TO...
* find circumference
* write and evaluate expressions

AS YOU...
* investigate a claim made in a tall tale

KEY TERMS
* circumference
* variable
* expression
* evaluate

Exploration

Finding Circumference

SET UP   *You will need: • metric ruler • coin • calculator*

▶ One version of the story says that 3 miles from camp Paul gave the griddle a push and it rolled to the spot where he wanted it. Is this an amazing feat? One measurement that can help you answer this question is the griddle's circumference. The **circumference** of a circle is the distance around it.

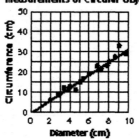

Measurements of Circular Objects

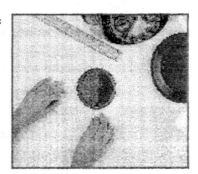

▸ **3** Some students measured the diameter and the circumference of several circular objects. Then they made a scatter plot and drew a fitted line. Does the line seem to fit the data? Explain.

40   **Module 1**   Amazing Feats, Facts, and Fiction

4 a. Copy the table. Then use the fitted line on page 40 to estimate the circumferences of circles with the given diameters.

b. Calculate the ratio of the *circumference* to the *diameter*. Round to the nearest tenth.

Diameter (cm)	Circumference (cm)	Circumference / diameter
11	?	?
1	?	?
8	?	?
14	?	?
4.5	?	?

c. Discussion The circumference of a circle appears to be about how many times its diameter?

5 Estimate the circumference of Paul Bunyan's pancake griddle. Explain how you made your estimate.

▶ **Using Variables** In mathematics, symbols are often used instead of words to express relationships. A **variable** is a symbol used to represent a quantity that is unknown or that can change.

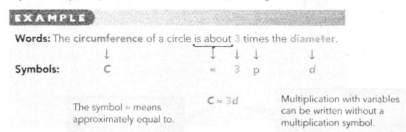

EXAMPLE

Words: The circumference of a circle is about 3 times the diameter.

Symbols: C $\approx$ 3   p   d

The symbol ≈ means approximately equal to.

$C \approx 3d$

Multiplication with variables can be written without a multiplication symbol.

▶ *C* and *3d* are *expressions*. An **expression** is a mathematical phrase that can contain numbers, variables, and operation symbols. Other examples of expressions are 8, *lw*, 5 + *n*, 9*x²*, and $\frac{a}{2}$.

6 **Discussion** The exact relationship between the circumference and the diameter of a circle is given below. Read π as *pi*.

$$\frac{C}{d} = \pi, \text{ or } C = \pi d$$

a. ▨ Calculator Press the ▮ key on a calculator. What number appears?

b. π is actually a letter from the Greek alphabet. Does this mean that it is a variable? Explain.

c. Which is a closer approximation for π, 3.14 or $\frac{22}{7}$? Explain.

Section 4   Circumference and Volume   **41**

2. Circumference of a Circle Using Calculus

Notice that since the circle has reflectional symmetry about the *x*-axis, its circumference is twice the length of the upper semicircle. Because the upper semicircle is the graph of a function, Theorem 7.1.1 applies.

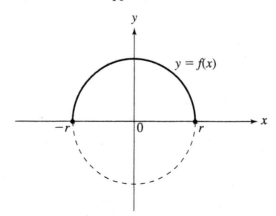

a. Under what conditions does a point (x, y) lie on the upper semicircle?

b. What is the equation of the upper semicircle?

c. Find the circumference $\mathcal{L}$ of the circle using Theorem 7.1.1. To answer this question, follow these steps:

    i. Evaluate the derivative $y'(x)$ using the square-root rule.

    ii. Evaluate and simplify the expression of $\sqrt{1 + [y'(x)]^2}$ as much as possible.

    iii. Use Theorem 7.1.1 to compute $\mathcal{L}$. You may use your calculator to integrate if needed. ◆

EXERCISES 7.1

In Exercises 1–4, do the following tasks: (a) Sketch the graph $y = f(x)$. (b) Find approximate values $\mathcal{L}_n$ for the length of the given graph using approximations of the graph by polygonal paths with n line segments for $n = 2, 4, \ldots, 20$ as in Classroom Discussion 7.1.3. Tabulate your results. (c) What is the value in your table that best approximates the length $\mathcal{L}$? Explain. (d) Use integral calculus to compute the length $\mathcal{L}$ of the given graph. You may use your calculator to compute the definite integral corresponding to the length $\mathcal{L}$, but you must first evaluate the integrand by hand.

1. $f(x) = x^3, \; 0 \le x \le 1$

2. $f(x) = 1/x, \; 1 \le x \le 3$

3. $f(x) = \sqrt{x}, \; 0 \le x \le 4$

4. $f(x) = e^x, \; -1 \le x \le 1$

In Exercises 5–10, sketch the graph $y = f(x)$ where $a \le x \le b$, then write down a formula for its length. Do not compute the integral but simplify the integrand as much as possible.

5. $f(x) = 1 - x^2, \; a = 0, \; b = 1$

6. $f(x) = x^3 - x, \; a = -5, \; b = 2$

7. $f(x) = 2x^2 + x, \; a = -3, \; b = 3$

8. $f(x) = x^2 e^x, \; a = -1, \; b = 2$

9. $f(x) = x(x^3 - 4), \; a = 1, \; b = 2$

10. $f(x) = x/(x + 1), \; a = 0, \; b = 3$

In Exercises 11–14, find a function whose graph has length equal to the given definite integral.

11. $\displaystyle\int_0^2 \sqrt{1 + x^2}\,dx$

12. $\displaystyle\int_0^\pi \sqrt{1 + e^{2x}}\,dx$

13. $\displaystyle\int_0^5 \sqrt{x^2 + 2x + 2}\,dx$

14. $\displaystyle\int_1^4 x^{-2}\sqrt{x^4 + 1}\,dx$

15. If the graph $y = f(x)$ has length $\mathcal{L}$, what is the length of the graph $y = f(x) + \alpha$? Here, $a \le x \le b$ and α is a constant that is given. Answer this question by giving two different arguments, one based on geometry only and the other on Theorem 7.1.1.

16. A rope will be hung between two poles of the same height and 50 meters apart. The rope takes the shape of the catenary $y = 25/2\left(e^{x/25} + e^{-x/25}\right)$, $-25 \le x \le 25$. Find the length of this rope.

17. Sketch the graph $y = \frac{1}{3}(x^2 + 2)^{3/2}$ for $0 \le x \le 2$, then compute its length by hand. You may use your calculator only to evaluate derivatives or to check your final answer.

18. Sketch the curve given by the equation $x^{2/3} + y^{2/3} = 1$, then compute its length by hand. You may use your calculator only to evaluate derivatives or to check your final answer.

19. Circumference of an Ellipse
  a. Recall the definition and the Cartesian equation of an ellipse.
  b. Can you adapt the method used in Classroom Connection 7.1.1 to the case of an ellipse?

  c. Use Theorem 7.1.1 to represent the circumference of an ellipse as an integral.
  d. Can you evaluate the integral in Problem c by hand? By using a calculator? Compare with evaluating the integral in the case of a circle.

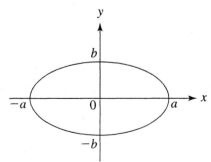

7.2   AREAS OF SURFACES OF REVOLUTION

Areas of cylinders, cones, and truncated cones • Surfaces of revolution • Areas of surfaces of revolution • Computing areas of spheres in middle-school and college

You have certainly heard and used the word *surface* on many occasions. Using your own words, how can you define the meaning of *surface*?

In real life, it is very important to be able to measure the area of a given surface. For example, if you were to tile the hemispherical roof of a mausoleum, then you would need to know how many tiles to order; if you were to paint the walls of a cylindrical room, you would want to know how much paint to buy. Our objective in this section is to find ways to measure the areas of surfaces. We will derive an explicit formula for computing the areas of certain surfaces called *surfaces of revolution*. In particular, we will use this formula to compute the area of a sphere.

Classroom Discussion 7.2.1: Generalities about Surfaces

1. We classified two-dimensional shapes as either polygons or irregular shapes. Can you think of a similar classification for the geometrical solids pictured here?

2. Explain how to compute the surface area of a polyhedron.

3. Ann asked Brian, "How can the surface area of an arbitrary solid be measured?" Brian answered "By measuring how much wallpaper it takes to cover the given solid." Do you agree with Brian? Explain. ◆

Classroom Discussion 7.2.2: Area of the Lateral Surface of a Cylinder

The goal of this Classroom Discussion is to find the area of the lateral surface of a right cylinder of radius $r > 0$ and height $h > 0$. We present two different approaches, both suitable for middle-school students: a simple one, based on cutting and unwrapping the lateral surface to make a rectangle, and one, more involved, based on approximating the lateral surface with the lateral faces of appropriate prisms.

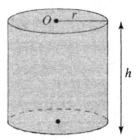

1. Classroom Connection 7.2.1: Transforming the Lateral Surface of a Cylinder into a Rectangle

The following exploration is taken from pages 318–320 in the eighth-grade textbook *Math Thematics, Book 3*. Discuss it in small groups. ◆

Section ① Working with Cylinders

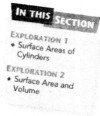

IN THIS SECTION

EXPLORATION 1
• Surface Areas of Cylinders

EXPLORATION 2
• Surface Area and Volume

Setting the Stage

The French general Napoleon Bonaparte once said that "an army marches on its stomach." He was not exaggerating. Hunger and poor nutrition caused more casualties in Napoleon's armies than actual combat. In 1795, the French government offered a prize of 12,000 francs to anyone who could invent a way to preserve food for the military.

Nicolas Appert, a candy maker from Paris, won the prize in 1809. Appert found that you can preserve food for months by sealing it in glass jars and heating the jars in boiling water. Glass jars break easily, however, and soldiers needed stronger containers.

This problem was solved by Peter Durand, an English inventor. Durand patented the use of metal cans for storing food. These cans were made of *tin plate* (iron coated with tin to prevent rusting) and came to be known as "tin cans." Tin cans were first used in 1813 to supply food to the British military.

Think About It

1 What two-dimensional shapes could you cut from a sheet of tin plate to make a tin can?

2 What factors might a manufacturer consider before designing a can?

▶ In this module, you'll learn about the history of inventions like the tin can. You'll also see how these inventions relate to mathematics.

Exploration

Surface Areas of Cylinders

GOAL

LEARN HOW TO...
* find the surface area of a cylinder

AS YOU...
* make a paper can

KEY TERM
* surface area

SET UP *You will need:* • *compass* • *metric ruler* • *scissors* • *tape* • *$8\frac{1}{2}$ in. by 11 in. sheet of paper*

▶ **A tin can is made by cutting two circles and a rectangle from a sheet of tin plate. The rectangle is rolled into a tube. The circles are added to the ends of the tube to form a cylinder.**

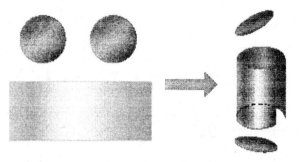

3 You can create your own can with paper.

a. Think about the dimensions of cans you see in a supermarket. Then choose a realistic radius and height for your paper can. Give the radius and the height in centimeters.

b. Use a compass to draw two circles having the radius you chose in part (a). Label the radius of each circle *r* as shown.

c. Draw a rectangle that you think you can use to make your can. Label the rectangle's length *l* and its width *w* as shown. Measure *l* and *w* to the nearest tenth of a centimeter.

4 Cut out the circles and the rectangle
you drew in Question 3. Tape the edges
of the rectangle together with little or
no overlap to form a tube. Hold a circle
over the top of the tube as shown.

a. In order for the tube and the circles
to form a can, how should the
length, *l*, of the rectangle be related
to the radius, *r*, of each circle?

b. How should the width, *w*, of the rectangle be related to the
height you chose in Question 3(a)?

c. If necessary, make a new rectangle whose length and width
have the properties you described in parts (a) and (b). Tape the
rectangle and circles together to form a paper can.

▶ A cylinder's *surface area* is the sum of the areas of the circles and the
rectangle that form the cylinder. In general, the **surface area** of a
space figure is the combined area of the figure's outer surfaces.

5 Use the paper can you made in Question 4.

a. How is the area of each circle related to the can's radius?

b. How is the area of the rectangle related to the rectangle's
length and width? to the can's radius and height?

c. Use your answers from parts (a) and (b) to write a formula for
the surface area, *S.A.*, of a cylinder in terms of its radius, *r*, and
height, *h*.

d. Find the surface area of your paper can.

✔ QUESTION 6

...checks that you
can find a cylinder's
surface area given its
radius and height.

6 **✔ CHECKPOINT** Find the surface area of each can.

a.   3.4 cm

10 cm

b.   3.9 cm

8.5 cm

7 Which of the cans in Question 6 uses more metal? Explain.

HOMEWORK EXERCISES ▶ See Exs. 1–9 on p. 325.

2. Approximating the Cylinder with Appropriate Prisms.

You can approximate the cylinder with regular n-hedrons, each of which is a right prism whose base is a regular n-gon as in the following figure.

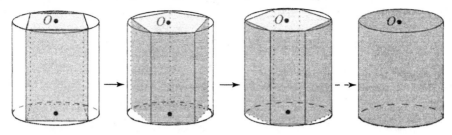

a. If p_n denotes the perimeter of the n-hedron's base, what is the area S_n of its lateral surface?

b. What happens to the prisms as n increases? What can you say about p_n and S_n?

c. Deduce from Problems a and b a formula for the area S of the lateral surface of the cylinder.   ◆

Classroom Discussion 7.2.3: Area of the Lateral Surface of a Cone

Our goal in this Classroom Discussion is to find the area of the lateral surface of a right cone with radius $r > 0$ and height $h > 0$. We present two approaches: one based on cutting the lateral surface into a sector, and one based on approximating the lateral surface with the lateral faces of appropriate pyramids; both approaches are suitable for high school students.

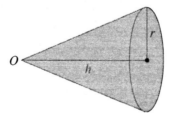

1. Cutting the Lateral Surface of a Cone into a Sector

By cutting the cone along segment OA, then unfolding, you get a sector centered at point O with radius $R = OA$ and angle α (in degrees). The radius R is called the *slant height* of the cone.

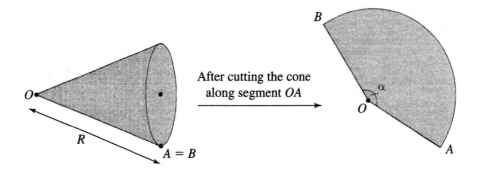

After cutting the cone
along segment OA

a. Express R in terms of r and h.

b. By identifying the circumference of the cone's base with the sector's arc length, find a relation between R, r, and α.

c. Express the area S of the lateral surface of the cone in terms of r and h.

2. Classroom Connection 7.2.2: Approximating the Cone with Appropriate Pyramids

The following exploration is taken from page 463 of the textbook *Discovering Geometry*, second edition, by Michael Serra. Discuss it in small groups. ◆

Finding the surface area of a right cone is related to how you found the surface area of a pyramid. To find the lateral surface of an octagonal pyramid, sum the areas of the eight triangles forming the lateral surface. To avoid confusing slant height with the height of the pyramid, use the variable l rather than h for slant height.

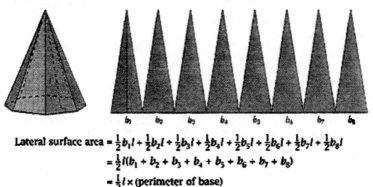

Lateral surface area $= \frac{1}{2}b_1 l + \frac{1}{2}b_2 l + \frac{1}{2}b_3 l + \frac{1}{2}b_4 l + \frac{1}{2}b_5 l + \frac{1}{2}b_6 l + \frac{1}{2}b_7 l + \frac{1}{2}b_8 l$

$\qquad\qquad\qquad = \frac{1}{2}l(b_1 + b_2 + b_3 + b_4 + b_5 + b_6 + b_7 + b_8)$

$\qquad\qquad\qquad = \frac{1}{2}l \times (\text{perimeter of base})$

Now imagine a pyramid whose polygonal base has more than eight sides. What does a regular polygon with 100 sides look like? With 1000 sides? What would a pyramid with 1000 lateral faces look like? As the number of sides of a regular polygon increases, the polygon approaches a circle. Its perimeter approaches the circumference of a circle. As the number of faces of a pyramid increases, it begins to look like a cone. Therefore we can take the formula for the lateral surface area of a pyramid and substitute circumference (C) for perimeter, which gives us a formula for the lateral surface area of a cone: lateral surface area $= \frac{1}{2}Cl = \frac{1}{2}(2\pi r)l = \pi r L$

Example D

Find the total surface area of the right cone with slant height 10 cm and radius 5 cm to the nearest centimeter.

Total surface area = lateral surface area + base area

$$= \pi r l + \pi r^2$$

$$= (\pi)(5)(10) + \pi(5)^2$$

$$= 75\pi \approx 235.6$$

The surface area of the cone is about 236 cm².

463

Classroom Discussion 7.2.4: Area of the Lateral Surface of a Truncated Cone

A cone of radius R and height $h + a$ was cut from the top as in the following figure.

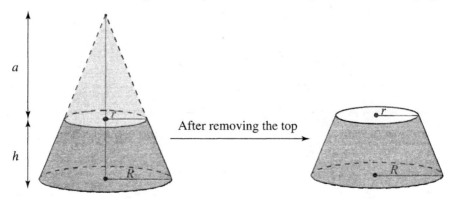

After removing the top

Follow the outline 1–5 to find the area of the lateral surface of the resulting truncated cone in terms of the radii r and R of its bases and of its height h.

1. What is the area S_1 of the original cone's lateral surface?
2. What is the area S_2 of the removed cone's lateral surface?
3. What is the area S of the lateral surface of the resulting truncated cone?
4. Using similarity, find a in terms of r, R, and h.
5. Using Problems 3 and 4, show that $S = \pi(R + r)\sqrt{h^2 + (R - r)^2}$. ◆

Summary

Using Euclidean geometry, you found a formula for the area S of the lateral surface of each of the following solids:

1. A cylinder of radius r and height h; $S = 2\pi rh$
2. A cone of radius r and height h; $S = \pi r\sqrt{h^2 + r^2}$
3. A truncated cone of radii r, R, and height h; $S = \pi(R + r)\sqrt{h^2 + (R - r)^2}$

The lateral surfaces of cylinders and cones can be obtained by revolving a line segment about a certain axis. For this reason, they are called **surfaces of revolution**.

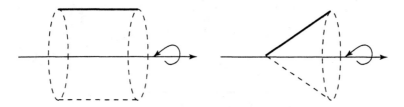

To understand this concept more fully, think about the surfaces generated by revolving the following graphs about the given axes.

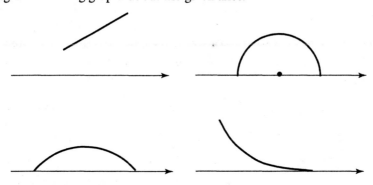

Can you think of any other examples of surfaces of revolution?

In what follows, we consider the general case of a surface obtained by revolving about the x-axis the graph of a given function $y = f(x)$ defined on the interval $[a, b]$. Our goal is to find, if possible, a formula for computing the area S of this surface.

Classroom Discussion 7.2.5: Area of an Arbitrary Surface of Revolution

Throughout this Classroom Discussion, the function f is assumed to take nonnegative values. Fix x in the interval $[a, b]$. Let $\mathcal{S}(x)$ be the area of the surface swept out by revolving about the x-axis the portion of the graph of f that lies between points $(a, f(a))$ and $(x, f(x))$.

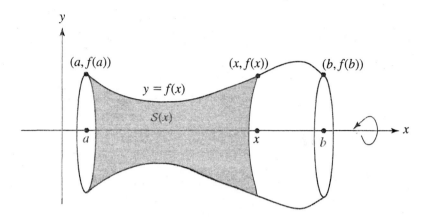

   In order to familiarize yourself with the function $\mathcal{S}(x)$, answer the following two questions.

1. What is the value of $\mathcal{S}(a)$, $\mathcal{S}(\frac{2a+b}{3})$, $\mathcal{S}(\frac{a+b}{2})$, and $\mathcal{S}(b)$?

2. Is the function $\mathcal{S}(x)$ increasing, decreasing, or neither? Explain.

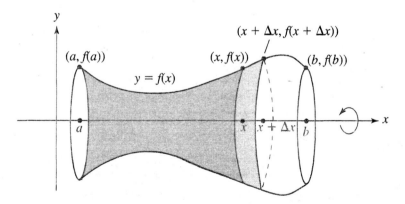

   Let x increase by a small amount $\triangle x$; then the area $\mathcal{S}(x)$ increases by an amount $\triangle \mathcal{S}$ corresponding to the additional surface swept out by revolving about the x-axis the portion of the graph of f that lies between points $(x, f(x))$ and $(x + \triangle x, f(x + \triangle x))$.

3. If Δx is very small, which common surface would you suggest using to approximate the additional surface?

4. Express the corresponding approximation for ΔS in terms of Δx, $f(x)$, and $f(x + \Delta x)$.

5. Using Problem 4, find an approximation for the rate of change $\dfrac{\Delta S}{\Delta x}$.

6. Assuming that the function f has a derivative $f'(x)$ on the interval $[a, b]$, evaluate the following two limits:

- $\displaystyle\lim_{\Delta x \to 0} f(x + \Delta x)$

- $\displaystyle\lim_{\Delta x \to 0} \dfrac{f(x + \Delta x) - f(x)}{\Delta x}$

7. Show that in this case,

$$\lim_{\Delta x \to 0} \frac{\Delta S}{\Delta x} = 2\pi f(x) \sqrt{1 + [f'(x)]^2}.$$

8. What can you say about the derivative $S'(x)$?

9. Express $S(x)$ as a definite integral. Here, assume that the function $f'(x)$ is continuous on the interval $[a, b]$. This implies that the function $\sqrt{1 + [f'(x)]^2}$ is also continuous. Consequently, the function $2\pi f(x) \sqrt{1 + [f'(x)]^2}$ is continuous and thus possesses an antiderivative.

10. Find an expression for the total surface area S. ◆

Theorem 7.2.1. *Let $f(x)$ be a nonnegative function defined on the interval $[a, b]$ whose derivative $f'(x)$ is continuous. The area S of the surface generated by revolving the graph $y = f(x)$ about the x-axis is given by the formula*

$$S = \int_a^b 2\pi f(x) \sqrt{1 + [f'(x)]^2}\, dx.$$

EXAMPLE Using Theorem 7.2.1, compute the area of the lateral surface of a right cylinder of radius $r > 0$ and height $h > 0$. Compare your result with the formula obtained in Classroom Discussion 7.2.2.

Solution   Consider the right cylinder of radius r and height h pictured here.

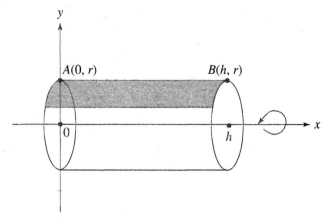

The lateral surface of this cylinder is generated by revolving $\overline{AB}$ about the x-axis, where $A(0, r)$ and $B(h, r)$. $\overline{AB}$ is the graph of the function $f(x) = r$ for all $0 \leq x \leq h$. The area of the cylinder's lateral surface is then given by Theorem 7.2.1. For $0 \leq x \leq h, f'(x) = 0$ and thus $\sqrt{1 + [f'(x)]^2} = 1$. Therefore,

$$S = \int_0^h 2\pi r \, dx = 2\pi r \int_0^h 1 \, dx = 2\pi r x \Big|_0^h = 2\pi r(h - 0) = 2\pi r h.$$

This is precisely the formula obtained earlier for the area of the lateral surface of a right cylinder of radius $r > 0$ and height $h > 0$. ∎

Practice Problems

1. Using Theorem 7.2.1, compute the area of the lateral surface of a right cone of radius $r > 0$ and height $h > 0$. To do so, first find the equation of $\overline{OA}$ (see the figure below). Compare your result with the formula obtained in Classroom Discussion 7.2.3.

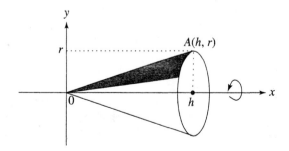

2. Compute the area of the surface generated by revolving the graph $y = x^3$ about the x-axis, where $0 \leq x \leq 1$.

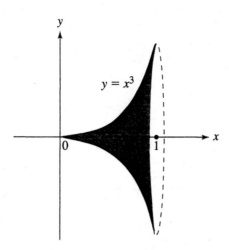

Classroom Discussion 7.2.6: Revolving about the x-axis the Graphs of Functions Taking Negative Values

Let $f(x)$ be an arbitrary function defined on the interval $[a, b]$ whose derivative $f'(x)$ is continuous over $[a, b]$.

1. Assume that $f(x) \leq 0$ for all $a \leq x \leq b$. Sketch the surface generated by revolving the graph of the function f about the x-axis. Then, express its area S as a definite integral.

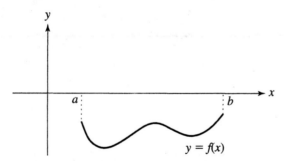

2. Assume that f takes both positive and negative values on the interval $[a, b]$. Sketch the surface generated by revolving the graph of the function f about the x-axis. Then, express its area S in terms of definite integrals. ◆

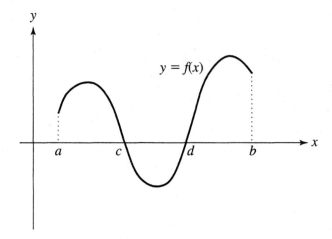

Classroom Discussion 7.2.7: Area of a Sphere from Different Points of View

The goal here is to find the area S of a sphere of radius $r > 0$. We present two different approaches: one for middle-school students based on experiments and one for college students based on calculus.

1. Experiment: Oranges and Areas of Spheres

Take an orange (or any spherical object that might apply for this experiment) and cut it in half. On a sheet of paper, trace the great circle of the orange. Peel the orange, then cut its skin into tiny flat pieces.

    a. How many times does the orange peel entirely fill the region inside the circle?

    b. Make a conjecture regarding the surface area S of the orange.

2. Areas of Spheres Using Calculus

The sphere of radius r centered at the origin is generated by revolving about the x-axis the upper semicircle of radius r centered at the origin.

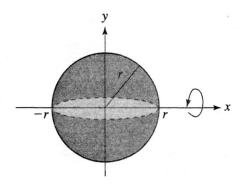

a. Define the function $f(x)$ whose graph is the upper semicircle.

b. Use the square-root rule to compute $f'(x)$.

c. Evaluate and simplify the function $\sqrt{1 + [f'(x)]^2}$ as much as possible.

d. Using Theorem 7.2.1, compute by hand the sphere's area S. ◆

EXERCISES 7.2

1. Each of the faces of the solid shown here is rectangular. Find its total surface area in terms of x.

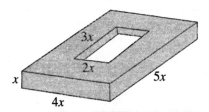

2. The top and bottom faces of the solid shown here are rectangular and are $4x$ meters apart. Find the solid's total surface area in terms of x.

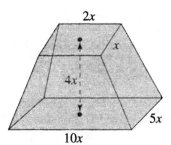

3. A solid has the shape of a right pyramid with a regular hexagonal base. The pyramid's height is h, and each side of its base has length x. Find the solid's total surface area in terms of h and x.

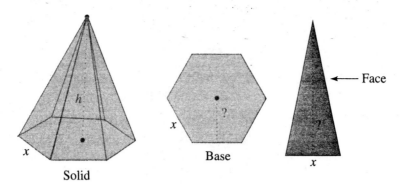

In Exercises 4–8, sketch the graph $y = f(x)$, where $a \le x \le b$, then sketch the surface generated by revolving this graph about the x-axis. Write down a formula for its area. Do not compute the integral but simplify the integrand as much as possible.

4. $f(x) = \sqrt{1 + x}, a = 3$, and $b = 8$

5. $f(x) = e^x, a = 0$, and $b = 2$

6. $f(x) = \sqrt{x}, a = 1$, and $b = 4$

7. $f(x) = 1/x, a = 1$, and $b = 5$

8. $f(x) = x^4, a = 0$, and $b = 2$

In Exercises 9–18, find the area of the surface obtained by revolving the graph $y = f(x)$ about the x-axis, where $a \le x \le b$. You may use your calculator to evaluate definite integrals when solving Exercises 13 and 16–18.

9. $f(x) = 3, a = -1$, and $b = 2$

10. $f(x) = x, a = 0$, and $b = 3$

11. $f(x) = 2x, a = 1$, and $b = 3$

12. $f(x) = x/3$ and $a = 1, b = 3$

13. $f(x) = x^3/3$ and $a = 0, b = 1$

14. $f(x) = \sqrt{1 - x^2}, a = 0$, and $b = 1$

15. $f(x) = \sqrt{2x - x^2}$ and $a = 0, b = 2$

16. $f(x) = \sqrt{x}$ and $a = 0, b = 1$

17. $f(x) = \sqrt{x - 1}$ and $a = 2, b = 5$

18. $f(x) = e^x$ and $a = 0, b = 2$

19. Fix a diameter in a sphere of radius $R > 0$ and divide it into twelve congruent segments. Then, using perpendicular slices, slice the sphere into twelve spherical bands as in the figure.

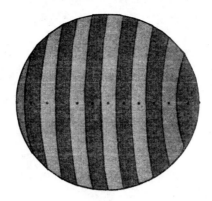

a. Do the twelve spherical bands each have the same area? Justify your answer.
b. Does your conclusion hold true if you replace twelve by an arbitrary whole number? Explain.

7.3 VOLUMES OF SOLIDS

Volumes of polyhedrons and cylinders • Volumes of solids • Computing volumes of cones in middle-school and college • Computing volumes of pyramids in middle-school and college • Computing volumes of spheres in middle-school, high school, and college

You have certainly heard and used the word *volume* on many occasions. Can you think of some real-life examples involving volumes? Using your own words, can you define the meaning of *volume*?

Our objective in this section is to develop an approach for measuring the volume of a solid, and to find an explicit formula for computing it. We use this formula to evaluate the volumes of cones, pyramids, and spheres.

Classroom Discussion 7.3.1: Volumes of Polyhedrons and Cylinders

Let us start with the simplest solid: a unit cube; that is, a cube whose length, height, and width are all equal to 1 unit. We define its volume to be 1 *cubic unit*. This definition should make sense to you since all unit cubes occupy the same amount of space.

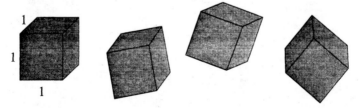

Thus, from now on, 1 cubic unit is the volume of any unit cube. Based only on this definition, explain how to compute the volumes of the solids described in Problems 1–5. At each step, you can use any results that you have proved earlier. Work in small groups.

1. A right rectangular prism with length ℓ, width w, and height h, where ℓ, w, and h are whole numbers.

2. A right rectangular prism with length ℓ, width w, and height h, where ℓ, w, or h is not a whole number; they are all rational numbers.

3. A right rectangular prism with length ℓ, width w, and height h, where ℓ, w, or h is an irrational number.

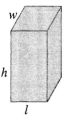

4. A right triangular prism whose base has an area A and whose height is h. Then, deduce a formula for the volume of an arbitrary right prism whose base has an area A and whose height is h.

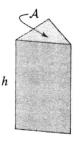
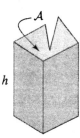

5. A generalized right cylinder whose height is h and whose base has an area A. Then, deduce the volume of a right cylinder with radius r and height h.

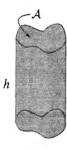
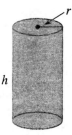

6. What would your answers be for Problems 4 and 5 if you replace the word *right* with the word *oblique* in each case? Hint: Imagine that each solid is a stack of sheets of paper.

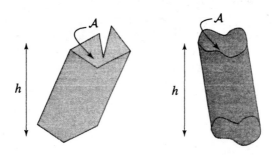

7. Can you find the volume of the following solids using the same techniques as in Problems 1–6?

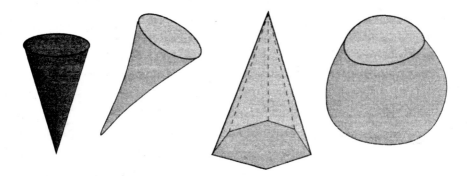

8. A cylindrical drinking cup has a height of 12 cm and a radius of 4 cm. The cup is filled with water, then the water is poured into an empty container, filling it completely. What is the volume of the container? What conclusion can you draw from this? How practical is this procedure for measuring the volume of arbitrary solids? ◆

Since solids have, in general, arbitrary shapes, an obvious question is: How can the volume of a solid with an arbitrary shape be measured?

One way to answer this question is to approximate the solid by shapes whose volumes are known, such as prisms, cylinders, etc. This leads to an approximation for the volume of the solid of interest. The smaller the gap between the given solid and the approximating shapes, the better the estimate will be for its volume. Thus, if the approximating shapes are chosen to approach the shape of the given solid, then their volumes will approach its volume. The exact value of the volume can then be obtained by passing to the limit. Note the analogy between this strategy and the one that uses Riemann sums to compute the areas of irregular shapes.

Another way to answer the question is to try to find the exact volume of a given solid using techniques similar to those employed in the proof of the Fundamental Theorem of Calculus. In the following Classroom Discussion, we describe this second approach in full detail.

Classroom Discussion 7.3.2: Formula for the Volumes of Solids

This Classroom Discussion's goal is to find, if possible, a formula for computing the volume $\mathcal{V}$ of the following solid S.

Fix $a \leq x \leq b$. The **cross-section** of the solid S corresponding to x is the plane region that contains all the points in S whose first coordinates are equal to x.

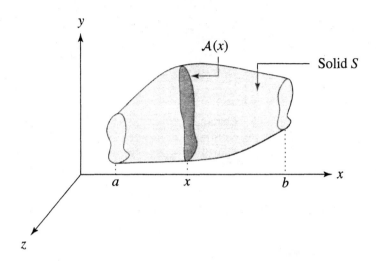

1. Describe the cross-sections of the solids listed here. In each case, consider different positions of the solid with respect to the x-axis.

   a. A rectangular prism
   b. A cylinder
   c. A cone
   d. A pentagonal pyramid
   e. A sphere
   f. A hemisphere

   Now, return to the solid S. Let $\mathcal{A}(x)$ be the area of the cross-section corresponding to x, and let $\mathcal{V}(x)$ be the volume-so-far function; that is, $\mathcal{V}(x)$ is the volume of the portion of the solid S that lies between the cross-sections corresponding to a and x. To familiarize yourself with the function $\mathcal{V}(x)$, answer the two following questions.

2. What are the values of $\mathcal{V}(a)$, $\mathcal{V}(\frac{2a+b}{3})$, $\mathcal{V}(\frac{a+b}{2})$, and $\mathcal{V}(b)$?
3. Is the function $\mathcal{V}(x)$ increasing, decreasing, or neither? Explain.

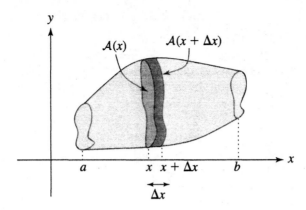

Let x increase by a small amount $\triangle x$; then the volume $\mathcal{V}(x)$ increases by an amount $\triangle \mathcal{V}$ corresponding to the additional portion of the solid S.

4. If $\triangle x$ is very small, how might you view this additional portion?

5. Express the corresponding approximation for $\triangle \mathcal{V}$ in terms of $\mathcal{A}(x)$ and $\triangle x$.

6. How would you approximate the rate of change $\dfrac{\triangle \mathcal{V}}{\triangle x}$?

7. Show that the derivative of $\mathcal{V}$ satisfies $\mathcal{V}'(x) = \mathcal{A}(x)$.

8. Assuming that the function $\mathcal{A}(x)$ has an antiderivative on the interval $[a, b]$, express $\mathcal{V}(x)$ as a definite integral.

9. Deduce a formula for the overall volume $\mathcal{V}$. ◆

Theorem 7.3.1. *For each fixed $a \leq x \leq b$, let $\mathcal{A}(x)$ denote the area of the cross-section corresponding to x of a given solid. If the function $\mathcal{A}(x)$ has an antiderivative on the interval $[a, b]$, then the solid's volume is given by the formula*

$$\mathcal{V} = \int_a^b \mathcal{A}(x)\, dx.$$

Historical Note: Johannes Kepler (1571–1630; from the Holy Roman Empire [now Germany])

Kepler is chiefly remembered for his work on astronomy: the first cosmological model of the solar system; his three laws of planetary motion; and his calculation of astronomical tables with extreme precision, which helped to establish the truth of heliocentric astronomy. In mathematics, Kepler studied close packing of spheres, and he gave the first proof of how logarithms work. Noticing the crude way in which volumes of wine casks were measured, Kepler was inspired to study volumes of solids of revolution, using ideas of Archimedes; the resulting treatise was later developed by Cavalieri and is part of the ancestry of infinitesimal calculus.

EXAMPLE Using calculus, find the volume of a prism whose height is h and whose base has area A; see the following figure. Then, compare your result with the formula obtained in Classroom Discussion 7.3.1.

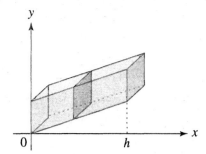

Solution   The prism's cross-sections are polygons that are congruent to the base, so for each $0 \le x \le h$, the cross section corresponding to x has area $\mathcal{A}(x) = A$. The prism's volume is given by Theorem 7.3.1:

$$\mathcal{V} = \int_0^h \mathcal{A}(x)\,dx = \int_0^h A\,dx = A\int_0^h 1\,dx = A\,x\Big|_0^h = A(h - 0) = Ah.$$

This formula is the same as the one obtained earlier.   ■

Practice Problem

Using calculus, find the volume of a generalized cylinder whose base has area A. Compare your result with the formula obtained in Classroom Discussion 7.3.1.

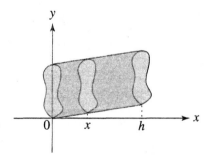

Classroom Discussion 7.3.3: Volumes of Cones

The goal of this Classroom Discussion is to find the volume of a cone with height $h > 0$ and radius $r > 0$. We present two different approaches: one for middle-school students based on comparing the volumes of cones and cylinders and one for college students based on calculus.

1. Classroom Connection 7.3.1: Comparing the Volumes of Cones and Cylinders

The following exploration is taken from pages 49–50 in the seventh-grade textbook *Connected Mathematics, Filling and Wrapping*. Discuss it in small groups.   ◆

Comparing Cones and Cylinders

In the last problem, you discovered the relationship between the volume of a sphere and the volume of a cylinder. In this problem, you will look for a relationship between the volume of a cone and the volume of a cylinder.

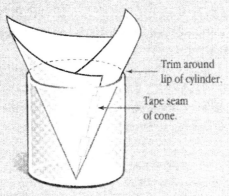
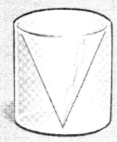

▦ Problem 5.2 Follow-Up

If a cone, a cylinder, and a sphere have the same radius and the same height, what is the relationship between the volumes of the three shapes?

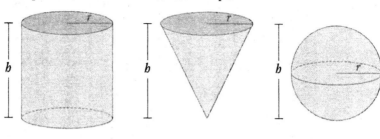

Melting Ice Cream

Olga and Serge buy ice cream from Chilly's Ice Cream Parlor. They think about buying an ice cream cone to bring back to Olga's little sister but decide the ice cream would melt before they got back home. Serge wonders, "If the ice cream all melts into the cone, will it fill the cone?"

Problem 5.3

Olga gets a scoop of ice cream in a cone, and Serge gets a scoop in a cylindrical cup. Each container has a height of 8 centimeters and a radius of 4 centimeters, and each scoop of ice cream is a sphere with a radius of 4 centimeters.

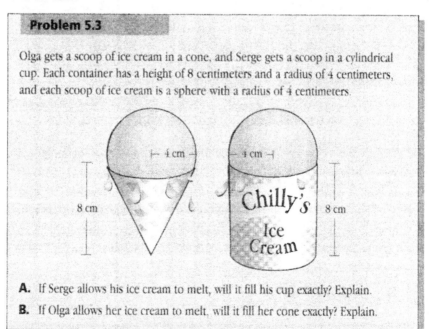

A. If Serge allows his ice cream to melt, will it fill his cup exactly? Explain.

B. If Olga allows her ice cream to melt, will it fill her cone exactly? Explain.

▦ Problem 5.3 Follow-Up

How many scoops of ice cream of the size above can be packed into each container?

2. Volumes of Cones Using Calculus

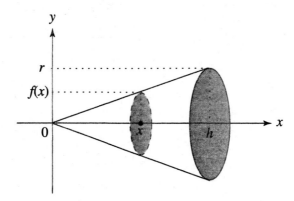

a. Explain why, when computing a cone's volume, the cone can be assumed to be right without loss of generality.

b. Describe the cross-sections of the cone in the preceding figure and compute their areas.

c. Use Theorem 7.3.1 to determine the volume of the cone. Compare this value with the volume of a cylinder that has the same radius and height as the cone. ◆

Classroom Discussion 7.3.4: Volumes of Pyramids

The goal is to find the volume of a pyramid whose height is $h > 0$ and whose base has an area $\mathcal{A}_0$. We present three different approaches: one for seventh-grade students based on combining pyramids to form prisms, one for eighth-grade students based on comparing the volumes of pyramids and prisms, and one for college students based on calculus.

1. Classroom Connection 7.3.2: Combining Pyramids to Form Prisms

The following exploration is taken from pages 544–545 and page 8-38 in the seventh-grade textbook *Math Thematics, Book 2*. Discuss it in small groups. ◆

GOAL

LEARN HOW TO...
• recognize a
  pyramid
• find the volume
  of a pyramid

AS YOU...
• build rectangular
  prisms using
  pyramid blocks

KEY TERM
• pyramid

Exploration 2

VOLUME
OF A PYRAMID

SET UP   *You will need: • Labsheet 2A • scissors • tape • metric ruler*

▶ A city skyline usually contains many different types of space figures.
To organize the space of a building efficiently, architects need to
consider the amount of space or volume a building will enclose.
In this exploration you'll see how the volume of a **pyramid** relates
to the volume of a rectangular prism. Several types of pyramids
are shown below.

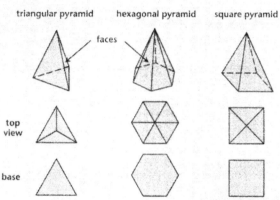

triangular pyramid   hexagonal pyramid   square pyramid

faces

top
view

base

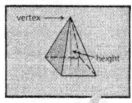

vertex ⟶

height

9 How are the three pyramids above
alike? How are they different?

10 **Use Labsheet 2A.**
Cut out the *Pyramid Nets*.
Crease each net along the
dashed lines. Then fold each
net and tape the edges
together to form two pyramids.

The *height* is the
perpendicular distance
from the base to the
vertex opposite it.

 Module 8   Heart of the City

For Questions 11–17, use the pyramids you made in Question 10.

11 The two pyramids you made are identical. Look at one of the pyramids.

   a. What type of pyramid did you form?

   b. Not including the base, how many faces does the pyramid have?

   c. Which faces of the pyramid are congruent? Explain.

12 Measure the height of one of your pyramids to the nearest centimeter.

height

13 The dashed lines below represent folds.

   a. Use your two pyramids to make a figure with the views shown.

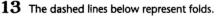

    front view   top view   right-side view

   b. Use your two pyramids to make a figure with the front and back views shown. Then draw the top view of the figure.

   front view   back view

14 Work with a partner.

   a. Make a cube by putting three of your pyramids together.

   b. Write an expression to represent the volume of the cube. Use B for the area of the base of the cube and h for the height of the cube.

   c. Compare the area of the base and the height of a single pyramid with the area of the base and the height of the cube you formed.

   d. Suppose you know the volume of a cube formed by putting together three identical pyramids as in part (a). How can you find the volume of each pyramid?

   e. Use your answers to parts (b)–(d) to write an expression for the volume of a pyramid with height h and base area B.

Name _____   Date _____

MODULE 8   **LABSHEET** **2A**

Pyramid Nets   (Use with Question 10 on page 544.)

Directions   Cut out each pyramid net. Crease each net along the dashed
lines. Then fold each net, tuck in the tabs, and tape the edges together to
form two pyramids.

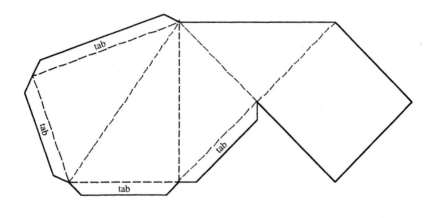

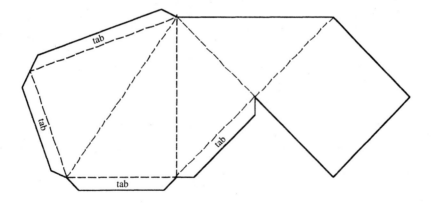

2. Classroom Connection 7.3.3: Comparing the Volumes of Pyramids and Prisms

The following exploration is taken from pages 422–423 and page 6-55 in the eighth-grade textbook *Math Thematics, Book 3*. Discuss it in small groups. ◆

...checks that you can find the surface area of a pyramid.

10 ✔ CHECKPOINT Find the surface area of each regular pyramid.

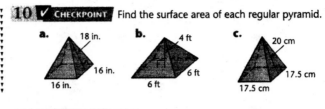

a. 18 in. 16 in. 16 in.

b. 4 ft 6 ft 6 ft

c. 20 cm 17.5 cm 17.5 cm

HOMEWORK EXERCISES ▶ See Exs. 1–8 on p. 427.

GOAL

LEARN HOW TO...
+ find volumes of prisms, pyramids, and cones

AS YOU...
+ look at models of block pyramids and prisms

KEY TERM
+ cone

Exploration 2

Volumes of PRISMS, PYRAMIDS, and CONES

SET UP *You will need Labsheet 3B.*

Winter *mat houses* were once used by people living in the Plateau region of the northwestern United States. These homes were usually occupied from mid-October to mid-March. A mat house was built roughly in the shape of a triangular prism.

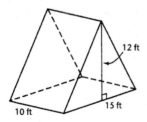

12 ft

10 ft   15 ft

11 In Module 1 you learned how to use the formula

Volume = area of base × height, or $V = Bh$

to find the volume of a cylinder or a rectangular prism. This formula can also be used to find volumes of prisms with other bases. How can you use this formula to find the volume of the triangular prism above?

▶ **Volume of a Pyramid** The Great Pyramid of Giza is made from blocks of stone. The sides are jagged, but they look smooth when viewed from afar, as if the block pyramid had straight edges and flat faces. Now you'll look at block pyramids and block prisms to discover the relationship between the volume of a pyramid and the volume of a prism with the same base and height.

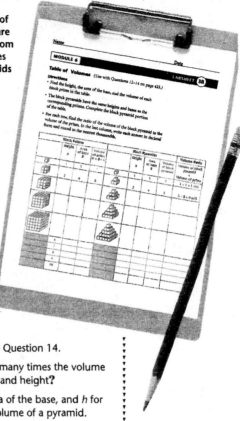

Use Labsheet 3B for Questions 12–14.

12 Follow the directions on the labsheet to complete the *Table of Volumes*.

13 Look at the volume ratio column of the table. What patterns do you notice in the values?

14 Discussion If the last column of the table were continued, the ratio for the 50th entry would be 0.343 and the 100th entry would be 0.338. What "nice" fraction do the ratios appear to be approaching?

15 Try This as a Class Use your answer to Question 14.

   a. The volume of a prism is about how many times the volume of a pyramid that has the same base and height?

   b. Using *V* for the volume, *B* for the area of the base, and *h* for the height, write a formula for the volume of a pyramid.

   c. Use your formula from part (b) to find the volume of a pyramid that has the same base and height as the triangular prism on page 422.

16 ✔ **CHECKPOINT** Find the volume of each space figure. Round decimal answers to the nearest tenth.

a.

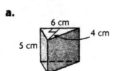

6 cm
4 cm
5 cm

b.

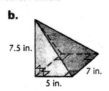

7.5 in.
7 in.
5 in.

✔ **QUESTION 16**

...checks that you can find volumes of prisms and pyramids.

Name _____   Date _____

| **MODULE 6** | **LABSHEET** 3B |

Table of Volumes (Use with Questions 12–14 on page 423.)

Directions
- Find the height, the area of the base, and the volume of each block prism in the table.

- The block pyramids have the same heights and bases as the corresponding prisms. Complete the block pyramid portion of the table.

- For each row, find the ratio of the volume of the block pyramid to the volume of the prism. In the last column, write each answer in decimal form and round to the nearest thousandth.

Block Prisms				Block Pyramids				Volume Ratio
	Height h	Area of base B	Volume of prism $V = B + h$		Height h	Area of base B	Volume of block pyramid	Volume of block pyramid ÷ Volume of prism
	1	1	1		1	1	1	$1 \div 1 = 1.000$
	2	4	8		2	4	5	$5 \div 8 = 0.625$
	6							
	7							
	8							
	9							
	10							

3. Volumes of Pyramids Using Calculus

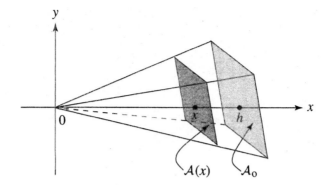

a. Explain why, when computing a pyramid's volume, the pyramid can be assumed to be right without loss of generality.

b. Using similarity, express the area $\mathcal{A}(x)$ of the cross-section corresponding to x in terms of x, the height h, and the base's area $\mathcal{A}_0$.

c. Use Theorem 7.3.1 to determine the volume of the given pyramid. Compare this value with the volume of a prism that has the same base area and height as the pyramid. ◆

Classroom Discussion 7.3.5: Volumes of Spheres

The goal is to find the volume of a sphere of radius $r > 0$. We present three different approaches: one for middle-school students based on comparing the volumes of spheres and cylinders, one for high school students based on approximating the sphere with tiny "pyramids," and one for college students based on calculus.

1. Classroom Connection 7.3.4: Volumes of Spheres and Cylinders

The following exploration is taken from pages 47–48 in the seventh-grade textbook *Connected Mathematics, Filling and Wrapping*. Discuss it in small groups. ◆

Although spheres may differ in size, they are all the same shape. We can describe a sphere by giving its radius.

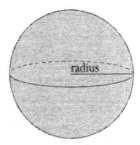

In this investigation, you will explore ways to determine the volume of cones and spheres.

Comparing Spheres and Cylinders

In this problem, you will make a sphere and a cylinder with the same radius and height and then compare their volumes. (The "height" of a sphere is just its diameter.) You can use the relationship you observe to help you develop a method for finding the volume of a sphere.

Did you know?

The Earth is nearly a sphere. You may have heard that, until Christopher Columbus's voyage in 1492, most people believed the Earth was flat. Actually, as early as the fourth century B.C., scientists in Greece and Egypt had figured out that the Earth was round. They observed the shadow of the Earth as it passed across the Moon during a lunar eclipse. It was clear that the shadow was round. Combining this observation with evidence gathered from observing constellations, these scientists concluded that the Earth was indeed spherical. In fact, in the third century B.C., Eratosthenes, a scientist from Alexandria, Egypt, was actually able to estimate the circumference of the Earth.

Problem 5.1

- Using modeling dough, make a sphere with a diameter between 2 inches and 3.5 inches.

- Using a strip of transparent plastic, make a cylinder with an open top and bottom that fits snugly around your sphere. Trim the height of the cylinder to match the height of the sphere. Tape the cylinder together so that it remains rigid.

- Now flatten the sphere so that it fits snugly in the bottom of the cylinder. Mark the height of the flattened sphere on the cylinder.

height of cylinder

height of empty space

height of flattened sphere

A. Measure and record the height of the cylinder, the height of the empty space, and the height of the flattened sphere.

B. What is the relationship between the volume of the sphere and the volume of the cylinder?

Remove the modeling dough from the cylinder, and save the cylinder for the next problem.

▇ Problem 5.1 Follow-Up

Compare your results with the results of a group that made a larger or smaller sphere. Did the other group find the same relationship between the volume of the sphere and the volume of the cylinder?

2. Classroom Connection 7.3.5: Volume of a Sphere Using Tiny "Pyramids"

The following exploration is taken from pages 546–547 in the textbook *Discovering Geometry*, third edition, by Michael Serra. In this exploration, the sphere's volume is known, and it is used to find its surface area. However, a similar reasoning works to find the sphere's volume if its surface area is known. Replace Steps 3 and 4 in the exploration with Step 3 here, and discuss it in small groups.

Step 3: The sphere's surface area is $S = 4\pi r^2$. Use the equation in Step 2 to find the sphere's volume V.   ◆

Investigation
The Formula for the Surface Area of a Sphere

In this investigation you'll visualize a sphere's surface covered by tiny shapes that are nearly flat. So the surface area, S, of the sphere is the sum of the areas of all the "nearly polygons." If you imagine radii connecting each of the vertices of the "nearly polygons" to the center of the sphere, you are mentally dividing the volume of the sphere into many "nearly pyramids." Each of the "nearly polygons" is a base for a pyramid, and the radius, r, of the sphere is the height of the pyramid. So the volume, V, of the sphere is the sum of the volumes of all the pyramids. Now get ready for some algebra.

Step 1   Divide the surface of the sphere into 1000 "nearly polygons" with areas $B_1, B_2, B_3, \ldots, B_{1000}$. Then you can write the surface area, S, of the sphere as the sum of the 1000 B's:

$$S = B_1 + B_2 + B_3 + \ldots + B_{1000}$$

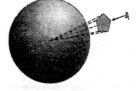

Step 2   The volume of the pyramid with base B_1 is $\frac{1}{3}(B_1)(r)$, so the total volume of the sphere, V, is the sum of the volumes of the 1000 pyramids:

$$V = \frac{1}{3}(B_1)(r) + \frac{1}{3}(B_2)(r) + \ldots + \frac{1}{3}(B_{1000})(r)$$

What common expression can you factor from each of the terms on the right side? Rewrite the last equation showing your factoring.

Step 3   But the volume of the sphere is $V = \frac{4}{3}\pi r^3$. Rewrite your equation from Step 2 by substituting $\frac{4}{3}\pi r^3$ for V and substituting for S the sum of the areas of all the "nearly polygons."

Step 4   Solve the equation from Step 3 for the surface area, S. You now have a formula for finding the surface area of a sphere in terms of its radius. State this as your next conjecture and add it to your conjecture list.

> **Sphere Surface Area Conjecture**
>
> The surface area, S, of a sphere with radius r is given by the formula _?_.

3. Volumes of Spheres Using Calculus

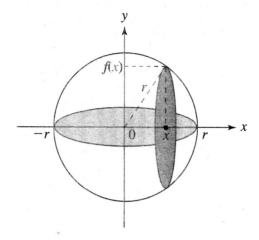

a. For each $-r \leq x \leq r$, describe the cross-section corresponding to x, then express its area $\mathcal{A}(x)$ in terms of x and r.

b. Use Theorem 7.3.1 to determine the volume of the given sphere. ◆

EXERCISES 7.3

1. A swimming pool is in the shape of a right prism as in the following figure. How many cubic feet of water can this swimming pool hold?

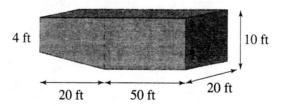

2. Classroom Connection 7.3.6: Revisiting Cylinders

The following exploration is taken from pages 502–503, page 7-56, and page 7-57 in the sixth-grade textbook *Math Thematics*, *Book 1*. Answer the questions therein. ◆

GOAL

LEARN HOW TO...
• recognize a cylinder
• find the volume of a cylinder

AS YOU...
• explore the size and shape of a kiva

KEY TERM
• cylinder

Exploration

Volume of a Cylinder

SET UP *Work in a group of three. You will need:*
• Labsheets 5B and 5C • scissors • tape • rice • ruler

In the summer of 1891, Gustaf Nordenskiöld of Sweden and his team began to uncover the ruins at Mesa Verde. Part of their task was to remove the layers of dust and rubbish that had piled up over the centuries. After digging to a depth of $\frac{1}{2}$ m at one location, they began to see a kiva take shape.

12 How do you think Nordenskiöld could have estimated the amount of dust and rubbish in the kiva without removing it?

A kiva is shaped like a circular **cylinder**. A circular cylinder is a space figure that has two circular bases that are parallel and congruent.

The bases are parallel and congruent.

13 **Use Labsheets 5B and 5C.** Cut out the nets for the open-topped *Prism A, Prism B and Cylinder.* Fold and tape each net.

14 How is the cylinder like a prism? How is it different?

15 Which has a larger volume, prism A or prism B? Explain.

16 Which do you think holds more, the cylinder or prism A? the cylinder or prism B? Explain your thinking in each case.

17 **a.** Fill prism B with rice and then pour the rice into the cylinder. Does the rice completely fill the cylinder, or is there too much or not enough rice?

b. Fill the cylinder with rice and then pour the rice into prism A. Does the rice completely fill prism A?

c. What can you conclude about the volume of the cylinder?

18 a. Place the cylinder inside the larger prism. Then place the smaller prism inside the cylinder.

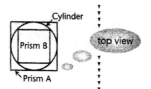

b. For each of the prisms and the cylinder, find the area of a base and the height. Make a table to record your results.

19 Discussion Add on to the table you completed in Question 18.

a. Find the volumes of prism A and prism B. Explain your method.

b. Use the same method you used in part (a) to find the volume of the cylinder.

c. Use your models and your results with rice to decide whether the volume you found for the cylinder is reasonable.

▶ You can find the volume *V* of a cylinder with height *h* and a base with area *B* in the same way you find the volume of a prism.

$$V = Bh, \text{ or } V = \pi r^2 h.$$

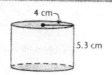

area of circular base

EXAMPLE

Find the volume of the cylinder shown to the nearest cubic centimeter. Use 3.14 for π.

4 cm

5.3 cm

SAMPLE RESPONSE

$$V = \pi r^2 h$$

$$\approx 3.14 \cdot 4^2 \cdot 5.3 = 266.272$$

Volume is measured in cubic units.

The volume is about **266 cm³**.

20 ✔ **CHECKPOINT** Find the volume of the cylinder to the nearest cubic meter. Use 3.14 for π.

10 m

7 m

✔ **QUESTION 20**

...checks that you can find the volume of a cylinder.

21 Gustaf Nordenskiöld reported that one of the kivas he uncovered had walls 2 m high with a diameter of 4.3 m. If this kiva was completely full of dust and rubbish, about how much material did Nordenskiöld have to remove?

HOMEWORK EXERCISES ▶ See Exs. 14–22 on p. 506.

Name _____   Date _____

Prism A   (Use with Questions 13–19 on page 502–503.)

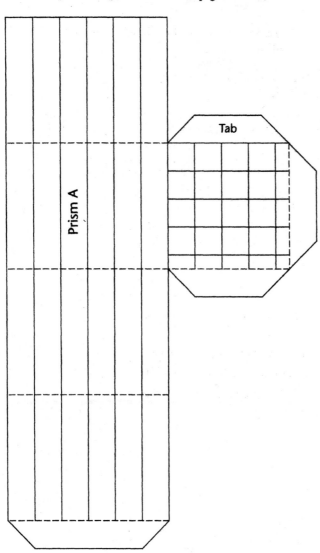

Prism A

Tab

Name _____   Date _____

Prism B and Cylinder   (Use with Question 13–19 on page 502–503.)

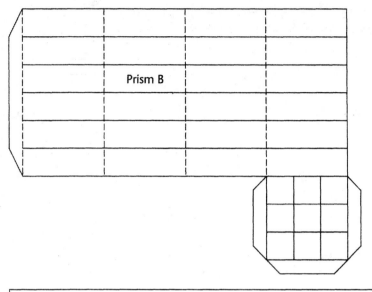

Prism B

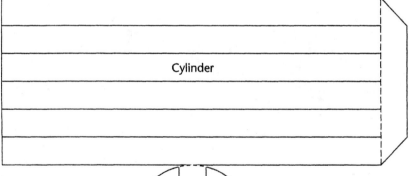

Cylinder

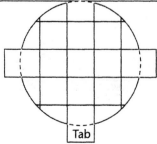

Tab

3. Classroom Connection 7.3.7: A Couple of Problems on Cylinders

The following extensions are from page 40 in the seventh-grade textbook *Connected Mathematics, Filling and Wrapping* . Answer the questions in Problems 12 and 13. ◆

Extensions

12. A cylindrical can is packed securely in a box as shown at right. The height of the box is 10 cm, and the sides of its square base measure 2 cm.

   a. Find the radius and height of the can.

   b. What is the volume of the empty space between the can and the box?

   c. Find the ratio of the volume of the can to the volume of the box.

   d. Make up a similar example with a different size can and box. What is the ratio of the volume of the can to the volume of the box for your example? How does the ratio compare to the ratio you got in part c?

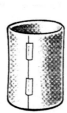

13. Start with two identical sheets of paper. Tape the long sides of one sheet together to form a cylinder. Form a cylinder from the second sheet by taping the short sides together. Imagine that each cylinder has a top and a bottom.

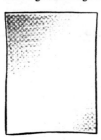

   a. Which cylinder has greater volume? Explain your reasoning.

   b. Which cylinder has greater surface area? Explain your reasoning.

In Exercises 4–14, sketch the graphs $y = f(x)$ and $y = g(x)$ for $a \le x \le b$. Then find the volume of the solid obtained by revolving about the x-axis the region of the plane that is bounded by the two graphs. You may use your calculator only to check your final answers.

4. $f(x) = 3, g(x) = x$ and $a = 0, b = 3$

5. $f(x) = x^2, g(x) = 0$ and $a = 0, b = \sqrt[5]{2}$

6. $f(x) = x, g(x) = 2x$ and $a = 0, b = 4$

7. $f(x) = -x^3, g(x) = 0$ and $a = 1, b = 3$

8. $f(x) = \sqrt{x}, g(x) = x$ and $a = 0, b = 1$

9. $f(x) = 1/\sqrt{x^3}, g(x) = 0$ and $a = 1, b = 4$

10. $f(x) = \sqrt{4 - x^2}, g(x) = 0$ and $a = -2, b = 2$

11. $f(x) = 2\sqrt{1 - \frac{x^2}{9}}, g(x) = 0$ and $a = -3, b = 3$

12. $f(x) = \sqrt{e^x}, g(x) = x^2$ and $a = 0, b = 1$

13. $f(x) = x\sqrt{x^3 + 1}, g(x) = 0$ and $a = 0, b = 1$

14. $f(x) = |x|, g(x) = 1$ and $a = -1, b = 1$

15. Find the volume of the solid obtained by revolving about the x-axis the region inside the triangle of vertices $A(0,0)$, $B(0,2)$, and $C(1,1)$.

16. Find the volume of the torus obtained by revolving about the x-axis the disc of radius 1 centered at point $(0,2)$.

17. Find the volume of the region in the space obtained by revolving about the x-axis the region of the plane that is bounded by the graphs $y = |x|$ and $y = 1$ for $-2 \le x \le 2$.

18. Find the volume of the region in the space obtained by revolving about the x-axis the region of the plane that is bounded by the graphs $y = \sqrt{x}$ and $y = x$ for $0 \le x \le 2$.

19. **I. The Disk Method**

    a. Sketch the graph of a nonnegative continuous function $f(x)$ defined on the interval $[a, b]$.

    b. Sketch the solid obtained by revolving about the x-axis the region of the plane that is between the x-axis and the graph $y = f(x)$.

    c. Give a few examples of solids that can be obtained as previously described.

    d. Using Theorem 7.3.1, find an explicit formula for the volume of this type of solids.

II. The Washer Method

    a. On the same coordinate system, sketch the graph of two continuous functions $f(x)$ and $g(x)$ defined on a given interval $[a, b]$ satisfying $0 \le g(x) \le f(x)$ for each $a \le x \le b$.

   b. Sketch the solid obtained by revolving about the x-axis the plane region that lies between the two preceding graphs.

   c. Give a few examples of solids that are obtained in this manner.

   d. How does the volume V of the solid in IIb compare to the volumes of the solids that are obtained by revolving about the x-axis the region below the graph of f and above the x-axis and the region below the graph of g and above the x-axis?

   e. Using Problem Id, find an explicit formula for the volume of the solid in Problem IIb.

20. The graph $y = x^2$, where $0 \le x \le h$, was rotated first about the x-axis to form Solid A, then about the y-axis to form Solid B.

   a. Find the volume $V_A(h)$ of Solid A.

   b. Find the volume $V_B(h)$ of Solid B.

   c. For which values of h do the two solids have the same volume?

   d. For which values of h is $V_A(h)$ larger than $V_B(h)$?

   e. For which values of h is $V_B(h)$ larger than $V_A(h)$?

PROJECTS AND EXTENSIONS 7.3

I. Relations between Length, Area, and Volume Using Derivatives

1. Let $C(r)$ denote the circumference of a circle of radius r, and let $\mathcal{A}(r)$ denote the area of a disc of radius r. Recall the expressions of $C(r)$ and $\mathcal{A}(r)$.

2. Using differential calculus, find a relation between the functions $C(r)$ and $\mathcal{A}(r)$.

3. Does the preceding relation also hold between the perimeters and areas of squares?

4. Is this only a coincidence or are there reasons for the relation you have found between $C(r)$ and $\mathcal{A}(r)$? To answer this question, follow the ideas in the proof of the Fundamental Theorem of Calculus given in Section 6.3.

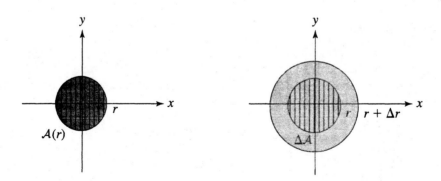

5. Let $S(r)$ and $V(r)$ be, respectively, the area and volume of a sphere of radius r. Recall the expressions of $S(r)$ and $V(r)$.

6. Using differential calculus, find a relation between the functions $S(r)$ and $V(r)$.

7. Consider a cylinder whose height equals its radius. Does the preceding relation also hold between the area and volume of the cylinder?

8. Is this simply a coincidence or are there reasons for the relation you have found between $S(r)$ and $V(r)$? To answer this question, follow the ideas in the proof of Theorem 7.3.1.

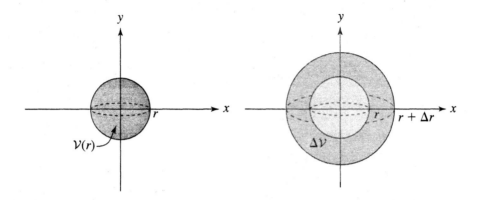

II. Gabriel's Trumpet

Consider the graph $y = 1/x$ for $x \geq 1$. The object obtained by revolving this graph about the x-axis is called *Gabriel's trumpet*.

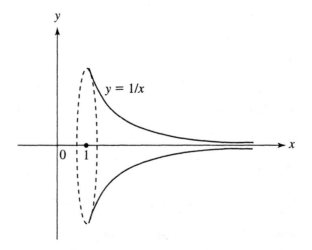

Notice that the trumpet is infinitely long, since its shank (small end) stretches along the x-axis forever. Suppose you wanted to paint the outside of this trumpet. How much paint would be required? If, instead, you poured the paint into the trumpet, how much paint would be needed to fill the trumpet? The answers may surprise you! Begin by fixing $a > 1$ and considering the portion of the trumpet corresponding to $1 \leq x \leq a$. Denote by $S(a)$ its area and by $V(a)$ its volume.

1. Using calculus, evaluate by hand $V(a)$.
2. What is the $\lim\limits_{a \to \infty} V(a)$?
3. How much paint can Gabriel's trumpet hold?
4. Using calculus, express $S(a)$ as a definite integral. Do not evaluate this integral, but simplify the integrand as much as possible.

⟨T⟩ 5. Using your calculator, fill in the following table.

n	$S(10^n)$
10	
10^2	
10^3	
10^4	
10^5	
10^6	
10^7	
10^8	
10^9	
10^{10}	

6. Make a conjecture about the $\lim\limits_{a \to \infty} S(a)$.
7. How much paint is needed to paint Gabriel's trumpet?
8. Can you resolve the paradox?

III. Displacement and Density

Research the concepts of displacement and density. Then, plan a lesson on how these quantities are used to find the volume of solids whose shapes are irregular.

CHAPTER 7 REVIEW

Roughly speaking, the length of a curve is the amount of string it takes to exactly cover the curve.

The length of the graph $y = f(x)$, where $a \le x \le b$, can be approximated by the lengths of polygonal paths approximating the curve. Let $n \ge 1$ be an integer, and denote by $x_i = a + i\frac{b-a}{n}$ for $0 \le i \le n$. The length $\mathcal{L}_n$ of the polygonal path obtained by joining the points whose x-coordinates are $x_0, x_1, ..., x_n$ is given by the formula

$$\mathcal{L}_n = \sum_{i=0}^{n-1} \sqrt{\frac{(b-a)^2}{n^2} + \left[f(x_{i+1}) - f(x_i)\right]^2}.$$

If $f(x)$ has derivative $f'(x)$ that is continuous on $[a, b]$, then the length $\mathcal{L}$ of the graph $y = f(x)$ is given by the formula

$$\mathcal{L} = \int_a^b \sqrt{1 + [f'(x)]^2} \, dx.$$

Roughly speaking, the area of a surface is the amount of paper needed to exactly cover the surface. Using geometry, we can find formulas for the areas of cylinders and cones. Using calculus, we can derive a formula for the areas of surfaces of revolution: The area S of the surface obtained by revolving the graph $y = f(x)$, where $a \le x \le b$, about the x-axis is given by

$$S = \int_a^b 2\pi f(x) \sqrt{1 + [f'(x)]^2} \, dx.$$

Here, $f(x)$ is a nonnegative function whose derivative $f'(x)$ is continuous.

The volume of a solid is the amount of space the solid occupies. By definition, the volume of a cube whose length, width, and height are all equal to 1 unit is 1 cubic unit. Based on this definition and by using geometry, we obtain formulas for the volumes of prisms and cylinders. Using calculus, we can derive a formula for the volumes of a variety of solids such as cones, pyramids, and spheres: For each fixed $a \le x \le b$, let $\mathcal{A}(x)$ denote the area of the cross-section corresponding to x in the given solid. If the function $\mathcal{A}(x)$ is integrable over the interval $[a, b]$, then the volume of the solid is given by

$$V = \int_a^b \mathcal{A}(x) \, dx.$$

CHAPTER 7 REVIEW EXERCISES

In Exercises 1–5, do the following: (a) sketch the graph $y = f(x)$ where $a \le x \le b$; (b) approximate its length using approximations by polygonal paths composed of n line segments for the given values of n. Tabulate your results; and (c) find the value in your table that best approximates the graph's length.

1. $f(x) = x^6, a = -1, b = 0, n = 5, 10, \cdots, 50$

2. $f(x) = x^5 - x^4, a = -1, b = 1, n = 15, 30, \cdots, 150$

3. $f(x) = x\sqrt{x}, a = 0, b = 4, n = 10, 20, \cdots, 100$

4. $f(x) = 3x/(x^2 + 1), a = 1, b = 3, n = 4, 8, \cdots, 40$

5. $f(x) = xe^x, a = 0, b = 1, n = 2, 4, \cdots, 20$

In Exercises 6–10, write down a formula for the length of the graph of the given function. Do not compute the integral but simplify the integrand as much as possible.

6. $f(x) = 1/x^2, 1 \le x \le 2$

7. $f(x) = x + e^x, 0 \le x \le 3$

8. $f(x) = x^2 + x + 1, -1 \le x \le 1$

9. $f(x) = \sqrt{x + 2}, -1 \le x \le 0$

10. $f(x) = (3 - 2x)^2, 0 \le x \le 3$

In Exercises 11–15, describe a graph whose length is represented by the given integral.

11. $\int_3^4 \sqrt{1 + \dfrac{1}{x^6}}\, dx$

12. $\int_2^7 \sqrt{1 + 5x^2}\, dx$

13. $\int_{-1}^1 \sqrt{3 + x}\, dx$

14. $\int_{-2}^{-1} \sqrt{6 + x}\, dx$

15. $\int_1^5 \left(x^2 + \dfrac{1}{4x^2}\right) dx$

In Exercises 16–20, use integral calculus to evaluate the length of the graph $y = f(x)$.

16. $f(x) = x\sqrt{2} - 1, -10 \le x \le 10$

17. $f(x) = x\sqrt{x}, 0 \le x \le 4$

18. $f(x) = \frac{1}{2}(e^x + e^{-x}), -1 \le x \le 1$

19. $f(x) = \dfrac{x^4}{4} + \dfrac{1}{8x^2}, 1 \le x \le 2$

20. $f(x) = \dfrac{x\sqrt{x}}{6} - 2\sqrt{x}, 4 \le x \le 9$

In Exercises 21–26, sketch the graph $y = f(x)$ where $a \le x \le b$; then sketch the surface generated by revolving this graph about the x-axis. Write down a formula for its area. Do not compute the integral but simplify the integrand as much as possible.

21. $f(x) = x^2, a = 0, b = 1$

22. $f(x) = 1/\sqrt{x}, a = 1, b = 4$

23. $f(x) = -xe^x, a = -1, b = 0$

24. $f(x) = x^3 - x, a = 1, b = 2$

25. $f(x) = 4x^2 - x^4, a = -2, b = 2$

26. $f(x) = \dfrac{x}{x^2+1}, a = 0, b = 1$

In Exercises 27–34, sketch the graph $y = f(x)$; then find the area of the surface obtained by revolving the graph about the x-axis, where $a \le x \le b$.

27. $f(x) = 9, a = -3, b = 5$

28. $f(x) = x + 3, a = 0, b = 1$

29. $f(x) = -4x + 5, a = 0, b = 1$

30. $f(x) = 2x + 1, a = 1, b = 2$

31. $f(x) = -3x + 5, a = 1, b = 2$

32. $f(x) = \sqrt{3x - 2}, a = \frac{2}{3}, b = 1$

33. $f(x) = \sqrt{1 - x}, a = -1, b = 0$

34. $f(x) = \sqrt{4 - x^2}, a = -1, b = 1$

In Exercises 35–43, sketch the graphs $y = f(x)$ and $y = g(x)$ for $a \leq x \leq b$. Then find the volume of the solid obtained by revolving about the x-axis the region of the plane that is bounded by the two graphs. You may use your calculator only to check your final answers.

35. $f(x) = 2, g(x) = 0$ and $a = 1, b = 2$

36. $f(x) = -x + 3, g(x) = 0$ and $a = 0, b = 3$

37. $f(x) = 2x + 1, g(x) = 0$ and $a = 0, b = 1$

38. $f(x) = x^2, g(x) = \sqrt{x}$ and $a = 0, b = 1$

39. $f(x) = x^2, g(x) = x^3$ and $a = 0, b = 1$

40. $f(x) = 2\sqrt{x}/(x^2 + 1), g(x) = 0$ and $a = 0, b = 2$

41. $f(x) = 2, g(x) = x$ and $a = 1, b = 2$

42. $f(x) = x^2\sqrt{x + 1}, g(x) = 0$ and $a = -1, b = 1$

43. $f(x) = \frac{1}{x^2}, g(x) = \frac{1}{x^3}$ and $a = 1/3, b = 1$

44. Sketch the triangle of vertices $A(1,1)$, $B(2,3)$, and $C(1,4)$. Then find the volume of the solid obtained by revolving about the x-axis the region inside the triangle.

45. Sketch the ellipse $\frac{(x-3)^2}{4} + \frac{y^2}{9} = 1$, then find the volume of the solid obtained by revolving about the x-axis the region inside the ellipse.

46. Sketch the triangle of vertices $A(2,2)$, $B(3,3)$, and $C(4,2)$. Then find the volume of the solid obtained by revolving about the x-axis the region inside the triangle.

47. Sketch the graphs $y = x^{3/2}$ and $y = x^2$ for $0 \leq x \leq 2$. Then find the volume of the solid obtained by revolving about the x-axis the region of the plane that is bounded by the two graphs.

48. Sketch the solid obtained by revolving about the x-axis the region of the plane that is bounded by the graphs $y = x^3 + x^2 + x$ and $y = -1$ for $-1 \leq x \leq 1$, then find its volume.

49. Sketch the solid obtained by revolving about the x-axis the region of the plane that is bounded by the graph $y = x\sqrt{16 - x^2}$ and the x-axis for $0 \leq x \leq 4$, then find its volume.

50. Sketch the solid obtained by revolving about the x-axis the region of the plane that is bounded by the graph $y = x^5\sqrt{9 - x^2}$ and the x-axis for $0 \leq x \leq 3$, then find its volume.

Appendix

Given a property P, the collection of all x's having the property P is denoted by $\{x : P\}$.

Sets of Numbers

The set of natural numbers $\mathbb{N} = \{1, 2, 3, 4, \ldots\}$
The set of whole numbers $\{0, 1, 2, 3, 4, \ldots\}$
The set of integers $\mathbb{Z} = \{\ldots, -4, -3, -2, -1, 0, 1, 2, 3, 4, \ldots\}$
The set of rational numbers $\mathbb{Q} = \{\frac{p}{q} : p, q \in \mathbb{Z}, q \neq 0\}$
The set of real numbers $\mathbb{R}$

Interval Notation

$\{x : a < x < b\} = (a, b)$
$\{x : a \leq x < b\} = [a, b)$
$\{x : a < x \leq b\} = (a, b]$
$\{x : a \leq x \leq b\} = [a, b]$
$\{x : x > a\} = (a, \infty)$
$\{x : x \geq a\} = [a, \infty)$
$\{x : x < b\} = (-\infty, b)$
$\{x : x \leq b\} = (-\infty, b]$

Factoring Special Polynomials

$x^2 - y^2 = (x - y)(x + y)$
$x^3 + y^3 = (x + y)(x^2 - xy + y^2)$
$x^3 - y^3 = (x - y)(x^2 + xy + y^2)$

Binomial Theorem

$(x + y)^2 = x^2 + 2xy + y^2$
$(x + y)^3 = x^3 + 3x^2y + 3xy^2 + y^3$
$(x - y)^3 = x^3 - 3x^2y + 3xy^2 - y^3$

$$(x + y)^n = x^n + nx^{n-1}y + \frac{n(n-1)}{2}x^{n-2}y^2 + \cdots$$
$$+ \binom{n}{k}x^{n-k}y^k + \cdots + nxy^{n-1} + y^n$$

where $\displaystyle \binom{n}{k} = \frac{n(n - 1)\cdots(n - k + 1)}{1 \cdot 2 \cdot 3 \cdots k}$

Quadratic Formula

If $ax^2 + bx + c = 0$ with $a \neq 0$ and $b^2 - 4ac \geq 0$, then $x = \dfrac{-b \pm \sqrt{b^2 - 4ac}}{2a}$.

Lines

Slope of the line passing through the points $P_1(x_1, y_1)$ and $P_2(x_2, y_2)$ for $x_1 \neq x_2$:

$$m = \frac{y_2 - y_1}{x_2 - x_1}$$

Point-slope equation of the line passing through the point $P_1(x_1, y_1)$ with slope m:

$$y - y_1 = m(x - x_1)$$

Slope-intercept equation of the line with slope m and y intercept b:

$$y = mx + b$$

Circles

Equation of the circle with radius r, centered at the point (a, b):

$$(x - a)^2 + (y - b)^2 = r^2$$

Right Triangle Trigonometry

$\sin(\theta) = \dfrac{\text{opp}}{\text{hyp}} \qquad \tan(\theta) = \dfrac{\text{opp}}{\text{adj}}$

$\cos(\theta) = \dfrac{\text{adj}}{\text{hyp}} \qquad \cot(\theta) = \dfrac{\text{adj}}{\text{opp}}$

$\sin^2(\theta) + \cos^2(\theta) = 1$

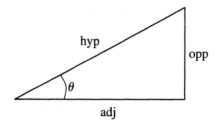

Answers to Classroom Discussions and Practice Problems

CHAPTER 1

SECTION 1.1

Practice Problems

1. A direct formula for the sequence is $x_n = \frac{n+1}{n}, n \geq 1$.
2. A recursive formula for the sequence is $y_{n+1} = y_n - n, n \geq 1, y_1 = 1$.
3. The first five terms of the sequence are $x_1 = 2, x_2 = 3, x_3 = 3x_2 - 2x_1 = 5$, $x_4 = 3x_3 - 2x_2 = 9$, and $x_5 = 3x_4 - 2x_3 = 17$.

Classroom Discussion 1.1.1: Direct Formula for Arithmetic Sequences

1. The recursive relations for the sequences in Classroom Connection 1.1.2 are as follows.

   Lab sheet 1: Since the lengths of the branches are $1, \frac{1}{2}, \frac{1}{4}, \frac{1}{8}, \ldots$, the recursive relation is $x_{n+1} = \frac{1}{2}x_n, n \geq 1$ with $x_1 = 1 \Rightarrow$ not arithmetic.

   EXAMPLE: $x_{n+1} = x_n + 4, n \geq 1, x_1 = 3 \Rightarrow$ arithmetic with common difference $d = 4$.

   6: $x_{n+1} = 4x_n, n \geq 1, x_1 = 3 \Rightarrow$ not arithmetic.
   8a: $x_{n+1} = 2x_n, n \geq 1, x_1 = 2 \Rightarrow$ not arithmetic.
   8b: $x_{n+1} = x \cdot x_n, n \geq 1, x_1 = 3x \Rightarrow$ not arithmetic.
   8c: $x_{n+1} = 0.1x_n, n \geq 1, x_1 = 0.3 \Rightarrow$ not arithmetic.
   8d: $x_{n+1} = x_n - 101, n \geq 1, x_1 = 400 \Rightarrow$ arithmetic with common difference $d = -101$.

2. Yes, I agree with Michelle. The second term in the sequence is obtained by adding the common difference to the first term. The third term in the sequence is obtained by adding the common difference to the second term, which is the same as adding 2 times the common difference to the first term. The fourth term in the sequence is obtained by adding the common difference to the third term, which, based on what we just obtained for the third term, is the same as adding 3 times the common difference to the first term, and so on.

3. If the sequence $\{x_n\}_{n \geq 1}$ is arithmetic with common difference d, then $x_2 = x_1 + d$, $x_3 = x_2 + d = (x_1 + d) + d = x_1 + 2d$, $x_4 = x_3 + d = (x_1 + 2d) + d = x_1 + 3d$. In general, if $x_n = x_1 + (n-1)d$ for some $n \geq 2$, then the next term will be

$$x_{n+1} = x_n + d = [x_1 + (n-1)d] + d = x_1 + nd.$$

4. **EXAMPLE:** $x_n = x_1 + (n - 1)4 = 3 + 4n - 4 = 4n - 1, n \geq 1$
   8d: $x_n = x_1 - (n - 1)101 = 400 - 101n + 101 = 501 - 101n, n \geq 1$

Classroom Discussion 1.1.2: Direct Formula for Geometric Sequences

1. **Lab sheet 1:** $x_{n+1} = \frac{1}{2}x_n, n \geq 1$ with $x_1 = 1 \Rightarrow$ geometric with common ratio
   $r = \frac{1}{2}$.
   6: $x_{n+1} = 4x_n, n \geq 1, x_1 = 3 \Rightarrow$ geometric with common ratio $r = 4$.
   8a: $x_{n+1} = 2x_n, n \geq 1, x_1 = 2 \Rightarrow$ geometric with common ratio $r = 2$.
   8b: $x_{n+1} = x \cdot x_n, n \geq 1, x_1 = 3x \Rightarrow$ geometric with common ratio $r = x$.
   8c: $x_{n+1} = 0.1x_n, n \geq 1, x_1 = 0.3 \Rightarrow$ geometric with common ratio $r = 0.1$.

2. Yes, I agree with Bob. The second term in the sequence is obtained by multiplying the first term by the common ratio. The third term in the sequence is obtained by multiplying the second term by the common ratio, which is the same as multiplying the first term by the square of the common ratio. The fourth term is obtained by multiplying the third term by the common ratio, which, based on what we just obtained for the third term, is the same as multiplying the first term by the cube of the common ratio, and so on.

3. If the sequence $\{y_n\}_{n \geq 1}$ is geometric with common ratio r, then $y_2 = y_1 \cdot r, y_3 = y_2 \cdot r = y_1 \cdot r \cdot r = y_1 \cdot r^2, y_4 = y_3 \cdot r = y_1 \cdot r^2 \cdot r = y_1 \cdot r^3$. So, in general, if $y_n = y_1 \cdot r^{n-1}$ for some $n \geq 2$, then the next term will be

$$y_{n+1} = y_n \cdot r = y_1 \cdot r^{n-1} \cdot r = y_1 \cdot r^n.$$

4. **Lab sheet 1:** $x_n = 1\left(\frac{1}{2}\right)^{n-1} = \left(\frac{1}{2}\right)^{n-1}, n \geq 1$
   6: $x_n = x_1 \cdot 4^{n-1} = 3 \cdot 4^{n-1}, n \geq 1$
   8a: $x_n = 2 \cdot 2^{n-1} = 2^n, n \geq 1$
   8b: $x_n = 3x \cdot x^{n-1} = 3x^n, n \geq 1$
   8c: $x_n = 0.3 \cdot 0.1^{n-1} = 3 \cdot 0.1^n, n \geq 1$

Classroom Discussion 1.1.3: The Circumference of a Circle and Perimeters of Inscribed Polygons

1. $L = 2\pi r$
2. $L = 25.1328$
3. As the number of sides of the inscribed polygons increases, the polygons will approximate increasingly better the circle. Consequently, the perimeters of these polygons will be closer and closer in value to the circumference of the circle C.
4. Let A, B be two adjacent vertices of a regular inscribed n-gon, and let O' be the point of intersection with the line AB of the line perpendicular to AB and passing through the center O. Then, the measure of angle $\widehat{AOB} = \frac{2\pi}{n}$ and the triangle $OO'B$ is right, with the measure of $\widehat{O'OB} = \frac{\pi}{n}$. Hence, $\sin\left(\frac{\pi}{n}\right) = \frac{O'B}{OB} = \frac{O'B}{4} \Rightarrow$
   $AB = 2 \cdot O'B = 8\sin\left(\frac{\pi}{n}\right) \Rightarrow p_n = n \cdot AB = 8n\sin\left(\frac{\pi}{n}\right)$.
5. Here is one set of approximations for the values of p_n for $n = 3, \ldots, 102$. As you expect, the last few digits of the p_n's will vary from calculator to calculator since they depend on the accuracy of the calculator used.

n	p_n		n	p_n		n	p_n		n	p_n
3	20.7846		28	25.0801		53	25.1175		78	25.1258
4	22.6275		29	25.0834		54	25.118		79	25.1259
5	23.5115		30	25.0869		55	25.1186		80	25.1261
6	24.0000		31	25.0893		56	25.1191		81	25.1263
7	24.2975		32	25.0912		57	25.1196		82	25.1264
8	24.4918		33	25.0948		58	25.1200		83	25.1266
9	24.6257		34	25.0970		59	25.1205		84	25.1267
10	24.7214		35	25.0991		60	25.1209		85	25.1269
11	24.7925		36	25.1019		61	25.1213		86	25.1270
12	24.8467		37	25.1032		62	25.1216		87	25.1272
13	24.8887		38	25.1031		63	25.1220		88	25.1273
14	24.9224		39	25.1044		64	25.1223		89	25.1274
15	24.9495		40	25.1070		65	25.1226		90	25.1275
16	24.9716		41	25.1082		66	25.1230		91	25.1276
17	24.9900		42	25.1094		67	25.1233		92	25.1277
18	25.0049		43	25.1103		68	25.1235		93	25.1279
19	25.0188		44	25.1114		69	25.1238		94	25.1280
20	25.0296		45	25.1120		70	25.1241		95	25.1281
21	25.0392		46	25.1132		71	25.1234		96	25.1282
22	25.0475		47	25.1140		72	25.1245		97	25.1283
23	25.0545		48	25.1149		73	25.1248		98	25.1284
24	25.0610		49	25.1155		74	25.1250		99	25.1285
25	25.0666		50	25.1162		75	25.1252		100	25.1286
26	25.0716		51	25.1169		76	25.1254		101	25.1287
27	25.0771		52	25.1194		77	25.1256		102	25.1288

Here is a scatter plot of the first forty terms in the table.

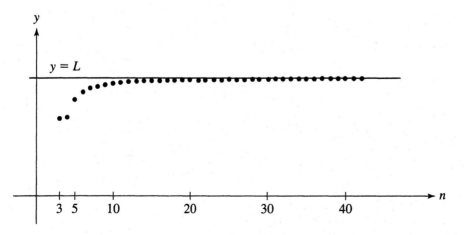

6. As the number of sides increases, the perimeters of the regular n-gons inscribed in the circle C also increase. The perimeters of the regular n-gons inscribed in the circle C are less than the circle's circumference. This conclusion is in line with the prediction made in Problem 3.

7. $L \approx 25.1288$ is the best approximation we obtain from the table in Problem 5. For an even better approximation one would have to consider a larger n.

Classroom Discussion 1.1.4: Limits of Sequences

1. **a.** As $n \to \infty$, the x_n's get smaller and smaller.

   b. All the terms of the sequence $\{x_n\}_n$ are contained in each one of the intervals $(-3, 5)$, $(-2, 4)$, $(-1, 3)$. No, we cannot conclude that $\lim_{n \to \infty} x_n = 1$ since we did not check that *all* open intervals centered at 1 contain all but finitely many terms of the sequence $\{x_n\}_n$.

   c. All of these intervals contain one term of the sequence $\{x_n\}_n$: $x_1 = 1$. All the other terms are outside. There is no contradiction with what we found in b.

   d. $\lim_{n \to \infty} \frac{1}{n} = 0$. To prove this, let $(-b, b)$ be an open interval centered at 0, with b being an arbitrary positive number. According to Definition 1.1.1, we need to show that all but finitely many terms of the sequence $\{\frac{1}{n}\}_n$ are contained in the interval $(-b, b)$. Let N_0 be the smallest positive integer verifying $\frac{1}{b} < N_0$. Then, if $n \geq N_0$, $x_n = \frac{1}{n} \leq \frac{1}{N_0} < b$, i.e., $x_n \in (-b, b)$ for all $n \geq N_0$. Thus, at most $N_0 - 1$ terms of the sequence are outside the interval $(-b, b)$.

2. **a.** As $n \to \infty$, the y_n's get larger and larger.

   b. The intervals contain 9, 99, and 999 terms of the sequence, respectively. Infinitely many terms are left outside each one of these intervals. No, we cannot conclude that $\lim_{n \to \infty} y_n = 0$.

   c. Suppose there exists a real number L such that $\lim_{n \to \infty} n = L$. Then, the interval $(L - 1, L + 1)$ should contain all but finitely many terms of the sequence. However, if N_0 is the smallest positive integer with $L + 1 \leq N_0$, then $y_n = n$ is outside $(L - 1, L + 1)$ for all $n \geq N_0$, so there are infinitely many terms outside $(L - 1, L + 1)$, leading to a contradiction. Hence our assumption that a number L with $\lim_{n \to \infty} n = L$ exists is false.

   d. Let M be an arbitrary fixed positive integer. Then the only terms less than or equal to M are $y_1, y_2, \ldots, y_{M-1}, y_M$.

   e. $\lim_{n \to \infty} (2n + 100) = \infty$ by the same reasoning as in d.

   f. Suppose there exists a real number L such that $\lim_{n \to \infty} (-n) = L$. Then, the interval $(L - 1, L + 1)$ should contain all but finitely many terms of the sequence. However, if N_0 is the smallest positive integer with $-N_0 \leq L - 1$, then $y_n = -n$ is outside $(L - 1, L + 1)$ for all $n \geq N_0$, so there are infinitely many terms outside $(L - 1, L + 1)$, leading to a contradiction. Hence our assumption that a number L exists with $\lim_{n \to \infty} (-n) = L$ is false.

   g. Let M be an arbitrary fixed negative integer. Then the only terms greater than or equal to M are $y_1, y_2, \ldots, y_{|M|-1}, y_{|M|}$.

   h. $\lim_{n \to \infty} (-n^2 + 1) = -\infty$

Practice Problem

Let a be an arbitrary positive real number. If $a \geq 2$, then all the terms of the sequence $x_n = \frac{1}{n+1}$, $n \geq 1$ are contained in the interval $(-a, a)$. If $0 < a < 2$, then $\frac{2}{a} - 1 > 0$, and for all natural numbers $n > \frac{2}{a} - 1$, we have $\frac{2}{n+1} \in (-a, a)$. Hence, only the x_n's with $n \leq \frac{2}{a} - 1$ are left outside the interval $(-a, a)$. Since $a > 0$ was arbitrary, it follows that any open interval centered at 0 contains all but finitely many x_n's. By Definition 1.1.1, we can conclude that $\lim_{n\to\infty} \frac{2}{n+1} = 0$.

Practice Problems

a. Since $0 \leq \frac{1}{n+2} \leq \frac{1}{n}$ for all n, and $\lim_{n\to\infty} \frac{1}{n} = 0$, by the sandwich theorem for Sequences it follows that $\lim_{n\to\infty} \frac{1}{n+2} = 0$.

b. We apply P5 after factoring and simplifying out n^2 from the numerator and denominator:

$$\lim_{n\to\infty} \frac{-n^2}{3n^2 + n + 5} = \lim_{n\to\infty} \frac{-n^2}{n^2(3 + \frac{1}{n} + \frac{5}{n^2})}$$

$$= \frac{\lim_{n\to\infty}(-1)}{\lim_{n\to\infty} 3 + \lim_{n\to\infty} \frac{1}{n} + \lim_{n\to\infty} 5 \cdot \lim_{n\to\infty} \frac{1}{n^2}} = -\frac{1}{3}.$$

c. We have

$$c_n = \begin{cases} -2 & \text{if } n \text{ is odd} \\ 2 & \text{if } n \text{ is even.} \end{cases}$$

Since all the terms of the sequence are outside the interval $(-\frac{1}{2}, \frac{1}{2})$, it follows that the sequence cannot converge to 0. If L is an arbitrary positive real number, the terms c_n will be outside the interval $(0, 2L)$ for $n = 1, 3, 5, 7, \ldots$ If L is an arbitrary negative real number, the terms c_n will be outside the interval $(2L, 0)$ for $n = 2, 4, 6, 8, \ldots$ Consequently, there is no real number L such that $\lim_{n\to\infty} c_n = L$, which means that $\{c_n\}_n$ is divergent.

d. By combining P5 with Example 2a, we obtain $\lim_{n\to\infty} \left| \frac{(-1)^n 2}{n^2} \right| = 2 \cdot \lim_{n\to\infty} \frac{1}{n^2} = 0$. Hence, based on the result of Example 3, it follows that $\lim_{n\to\infty} \frac{(-1)^n 2}{n^2} = 0$.

Classroom Discussion 1.1.5: Convergence Criteria for Arithmetic and Geometric Sequences

1. Suppose $\{x_n\}_{n\geq 1}$ is an arithmetic sequence with common difference d. Then, $x_n = x_1 + (n-1)d$ for $n \geq 1$. Examples of such sequences include

$$3, \quad 3, \quad 3, \quad 3 \ldots \quad (x_1 = 3, d = 0),$$

$$1, \quad \tfrac{4}{3}, \quad \tfrac{5}{3}, \quad 2, \ldots \quad (x_1 = 1, d = \tfrac{1}{3}),$$

$$1, \quad -4, \quad -9, \quad -14, \quad \ldots \quad (x_1 = 1, d = -5).$$

2. The sequence $x_n = 3, n \geq 1$ is convergent and $\lim_{n \to \infty} 3 = 3$. The sequence

$$x_n = 1 + (n - 1)\frac{1}{3} = \frac{n + 2}{3}, \quad n \geq 1$$

is divergent and $\lim_{n \to \infty} \frac{n + 2}{3} = \infty$. The sequence

$$x_n = 1 + (n - 1)(-5) = -5n + 6, \quad n \geq 1$$

is divergent and $\lim_{n \to \infty} (-5n + 6) = -\infty$.

3. If $d = 0$, the sequence is constant, $x_n = x_1$ for $n \geq 1$, and clearly convergent to x_1. On the other hand, if $d \neq 0$, the sequence is unbounded and, by P3, divergent.

a. An arithmetic sequence $\{x_n\}_n$ with positive common difference is divergent and $\lim_{n \to \infty} x_n = \infty$.

b. An arithmetic sequence $\{x_n\}_n$ with negative common difference is divergent and $\lim_{n \to \infty} x_n = -\infty$.

c. An arithmetic sequence $\{x_n\}_n$ with common difference equal to zero is convergent and $\lim_{n \to \infty} x_n = x_1$.

4. Suppose $\{y_n\}_{n \geq 1}$ is a geometric sequence with common ratio r. Then, $y_n = y_1 \cdot r^{n-1}$ for $n \geq 1$. Examples of such sequences include:

a. $r = 1$

$$2, \quad 2, \quad 2, \quad 2, \quad \ldots \quad (y_1 = 2),$$
$$0.1, \quad 0.1, \quad 0.1, \quad 0.1, \quad \ldots \quad (y_1 = 0.1),$$
$$-1, \quad -1, \quad -1, \quad -1, \quad \ldots \quad (y_1 = -1).$$

b. $r = -1$

$$2, \quad -2, \quad 2, \quad -2, \quad \ldots \quad (y_1 = 2),$$
$$0.1, \quad -0.1, \quad 0.1, \quad -0.1, \quad \ldots \quad (y_1 = 0.1),$$
$$-1, \quad 1, \quad -1, \quad 1, \quad \ldots \quad (y_1 = -1).$$

c. $|r| > 1$

$$2, \quad 4, \quad 8, \quad 16, \quad \ldots \quad (y_1 = 2, r = 2),$$
$$2, \quad -4, \quad 8, \quad -16, \quad \ldots \quad (y_1 = 2, r = -2),$$
$$3, \quad 15, 75, \quad 375, \quad \ldots \quad (y_1 = 3, r = 5),$$
$$1, \quad -\frac{4}{3}, \frac{16}{9}, \quad -\frac{64}{27}, \quad \ldots \quad (y_1 = 1, r = -\frac{4}{3}).$$

d. $|r| < 1$

$$1, \quad 0.1, \quad 0.01, \quad 0.001, \quad 0.0001, \quad \ldots \quad (y_1 = 1, r = 0.1),$$
$$1, \quad \frac{1}{2}, \quad \frac{1}{4}, \quad \frac{1}{8}, \quad \frac{1}{16}, \quad \ldots \quad (y_1 = 1, r = \frac{1}{2}),$$
$$4, \quad \frac{4}{3}, \quad \frac{4}{9}, \quad \frac{4}{27}, \quad \frac{4}{81}, \quad \ldots \quad (y_1 = 4, r = \frac{1}{3}),$$
$$1, \quad -\frac{1}{2}, \quad \frac{1}{4}, \quad -\frac{1}{8}, \quad \frac{1}{16}, \quad \ldots \quad (y_1 = 1, r = -\frac{1}{2}),$$
$$-3, \quad \frac{3}{5}, \quad -\frac{3}{25}, \quad \frac{3}{125}, \quad -\frac{3}{625}, \quad \ldots \quad (y_1 = -3, r = -\frac{1}{5}).$$

5. **a.** The sequences are convergent.
   b. The sequences are divergent.
   c. The sequences are divergent.
   d. The sequences are convergent.

6. A geometric sequence converges if and only if the common ratio r is equal to 1 or $|r| < 1$.

   For a proof of this statement, we need to treat separately each case a–d.
   a. For $r = 1$, the sequence becomes $y_n = y_1$ for $n \geq 1$; hence, it is constant and convergent to y_1.
   b. For $r = -1$, the sequence is of the form $y_n = (-1)^{n-1} y_1, n \geq 1$, so

   $$ y_n = \begin{cases} y_1 & \text{if } n \text{ is odd} \\ -y_1 & \text{if } n \text{ is even.} \end{cases} $$

   If $y_1 = 0$, $\{y_n\}_n$ is convergent and $\lim_{n \to \infty} y_n = 0$. If $y_1 \neq 0$, we use Definition 1.1.1 to prove that it is divergent. Since in this case $y_n \notin (-\frac{3}{2}y_1, -\frac{1}{2}y_1)$ for $n = 1, 3, 5, 7, \ldots$, by Definition 1.1.1, we conclude that $\{(-1)^{n-1} y_1\}_n$ does not converge to $-y_1$. On the other hand, if L is an arbitrary real number with $L \neq 0$ and $L \neq -y_1$, we can choose an open interval I centered at L such that $-y_1 \notin I$. In this case, $y_n \notin I$ for $n = 2, 4, 6, 8, \ldots$, so, again, infinitely many terms of the sequence will be outside the interval I; hence, the sequence does not converge to L. Thus, there is no real number L such that $\lim_{n \to \infty} (-1)^{n-1} y_1 = L$. The sequence $\{(-1)^{n-1} y_1\}_n$ is therefore divergent.
   c. For $|r| > 1$, all sequences $\{y_1 \cdot r^n\}_n$ are unbounded and hence, by P3, divergent.
   d. For $|r| < 1$, all sequences $\{y_1 \cdot r^n\}_n$ converge to 0. If we can prove that $\lim_{n \to \infty} r^n = 0$ whenever $|r| < 1$, then $\lim_{n \to \infty} (y_1 \cdot r^n) = y_1 \lim_{n \to \infty} r^n = 0$ whenever $|r| < 1$. Thus, we are left with showing that $\lim_{n \to \infty} r^n = 0$ whenever $|r| < 1$. When $0 < r < 1$, for each $a > 0$ we can find a whole number m such that $r^m < a$. Since $r^n < r^m$ for all $n \geq m$, we can conclude that $r^n \in (-a, a)$ for all $n \geq m$, which, by Definition 1.1.1, is equivalent to $\lim_{n \to \infty} r^n = 0$. If we now assume that $-1 < r < 0$, the sequence $x_n = r^n, n \geq 1$, verifies $\lim_{n \to \infty} |x_n| = \lim_{n \to \infty} |r|^n = 0$; according to what we proved in Example 3, $\lim_{n \to \infty} x_n = 0$. Hence, we can conclude that $\lim_{n \to \infty} r^n = 0$ for $|r| < 1$.

Practice Problems

a. The sequence $x_n = -2, n \geq 1$ is arithmetic with first term $x_1 = -2$ and common difference $d = 0$ and is thus convergent with $\lim_{n \to \infty} x_n = -2$.

b. The sequence $x_n = -\frac{3}{4} - \frac{2n-2}{5}, n \geq 1$ is arithmetic with first term $x_1 = -\frac{3}{4}$ and common difference $d = -\frac{2}{5}$. Since $d \neq 0$, the sequence is not convergent, and we also have $\lim_{n \to \infty} x_n = -\infty$.

c. The sequence $x_n = 1 + 2(n - 1), n \geq 1$ is arithmetic with first term $x_1 = 1$ and common difference $d = 2$. Since $d \neq 0$, the sequence is not convergent, and we also have $\lim_{n \to \infty} x_n = \infty$.

d. The sequence $x_n = \frac{(-5)^{n+1}}{6^{n-1}}$, $n \geq 1$ is geometric with first term $x_1 = 25$ and common ratio $r = -\frac{5}{6}$. Since $|r| < 1$, the sequence is convergent and $\lim\limits_{n \to \infty} x_n = 0$.

e. The sequence $x_n = \frac{(-7)^n}{4^{n+2}}$, $n \geq 1$ is geometric with first term $x_1 = -\frac{7}{64}$ and common ratio $r = -\frac{7}{4}$. Since $r < -1$, the sequence is not convergent, and $\lim\limits_{n \to \infty} x_n$ does not exist.

f. The sequence $x_n = \frac{-4^n}{3^n}$, $n \geq 1$ is geometric with first term $x_1 = -\frac{4}{3}$ and common ratio $r = \frac{4}{3}$. Since $r > 1$, the sequence is not convergent and $\lim\limits_{n \to \infty} x_n = -\infty$.

Classroom Discussion 1.1.6: Adding Consecutive Terms of an Arithmetic Sequence

1. Yes, the arithmetic sequence is $x_n = n$, $n \geq 1$ with first term $x_1 = 1$ and common difference $d = 1$.

2. All are equal to 1,001.

3. Let $S = 1 + 2 + 3 + 4 + \cdots + 999 + 1{,}000$. Then, from the diagram, it follows that $2S = 1{,}000 \cdot 1{,}001 \Rightarrow S = 500{,}500$.

4. Let $S = 1 + 3 + 5 + \cdots + 997 + 999$. The numbers we want to add are consecutive terms of the arithmetic sequence $x_n = 2n - 1$, $n \geq 1$. To determine the number of terms added, it suffices to solve the equation $2n - 1 = 999$, which in turn yields $n = 500$.
   Since $1 + 999 = 1{,}000$, $3 + 997 = 1{,}000$, $\ldots$, we have $2S = 500 \cdot 1{,}000 \Rightarrow S = 250{,}000$.

5. Let $S = x_1 + x_2 + \cdots + x_n$. Then $2S = n[2a + (n - 1)d] \Rightarrow S = \frac{n[2a + (n - 1)d]}{2} = na + \frac{n(n - 1)}{2}d$.

6. $2 + 4 + 6 + 8 + 10 + \cdots + 1{,}998 + 2{,}000 = \frac{1{,}000 \cdot 2{,}002}{2} = 1{,}001{,}000$
   $4 + 7 + 10 + 13 + 16 + \cdots + 3{,}001 + 3{,}004 + 3{,}007 = \frac{1{,}002 \cdot 3{,}011}{2} = 1{,}508{,}511$
   $97 + 95 + 93 + 91 + \cdots + 9 + 7 + 5 = \frac{47 \cdot 102}{2} = 2{,}397$

Classroom Discussion 1.1.7: Adding Consecutive Terms of a Geometric Sequence

1. Yes, the geometric sequence is $x_n = 3 \cdot 2^{n-1}$, $n \geq 1$ with first term $x_1 = 3$ and common ratio $r = 2$.

2. $2S = 3 \cdot 2 + 3 \cdot 2^2 + 3 \cdot 2^3 + \cdots + 3 \cdot 2^{2{,}003} + 3 \cdot 2^{2{,}004}$. After adding 3 to both sides, it follows that $2S + 3 = 3 + 3 \cdot 2 + 3 \cdot 2^2 + 3 \cdot 2^3 + \cdots + 3 \cdot 2^{2{,}003} + 3 \cdot 2^{2{,}004} = S + 3 \cdot 2^{2{,}004}$.

3. $S = 3 \cdot 2^{2{,}004} - 3$.

4. $\frac{1}{2}S + 2 = S + \frac{1}{2^{102}} \Rightarrow S = 4 - \frac{1}{2^{101}}$.

5. $r(y_1 + y_2 + \cdots + y_n) = r(a + ar + \cdots + ar^{n-1}) = ar + ar^2 + \cdots + ar^n = y_2 + y_3 + \cdots + y_n + y_{n+1}$.

6. $rS = y_2 + y_3 + \cdots + y_n + y_{n+1}$, adding $a = y_1$ to both sides yields $rS + a = y_1 + y_2 + y_3 + \cdots + y_n + y_{n+1} = S + y_{n+1} = S + ar^n$.

7. If $r \neq 1$, $S = \frac{ar^n - a}{r - 1} = a\frac{r^n - 1}{r - 1}$.

8. If $r = 1$, we obtain $S = \sum\limits_{i=1}^{n} y_i = \sum\limits_{i=1}^{n} y_1 = n\,a.$

9. $a = 2, r = 3, n = 2{,}002 \Rightarrow \sum\limits_{i=1}^{2{,}002} 2 \cdot 3^{i-1} = \dfrac{2 \cdot 3^{2{,}002} - 2}{3 - 1} = 3^{2{,}002} - 1.$

SECTION 1.2

Classroom Discussion 1.2.1: The Intriguing Number $0.\overline{9}$

1. The sequence $\{a_n\}_n$ is geometric with common ratio $r = \dfrac{1}{10}$.

2. We observe that $s_1 < s_2 < \cdots < s_{100}, \ldots$, their values being closer and closer to 1 as the subscript increases.

3.

$$s_n = \sum_{i=1}^{n} \frac{9}{10}\left(\frac{1}{10}\right)^{i-1} = \frac{\frac{9}{10}\left[\left(\frac{1}{10}\right)^n - 1\right]}{\frac{1}{10} - 1} = 1 - \left(\frac{1}{10}\right)^n$$

4. Yes, the sequence $\{s_n\}_n$ is convergent. Since $\lim\limits_{n \to \infty}\left(\dfrac{1}{10}\right)^n = 0$, from the expression of s_n it follows that $\lim\limits_{n \to \infty} s_n = 1$.

5. This conclusion is in agreement with the computation in (1).

Classroom Discussion 1.2.2: The Mysterious Series $1 - 1 + 1 - 1 + 1 - 1 + \ldots$

Neither Dan nor John is right. The series $1 - 1 + 1 - 1 + 1 - 1 + \ldots$ is divergent. The ad-hoc ways of adding the terms of the series used by Dan and John are not consistent with the definition of a sum of series. The sequence of partial sums for the series under discussion is $s_1 = 1, s_2 = 0, s_3 = 1, s_4 = 0, s_5 = 1, \ldots$, which is divergent.

Classroom Discussion 1.2.3: Convergence of Geometric Series

1. If $|r| < 1$, we have $\lim\limits_{n \to \infty} r^n = 0$, thus, $\lim\limits_{n \to \infty} s_n = \dfrac{a}{1 - r}$. In this case, the sequence $\{s_n\}_n$ converges to $\dfrac{a}{1 - r}$, thus, the series converges to $\dfrac{a}{1 - r}$.

2. If $r > 1$, $\lim\limits_{n \to \infty} r^n = \infty$, thus the sequence $\{s_n\}_n$ diverges.

3. If $r < -1$, $\lim\limits_{n \to \infty} r^n$ does not exist. Consequently, $\lim\limits_{n \to \infty} s_n$ also does not exist, and the series diverges for $r < -1$.

4. If $r = 1$, then $s_n = na$, and since $\lim\limits_{n \to \infty} n = \infty$, it follows that $\lim\limits_{n \to \infty} s_n = \infty$ if $a > 0$ and $\lim\limits_{n \to \infty} s_n = -\infty$ if $a < 0$. Consequently, the series diverges for $r = 1$.

5. For $r = -1$, $s_n = a$ if n is odd and $s_n = 0$ if n is even, thus the sequence $\{s_n\}_n$ is divergent. Consequently, the series diverges for $r = -1$.

Practice Problems

1. $0.\overline{41} = \dfrac{41}{100} + \dfrac{41}{10{,}000} + \dfrac{41}{1{,}000{,}000} + \ldots$ is a geometric series with ratio $r = \dfrac{1}{100}$ and first term $a = \dfrac{41}{100}$. Since $|r| < 1$, by Theorem 1.2.2 it follows that the series converges to $\dfrac{\frac{41}{100}}{1 - \frac{1}{100}} = \dfrac{41}{99}$.

Reasoning as in (1), it follows that

$$\begin{aligned} 100x &= 41.\overline{41} \\ - x &= -0.\overline{41} \\ \hline 99x &= 41 \end{aligned}.$$

Therefore, $x = \frac{41}{99}$.

2. This is a geometric series of ratio $r = 2$. Since $r > 1$, it follows that the series diverges.

CHAPTER 2

SECTION 2.1

Classroom Discussion 2.1.1: More on Functions

1. **a.** The domain of g is $\mathbb{R}$ since $g(x)$ is well defined for all real numbers x.
   b. Image$(g) = \mathbb{R}$.
   c. $g(32) = 0$, $g(65) = \frac{55}{3}$, $g(99) = \frac{335}{9}$, and they represent the temperatures in degrees Celsius when the temperatures in degrees Fahrenheit are $32, 65$, and 99, respectively.
   d. $g(x + 1) = \frac{5x - 155}{9}$, $g(2x) = \frac{10x - 160}{9}$, $g(3x - 5) = \frac{15x - 185}{9}$, and they represent the temperatures in degrees Celsius when the temperatures in degrees Fahrenheit are $x + 1, 2x$, and $3x - 5$, respectively.
   e. The statement is equivalent to the algebraic identity $g(x + 1) = g(x) + 1$. This statement is false since $g(x + 1) = \frac{5(x - 32 + 1)}{9} = g(x) + \frac{5}{9} \neq g(x) + 1$.

2. **a.** The domain of f is $(0,13]$, and the image is
   $$\{0.37, 0.60, 0.83, 1.06, 1.29, 1.52, 1.75, 1.98, 2.21, 2.44, 2.67, 2.90, 3.13\}.$$
   b. $f(x) = 0.37 + 0.23\lceil x - 1 \rceil$.

Practice Problems

a. The expression of $f(x)$ is meaningful only if $x + 1 \geq 0$, which is equivalent to $x \geq -1$. Thus, the domain of $f(x)$ is $[-1, \infty)$.

b. The expression of $g(x)$ is well defined for all real numbers for which $x + 3 \neq 0$. Hence, Dom$(g) = \mathbb{R}\backslash\{3\}$.

c. For $h(x)$ to make sense, we need to impose the condition $2 - x > 0$, which gives $x < 2$. Consequently, Dom$(h) = (-\infty, 2)$.

Classroom Discussion 2.1.2: The Vertical Line Test

1. Suppose the line $x = c$ intersects a given curve at two points (c, y_1) and (c, y_2). If the curve were the graph of a function f, then $y_1 = f(c)$ and $y_2 = f(c)$. Hence, the law f would associate to the point $x = c$ two different values, which, according to the definition of a function, is not possible. The curve cannot represent the graph of a function.

2. Yes, I agree. According to what we have proved in Problem 1, if a given curve is the graph of a function, then any vertical line intersects the curve at most one point. On the other hand, if we know that any vertical line intersects a curve in the plane at most once, then we can define a real valued function f whose graph is

the curve. The domain of this function f is the set of all real numbers x for which there exists y such that (x, y) is on the curve and the correspondence is $f(x) = y$.

3. The graphs in a and b fail the vertical line test; more than one temperature reading corresponds to one time in certain cases. The graphs in c and d are not likely to occur because of the way temperatures vary in general. In c there are times when no temperature is marked, while in d, a sudden drop from 90 degrees to freezing and back to 90 degrees is graphed within 1 hour.

4. Solving for y in terms of x, we obtain the following:
   a. $y = \pm\sqrt{1 - x^2}$ is not a function of x since, for example, when $x = 0$, the equation yields two values for y: -1 and 1. As such, the vertical line test fails: the line $x = 0$ intersects the curve $x^2 + y^2 = 1$ at two points.
   b. $y = \frac{3}{x}$ is a function. For each $x \neq 0$, the equation yields one value for y, hence the vertical line test holds.
   c. $y = 1$ is a function. For each real number x, the equation yields one value for y, hence the vertical line test holds.
   d. $y = \pm\sqrt{2x}$ is not a function of x since, for example, when $x = 2$, the equation yields two values for y: -1 and 1. As such the vertical line test fails: the line $x = 2$ intersects the curve $y^2 = 2x$ at two points.
   e. $y = \frac{1}{2}x + \frac{5}{2}$ is a function. For each real number x, the equation yields one value for y, hence the vertical line test holds.

Classroom Discussion 2.1.3: Examples of Composite Functions

1. a. $f(g(x)) = f(2x + 5) = \sqrt{2x + 5}$
   b. Since $g(-3) = 2(-3) + 5 = -1 < 0$, and the domain of f is $[0, \infty)$, we conclude that $g(-3)$ is not an element of the domain of f.
   c. We need to impose the condition $g(x) \geq 0$ since the domain of f is $[0, \infty)$.
   d. The condition $g(x) \geq 0$ is equivalent to $2x + 5 \geq 0$, which in turn is equivalent to $x \geq -\frac{5}{2}$. If we select $A = [-\frac{5}{2}, \infty)$, then $(f \circ g)(x)$ makes sense for all $x \in A$. This is the largest set with this property.

2. a. Since $g(x) \in (0, \infty) = \text{Domain}(f)$ for all x, we can conclude that $f \circ g$ makes sense for any $x \in (0, \infty)$.
   b. Since $f(x) \in (0, \infty) = \text{Domain}(g)$ for all $x \in (0, \infty)$, we can conclude that $g \circ f$ makes sense for any $x \in (0, \infty)$.
   c. $(f \circ g)(x) - (g \circ f)(x) = f(x^2) - g\left(x + \frac{1}{x}\right) = x^2 + \frac{1}{x^2} - $
   $\left(x + \frac{1}{x}\right)^2 = x^2 + \frac{1}{x^2} - x^2 - 2 - \frac{1}{x^2} = -2$; thus $f \circ g - g \circ f = -2$.
   d. Yes. We saw that if the order of composition is changed, the resulting function is different.
   e. If g acts first, then f needs to be applied to $g(x)$, yielding $f(g(x))$, which is denoted by $(f \circ g)(x)$.

Practice Problem

Since the domain of f is $[-2, \infty)$, we have that $(f \circ g)(x)$ is well defined for all x in the domain of g for which $g(x) \geq -2$; that is, for all $x \in \mathbb{R}$ for which $\frac{1}{x^2 + 1} \geq -2$. The last inequality is satisfied for all $x \in \mathbb{R}$. Hence, for all $x \in \mathbb{R}$, we have

$$(f \circ g)(x) = f\left(\frac{1}{x^2 + 1}\right) = \sqrt{\frac{1}{x^2 + 1} + 2} = \sqrt{\frac{2x^2 + 3}{x^2 + 1}}.$$

Since the domain of g is $\mathbb{R}$, it follows that $(g \circ f)(x)$ is well defined for all x in the domain of f, which is $[-2, \infty)$. In addition, for all $x \in [-2, \infty)$, we have

$$g(f(x)) = g(\sqrt{x + 2}) = \frac{1}{(\sqrt{x + 2})^2 + 1} = \frac{1}{x + 3}.$$

SECTION 2.2

Classroom Discussion 2.2.1: Graphical Approach for Finding Limits

1. **a.** As the length of the femur increases from 16.5 inches to 17 inches, the values of the height increase from 62.28 to 63.44.
   b. Yes, $L = 63.44$.
   c. As the femur's length decreases from 17.5 inches to 17 inches, the values of the height decrease from 64.6 to 63.44.
   d. Yes, $L' = 63.44$.
   e. $L = L'$
   f. As $x \neq 17$ gets closer and closer to 17 from either side of 17, the values of $h(x)$ get closer and closer to 63.44.

2. **The Function** $f(x) = \frac{x^2 - 4}{x - 2}$
   a. The domain of $f(x) = \frac{x^2 - 4}{x - 2}$ is $\mathbb{R} \setminus \{2\}$.
   b. The equality $f(x) = x + 2$ is true for all real values x different from 2.
   c.

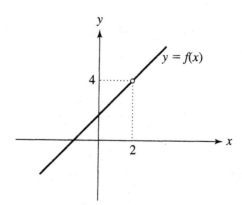

   d. $\lim\limits_{x \to 2^-} f(x) = 4$, $\lim\limits_{x \to 2^+} f(x) = 4$.
   e. $\lim\limits_{x \to 2} f(x) = 4$.
   f. The value of $\lim\limits_{x \to 2} f(x)$ is independent on whether f is defined at 2. In the study of the limit of f as x approaches 2, we are interested only in the values of f as x gets closer and closer to 2 while keeping $x \neq 2$.

3. USPS Prices

a. $\lim\limits_{x \to 4^-} f(x) = 1.06$, $\lim\limits_{x \to 4^+} f(x) = 1.29$.

b. The two limits are different.

c. Since $\lim\limits_{x \to 4^-} f(x) = 1.06 \neq \lim\limits_{x \to 4^+} f(x) = 1.29$, we have that $\lim\limits_{x \to 4} f(x)$ does not exist.

d. It does not make sense to discuss $\lim\limits_{x \to 0^-} f(x)$ since the function is not even defined for $x < 0$; the weight of a letter cannot be a negative number. It makes sense to discuss $\lim\limits_{x \to 0^+} f(x)$. We know the shipping cost of letters of arbitrarily small weight, and it makes sense to look at what happens to the shipping cost of letters if their corresponding weights approach zero. As seen from the graph, $\lim\limits_{x \to 0^+} f(x) = 0.37$.

e. It does not make sense to discuss $\lim\limits_{x \to 0} f(x)$. As seen before, we make sense of $\lim\limits_{x \to 0} f(x)$ only if $\lim\limits_{x \to 0^-} f(x) = \lim\limits_{x \to 0^+} f(x)$. In our case, $\lim\limits_{x \to 0^-} f(x)$ is not even defined.

4. a.
$\lim\limits_{x \to -9^-} f(x) = 0$, $\lim\limits_{x \to -9^+} f(x) = 0$. Since $\lim\limits_{x \to -9^-} f(x) = \lim\limits_{x \to -9^+} f(x) = 0$, it follows that $\lim\limits_{x \to -9} f(x) = 0$.

b. $\lim\limits_{x \to 3^-} f(x) = 2$, $\lim\limits_{x \to 3^+} f(x) = 8$. Since $\lim\limits_{x \to 3^-} f(x) \neq \lim\limits_{x \to 3^+} f(x)$, it follows that $\lim\limits_{x \to 3} f(x)$ does not exist.

c. $\lim\limits_{x \to 13^-} f(x) = 14 = \lim\limits_{x \to 13^+} f(x)$. Hence, $\lim\limits_{x \to 13} f(x) = 14$. Observe that even though f is not defined at $x = 13$, the limit $\lim\limits_{x \to 13} f(x)$ makes sense.

d. $\lim\limits_{x \to 15^-} f(x) = 9$, $\lim\limits_{x \to 15^+} f(x) = 9$. Since $\lim\limits_{x \to 15^-} f(x) = \lim\limits_{x \to 15^+} f(x) = 9$, it follows that $\lim\limits_{x \to 15} f(x) = 9$.

Classroom Discussion 2.2.2: Numerical Approach for Finding Limits

1. a.

x	$f(x)$
1.999	0.59992
1.9991	0.59993
1.9992	0.59994
1.9993	0.59994
1.9994	0.59995
1.9995	0.59996
1.9996	0.59997
1.9997	0.59998
1.9998	0.59998
1.9999	0.59999

We observe that the closer x gets to 2, the closer $f(x)$ gets to 0.6.

b.

x	$f(x)$
2.001	0.60008
2.0009	0.60007
2.0008	0.60006
2.0007	0.60006
2.0006	0.60005
2.0005	0.60004
2.0004	0.60003
2.0003	0.60002
2.0002	0.60002
2.0001	0.60001

We observe that the closer x gets to 2, the closer $f(x)$ gets to 0.6.

c.

$$\lim_{x \to 2^-} f(x) = 0.6 \qquad \lim_{x \to 2^+} f(x) = 0.6 \qquad \lim_{x \to 2} f(x) = 0.6$$

2. a. As $x > 0$ gets smaller and smaller, $g(x)$ gets larger and larger.

x	$g(x)$
0.1	10
0.01	100
0.001	1,000
0.0001	10,000
0.00001	100,000
0.000001	1,000,000
0.0000001	10,000,000

b. The notation $\lim\limits_{x \to 0^+} \dfrac{1}{x} = \infty$ is appropriate since the function $\dfrac{1}{x}$ increases without bound as $x \to 0^+$.

c. As $x < 0$ gets smaller and smaller in absolute value, $g(x)$, which is negative, gets larger and larger in absolute value.

x	$g(x)$
-0.1	-10
-0.01	-100
-0.001	$-1,000$
-0.0001	$-10,000$
-0.00001	$-100,000$
-0.000001	$-1,000,000$
-0.0000001	$-10,000,000$

d. The notation $\lim\limits_{x \to 0^-} \dfrac{1}{x} = -\infty$ is appropriate: as $x \to 0^-$, the function $\dfrac{1}{x}$ takes negative values that are larger and larger in absolute value.

e. No, the limit does not exist since for the limit of a function at a point to exist, both the right-hand and left-hand limits should exist and be equal.

f. Since $\lim\limits_{x \to 0^-} \frac{1}{x} = -\infty$, the portion of the graph of $\frac{1}{x}$ corresponding to values $x \to 0^-$ is traced from left to right, "closer and closer" to the y-axis and "lower and lower" in the negative direction of the y-axis. Similarly, since $\lim\limits_{x \to 0^+} \frac{1}{x} = \infty$, the portion of the graph of $\frac{1}{x}$ corresponding to values $x \to 0^+$ is traced from right to left, "closer and closer" to the y-axis and "higher and higher" in the positive direction of the y-axis.

Practice Problem

Suppose there exists $M > 0$ such that $|g(x)| \leq M$ for all $x < 0$. Since $g(x) < 0$ for $x < 0$, the condition $|g(x)| \leq M$ for all $x < 0$ is equivalent to $-\frac{1}{x} \leq M$ for all $x < 0$. However, if $-\frac{1}{M} < x < 0$, then $-\frac{1}{x} > M$. Hence, the assumption that there exists $M > 0$ such that $|g(x)| \leq M$ for all $x < 0$ is false, and we can conclude that M with the required property does not exist.

Classroom Discussion 2.2.3: Limit of Rational Functions

1. As $x \to a$, the values of the function $f(x) = x$ also get closer and closer to a; thus $\lim\limits_{x \to a} x = a$. As $x \to a$, the values of the function $g(x) = c$ stay equal to c; hence $\lim\limits_{x \to a} c = c$.

2. $\lim\limits_{x \to a} c \cdot x^2 = c \cdot \lim\limits_{x \to a} x^2 = c(\lim\limits_{x \to a} x)(\lim\limits_{x \to a} x) = c a^2$

   $\lim\limits_{x \to a} c \cdot x^3 = c \cdot \lim\limits_{x \to a} x^3 = c(\lim\limits_{x \to a} x)(\lim\limits_{x \to a} x^2) = c a^3$

   $\lim\limits_{x \to a} c \cdot x^4 = c(\lim\limits_{x \to a} x)(\lim\limits_{x \to a} x^3) = c a^4$

3. $\lim\limits_{x \to a} c \cdot x^n = c a^n$

4. $\lim\limits_{x \to a} (2x^3 - \frac{4}{3}x^2 + x + \frac{1}{2}) = \lim\limits_{x \to a} (2x^3) - (\frac{4}{3} \lim\limits_{x \to a} x^2) + \lim\limits_{x \to a} x + \lim\limits_{x \to a} \frac{1}{2}$

   $= 2a^3 - \frac{4}{3}a^2 + a + \frac{1}{2}$

5. $\lim\limits_{x \to a} f(x) = \lim\limits_{x \to a} (b_n x^n) + \cdots + \lim\limits_{x \to a} (b_1 x) + \lim\limits_{x \to a} b_0 = b_n a^n + \cdots + b_1 a + b_0 = f(a)$

   For the first equality, we used P1, for the second equality, we used Problem 3. If $f(x)$ is a polynomial function, then for any real number a, $\lim\limits_{x \to a} f(x) = f(a)$.

6. By Problem 5, we have that $\lim\limits_{x \to a} p(x) = p(a)$, $\lim\limits_{x \to a} q(x) = q(a)$, so by P4 it follows that

$$\lim_{x \to a} \frac{p(x)}{q(x)} = \frac{\lim\limits_{x \to a} p(x)}{\lim\limits_{x \to a} q(x)} = \frac{p(a)}{q(a)}.$$

If $p(x)$ and $q(x)$ are polynomial functions, then $\lim\limits_{x \to a} \frac{p(x)}{q(x)} = \frac{p(a)}{q(a)}$ for any real number a satisfying $q(a) \neq 0$.

Practice Problems

1. The function $f(x) = 5x^4 + x^3 - \frac{1}{2}x + 5$ is a polynomial function. Applying the results from Classroom Discussion 2.2.3 we obtain

$$\lim_{x \to -1} (5x^4 + x^3 - \frac{1}{2}x + 5) = f(-1) = \frac{19}{2}.$$

2. Using P5, we have that $\lim_{x \to 0} \sqrt{1 + 6x - 5x^2} = \sqrt{\lim_{x \to 0}(1 + 6x - 5x^2)}$. In addition, since $1 + 6x - 5x^2$ is a polynomial function, we also have $\lim_{x \to 0}(1 + 6x - 5x^2) = 1 + 6 \cdot 0 - 5 \cdot 0 = 1$. Thus, $\lim_{x \to 0} \sqrt{1 + 6x - 5x^2} = \sqrt{1} = 1$.

3. The functions $p(x) = x^2 - 4$ and $q(x) = x^2 + 2$ are polynomial functions, and since $q(3) = 11 \neq 0$, the results from Classroom Discussion 2.2.3 imply that

$$\lim_{x \to 3} \frac{x^2 - 4}{x^2 + 2} = \frac{p(3)}{q(3)} = \frac{5}{11}.$$

Classroom Discussion 2.2.4: Limits and Intersecting Streets

1. I agree with the statement, since for $x \neq 17$ and $|x - 17| < 0.21551724 \Rightarrow$ $16.784483 < x < 17.21551724 \Rightarrow 38.94 < 2.32 x < 39.94 \Rightarrow 62.94 <$ $2.32 x + 24 < 63.94 \Rightarrow -0.5 < 2.32 x + 24 - 63.44 < 0.5 \Rightarrow |h(x) -$ $63.44| < 0.5$.

2. No. Take $x = 17.009$. Then x is within 0.01 of 17, $h(x) = 63.46088$, and $h(x) - 63.44 = 0.02088$. Thus, $h(x) - 63.44 > 0.001$.

3. If $|x - 17| < \delta \Rightarrow -\delta + 17 < x < \delta + 17 \Rightarrow -2.32 \delta + 39.44 < 2.32 x <$ $2.32 \delta + 39.44 \Rightarrow -2.32 \delta + 63.44 < 2.32 x + 24 < 2.32 \delta + 63.44 \Rightarrow$ $-2.32 \delta < 2.32 x + 24 - 63.44 < 2.32 \delta \Rightarrow |h(x) - 63.44| < 2.32 \delta$. To ensure that $|h(x) - 63.44| < 0.001$, it is enough to select δ such that $2.32 \delta < 0.001$, which is equivalent to $\delta < \frac{0.001}{2.32} = 0.0004310344$. Thus, for $|x - 17| < \delta < \frac{0.001}{2.32} = 0.0004310344$, we have $|h(x) - 63.44| < 2.32 \delta < 2.32 \cdot \frac{0.001}{2.32} = 0.001$.

4. Using the computations in Problem 3, we see that if $\delta < \frac{\varepsilon}{2.32}$, then $|h(x) - 63.44| < 2.32 \delta < 2.32 \cdot \frac{\varepsilon}{2.32} = \varepsilon$.

5. As seen from the following figure, for an arbitrary open interval I on the y-axis centered at 9, we can find an interval J on the x-axis centered around 11 such that, if $x \in J \setminus \{11\}$, then the point $(x, g(x))$ is in the portion of the graph of g contained in the horizontal band whose intersection with the y-axis is I.

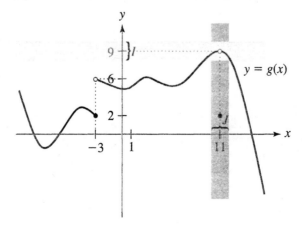

Thus, $\lim_{x \to 11} g(x) = 9$. On the other hand, for the interval I on the y-axis centered at 6 as seen in the following figure, no matter how small we select an interval

J on the x-axis centered around -3, there are points $x \in J\backslash\{-3\}$, such that the corresponding points $(x, g(x))$ are left outside the portion of the graph of g contained in the horizontal band whose intersection with the y-axis is I. This is the case for all $x \in J\backslash\{-3\}, x < -3$.

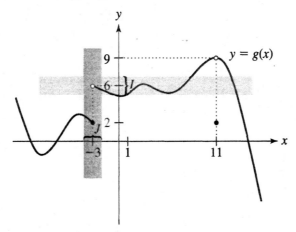

Consequently, $\lim\limits_{x \to -3} g(x) \ne 6$.

Practice Problem

Fix an open interval I centered at $-\frac{1}{32}$. We want to find an open interval J containing $-\frac{1}{2}$, with $J\backslash\{-\frac{1}{2}\}$ contained in the domain of f, such that for every $x \in J\backslash\{-\frac{1}{2}\}$ we have $f(x) \in I$. Let $2l$ be the length of the interval I; that is, $I = (-\frac{1}{32} - l, -\frac{1}{32} + l)$. Select $J = \left(\sqrt[5]{-\frac{1}{32} - l}, \sqrt[5]{-\frac{1}{32} + l}\right)$. Then, $-\frac{1}{2} \in J, J\backslash\{-\frac{1}{2}\}$ is contained in the domain of f and, for all $x \in J\backslash\{-\frac{1}{2}\}$, the inequality $|x^5 - (-\frac{1}{32})| < l$ is satisfied. Hence, $f(x) \in I$ for all $x \in J\backslash\{-\frac{1}{2}\}$, which completes the proof.

SECTION 2.3

Classroom Discussion 2.3.1: Setting the Stage

1. Only the graph of f_4 can be traced without lifting the pen. The functions f_1 and f_2 have a jump at $x = 1$, while f_3 is not defined at $x = 1$.
   $\lim\limits_{x \to 1^-} f_1(x) = 1.5$, $\lim\limits_{x \to 1^+} f_1(x) = 3$; thus $\lim\limits_{x \to 1} f_1(x)$ does not exist, while $f_1(1) = 1.5$.
   $\lim\limits_{x \to 1} f_2(x) = 2, f_2(1) = 3.5$.
   $\lim\limits_{x \to 1} f_3(x) = 2$; f_3 is not defined at $x = 1$.
   $\lim\limits_{x \to 1} f_4(x) = 0.5 = f_4(1)$.
2. **a.** The graph of h can be traced without lifting the pen, while the graph of f cannot.
   b. The graph of the function h has no jumps, while the graph of the function f has jumps at 1, 2, 3, 4, 5, 6, 7, 8, 9, 10, 11, and 12.

c. The function h models a process without gaps or sudden changes; f models a process with sudden changes.

Practice Problem

The function g is continuous on $(-\infty, 0)$ since it coincides with a polynomial function on this interval. On the interval $(0, \infty)$, the rational function $\dfrac{1}{x+1}$ is continuous; thus g is also continuous on $(0, \infty)$. We are left with analyzing the continuity of g at $x = 0$. Because

$$\lim_{x \to 0^-} g(x) = \lim_{x \to 0^-} (3x + 2) = (3x + 2)|_{x=0} = 2$$

and

$$\lim_{x \to 0^+} g(x) = \lim_{x \to 0^+} \frac{1}{x+1} = \frac{1}{x+1}\Big|_{x=0} = 1,$$

we can conclude that $\lim_{x \to 0} g(x)$ does not exist. Since condition b fails, we have that g is discontinuous at 0. The conclusion is that g is continuous on $\mathbb{R} \backslash \{0\}$.

CHAPTER 3

SECTION 3.1

Classroom Discussion 3.1.1: Rates of Change in Real Life

1. $\text{heart rate} = \dfrac{\text{number of heartbeats}}{\text{time in minutes}}$
2. The area of the surface to be painted is equal to $15 \cdot 3 \cdot 5 \cdot 2 = 450$ square feet. The two painters paint together at a rate of 45 square feet per minute. It will take $\dfrac{450}{45} = 10$ minutes for the two painters to finish painting both sides of the wooden fence.
3. Adrian's walking rate is $\dfrac{7 \cdot 1000}{1.5 \cdot 3600} = \dfrac{35}{27}$ meters per second.

Classroom Discussion 3.1.2: Rates of Change of Linear and Nonlinear Functions

1. **Water levels**
   a. $\dfrac{f(2) - f(1)}{2 - 1} = 5 - 4.5 = 0.5$
   b. $\dfrac{f(5) - f(2)}{5 - 2} = \dfrac{5 - 5}{5 - 2} = 0$
   c. $\dfrac{f(15) - f(8)}{15 - 8} = \dfrac{4 - 6}{7} = -\dfrac{2}{7}, \dfrac{f(30) - f(15)}{30 - 15} = \dfrac{7 - 4}{15} = \dfrac{1}{5}$.
   The average rates of change computed are different.
2. **Walking Rates**
   a. Diana's average walking rates on the given intervals are: $\dfrac{d(4) - d(1)}{4 - 1} = \dfrac{240 - 60}{3} = 60$ on $[1, 4]$, $\dfrac{d(7) - d(6)}{7 - 6} = 60$ on $[6, 7]$, and $\dfrac{d(25) - d(0)}{25 - 0} = 60$ on $[0, 25]$. All the average walking rates we computed are equal to 60 meters per minute; they are independent of the interval of time. This means that Diana is walking at a constant rate over these intervals.

b. The average rates of change of d computed over two different intervals are equal.

Let $[t_1, t_2]$ and $[t_3, t_4]$ be two arbitrary intervals. Diana's average walking rates on these intervals are:
$\frac{d(t_2) - d(t_1)}{t_2 - t_1} = \frac{60 \cdot t_2 - 60 \cdot t_1}{t_2 - t_1} = 60$ and $\frac{d(t_4) - d(t_3)}{t_4 - t_3} = \frac{60 \cdot t_4 - 60 \cdot t_3}{t_4 - t_3} = 60$. They are equal.

3. **Linear Functions versus Nonlinear Functions**

a. The average rate of change of g over an interval $[a, b]$ is $\frac{g(b) - g(a)}{b - a} = \frac{mb + c - ma - c}{b - a} = m$. It is independent of the interval.

b. Suppose that the average rate of change of a function g over any interval is equal to m. We fix a real number a and let x be an arbitrary real number. Since the average rate of change of g over any interval is equal to m, it follows that $\frac{g(x) - g(a)}{x - a} = m$; thus, $g(x) = mx + g(a) - ma$. Letting $c = g(a) - ma$, we see that g is indeed of the form, $g(x) = mx + c$; that is, g is a linear function.

c. The function f from Problem 1 is not linear. This can be seen either from the graph—the graph is not a line—or by observing that the rates of change of f over the intervals $[1, 2]$ and $[3, 5]$ are different.

d. The average rate of change of a function f over an interval $[a, b]$ is equal to the slope of the line passing through the points $(a, f(a))$ and $(b, f(b))$.

SECTION 3.2

Classroom Discussion 3.2.1: Instantaneous Rate of Change of the Function $f(x) = x^2$

2. $\frac{f(1.1) - f(1)}{1.1 - 1} = \frac{1.1^2 - 1}{0.1} = 2.1$

3.

h	$\dfrac{f(1 + h) - f(1)}{h}$
10^{-1}	2.1
10^{-2}	2.01
10^{-3}	2.0009999999997
10^{-4}	2.00009999999917
10^{-5}	2.00001000001393
10^{-6}	2.00000099992437

As the values of h get smaller and smaller, the average rates of change of f over the intervals $[1, 1 + h]$ get closer and closer to 2. As such, as $h \to 0^+$, the slope of the line passing through the points $(1, f(1))$ and $(1 + h, f(1 + h))$ gets closer and closer to 2.

4.

h	$\dfrac{f(1) - f(1 + h)}{-h}$
10^{-1}	1.9
10^{-2}	1.99
10^{-3}	1.99899999999997
10^{-4}	1.99989999999928
10^{-5}	1.99998999999007
10^{-6}	1.99999900007963

As $h < 0$ decreases in absolute value, the average rates of change of f over the intervals $[1 + h, 1]$ get closer and closer to 2. The slope of the line passing through the points $(1, f(1))$ and $(1 + h, f(1 + h))$ gets closer and closer to 2.

5. As $h \to 0$, we expect that the values of the average rates of change of f over the intervals with endpoints 1 and $1 + h$ will get closer and closer to 2. Moreover, as $h \to 0$, the point $(1 + h, f(1 + h))$ moves on the graph of f closer and closer to the point $(1, f(1))$. In the process, the slope of the line passing through $(1, f(1))$ and $(1 + h, f(1 + h))$ gets closer and closer to 2. Hence, we expect that a tangent line to the graph of $f(x)$ at $(1, 1)$ exists and that it coincides with the line of slope 2 passing through the point $(1, 1)$.

6. $\dfrac{f(1 + h) - f(1)}{1 + h - 1} = \dfrac{(1 + h)^2 - 1}{h} = \dfrac{1 + 2h + h^2 - 1}{h} = 2 + h$

$\lim\limits_{h \to 0} (2 + h) = 2$

Since the value $2 + h$ is the slope of the line passing through the points $(1, f(1))$ and $(1 + h, f(1 + h))$, the limit of this expression as $h \to 0$ should be the same as the slope of the tangent line to the graph of f at $(1, f(1))$. Thus, the slope of the tangent line to the graph of f at $(1, f(1))$ is equal to 2.

7. Since it is the result of taking the limit as $h \to 0$ of the average rates of change of f over the intervals with endpoints 1 and $1 + h$, these intervals shrink to the point $x = 1$ as $h \to 0$.

8. The slope of the tangent line to the graph $y = x^2$ at the point $(1,1)$ is equal to 2.

9. $\lim\limits_{h \to 0} \dfrac{f(x + h) - f(x)}{h} = \lim\limits_{h \to 0} \dfrac{(x + h)^2 - x^2}{x + h - x} = \lim\limits_{h \to 0} \dfrac{x^2 + 2xh + h^2 - x^2}{h}$
$= \lim\limits_{h \to 0} (2x + h) = 2x.$

10. The instantaneous rate of change of f at x is $\lim\limits_{h \to 0} \dfrac{f(x + h) - f(x)}{h}$. The value of this limit coincides with the slope of the tangent line to the graph $y = x^2$ at the point $(x, f(x))$.

Practice Problem

Let $f(x) = x^2$. Then the slope of the tangent line to the graph of f at $(3, 9)$ has slope $m = f'(3) = 2 \cdot 3 = 6$. Thus, the tangent line has the equation $y - 9 = 6(x - 3)$, or equivalently, $y = 6x - 9$.

Classroom Discussion 3.2.2: Derivatives of Linear and Quadratic Functions

1. The tangent lines to the graph of $f(x) = 3$ have slopes equal to 0 since they all coincide with the line $y = 3$. Applying Definition 3.2.1 we have $f'(x) =$
$\lim\limits_{h \to 0} \dfrac{f(x+h) - f(x)}{h} = \lim\limits_{h \to 0} \dfrac{3-3}{h} = 0.$

2. The lines tangent to the graph of $f(x) = c$ are all of the form $y = c$. As in Problem 1, their slopes are equal to 0.
$$f'(x) = \lim_{h \to 0} \frac{f(x+h) - f(x)}{h} = \lim_{h \to 0} \frac{c-c}{h} = 0.$$
If $f(x) = c$ for every x, then $f'(x) = 0$.

3. The graph of g is the line $y = 3x + 1$ whose slope is 3. Hence, any tangent line to this graph coincides with the line $y = 3x + 1$ and has slope equal to 3.
$$g'(x) = \lim_{h \to 0} \frac{g(x+h) - g(x)}{h} = \lim_{h \to 0} \frac{3(x+h) + 1 - (3x + 1)}{h} = \lim_{h \to 0} 3 = 3.$$

4. The graph of f is a line, hence any tangent line to its graph coincides with the line $y = mx + b$ of slope m.
$$f'(x) = \lim_{h \to 0} \frac{f(x+h) - f(x)}{h} = \lim_{h \to 0} \frac{m(x+h) + b - (mx + b)}{h} = \lim_{h \to 0} m = m.$$
If $f(x) = mx + b$ for every x, then $f'(x) = m$.

5.

$$\lim_{h \to 0} \frac{f(x+h) - f(x)}{h} = \lim_{h \to 0} \frac{a(x+h)^2 + b(x+h) + c - (ax^2 + bx + c)}{h}$$

$$= \lim_{h \to 0} \frac{ax^2 + 2axh + ah^2 + bx + bh + c - ax^2 - bx - c}{h}$$

$$= \lim_{h \to 0} (2ax + ah + b) = 2ax + b.$$

Hence, if $f(x) = ax^2 + bx + c$ for every x, then $f'(x) = 2ax + b$.

Classroom Discussion 3.2.3: Rising or Falling Graphs and Derivatives of Functions

1.

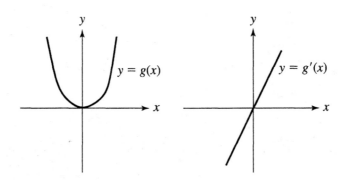

2. g is nonincreasing on the interval $(-\infty, 0)$, and $g'(x)$ is negative on this interval.
3. g is nondecreasing on the interval $(0, \infty)$, and $g'(x)$ is positive on this interval.
4. The derivative is zero at $x_0 = 0$. The tangent line to the graph of g at $(0,0)$ is horizontal.

5. If f is nonincreasing on an interval (a, b), then the slopes of the tangent lines to the graph of f corresponding to points $x \in (a, b)$ are nonpositive. Hence, we expect that $f' \leq 0$ on the interval (a, b).

Reasoning based on the definition of the derivative, we have the following, If f is differentiable and verifies $f(x_1) \geq f(x_2)$ for all $a < x_1 < x_2 < b$, then $\frac{f(x_2) - f(x_1)}{x_2 - x_1} \leq 0$, which in turn implies that $\lim_{x_2 \to x_1} \frac{f(x_2) - f(x_1)}{x_2 - x_1} \leq 0$. Hence, $f'(x_1) \leq 0$.

6. If f is nondecreasing on an interval (a, b), then the slopes of the tangent lines to the graph of f corresponding to points $x \in (a, b)$ are nonnegative. Hence, we expect that $f' \geq 0$ on the interval (a, b).

Reasoning based on the definition of the derivative, we have the following: If f is differentiable and verifies $f(x_1) \leq f(x_2)$ for all $a < x_1 < x_2 < b$, then $\frac{f(x_2) - f(x_1)}{x_2 - x_1} \geq 0$, which in turn implies that $\lim_{x_2 \to x_1} \frac{f(x_2) - f(x_1)}{x_2 - x_1} \geq 0$. Hence, $f'(x_1) \geq 0$.

7. We have seen that for a differentiable function f, the slope of the tangent line to the graph of f at a point $(c, f(c))$ is equal to $f'(c)$. As such, the tangent line at $(c, f(c))$ will be horizontal if and only if $f'(c) = 0$.

Classroom Discussion 3.2.4: Sketching Graphs of Derivatives

1. The derivative is constant, positive, and defined everywhere. A tangent line exists at each point on the graph.

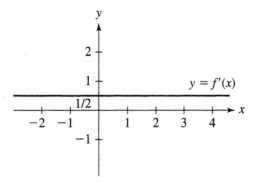

2. The derivative is constant, negative, and defined everywhere. A tangent line exists at each point on the graph.

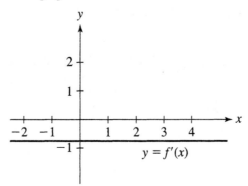

3. The derivative is not defined at $x = 2$ and equals zero everywhere else on the real line. The graph does not have a tangent line at $x = 2$.

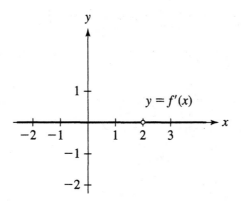

4. The derivative is not defined at $x = 0$, $x = 1$, $x = 2$, and $x = 3$. It is zero on the intervals $(0, 1)$ and $(3, \infty)$, constant negative on each of the intervals $(-\infty, 0)$ and $(1, 2)$, and constant positive on the interval $(2, 3)$. The graph does not have a tangent line at $x = 0$, $x = 1$, $x = 2$, and $x = 3$.

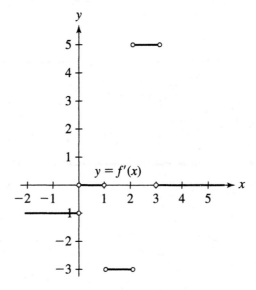

5. The graph looks like the graph of a quadratic function, so the derivative is a linear function. The derivative is defined everywhere, being zero at $x = 1$, positive on the interval $(-\infty, 1)$, and negative on the interval $(1, \infty)$. A tangent line exists at each point on the graph.

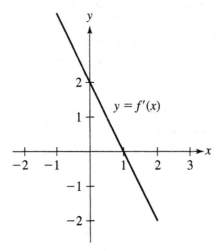

6. The graph looks like the graph of a quadratic function, so the derivative is a linear function. The derivative is defined everywhere, being zero at $x = 1$, negative on the interval $(-\infty, 1)$, and positive on the interval $(1, \infty)$. A tangent line exists at each point on the graph.

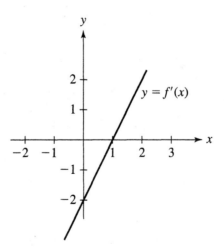

7. The derivative is defined everywhere, being negative on the whole real line. A tangent line exists at each point on the graph. The graph of f suggests that as x increases, the slopes of the tangent lines at $(x, f(x))$ decrease. If we let $x \to -\infty$, then the slopes of the corresponding tangent lines seem to approach zero from the left.

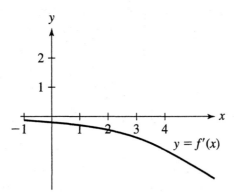

8. The derivative is defined everywhere, being negative on the whole real line. A tangent line exists at each point on the graph. The graph of f suggests that as x increases, the slopes of the tangent lines at $(x, f(x))$ increase, and as $x \to \infty$, the slopes approach zero from the left.

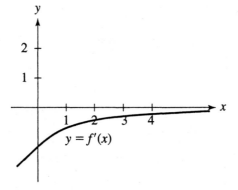

9. The derivative is defined everywhere, being positive on the whole real line. A tangent line exists at each point on the graph. The graph of f suggests that as x increases, the slopes of the tangent lines at $(x, f(x))$ decrease and approach zero from the right as $x \to \infty$.

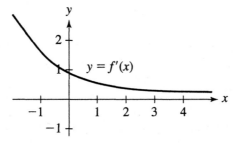

10. The derivative is defined everywhere, being positive on the whole real line. A tangent line exists at each point on the graph. The graph of f suggests that as x

increases, the slopes of the tangent lines at $(x, f(x))$ increase. If we let $x \to -\infty$, then the slopes of the corresponding tangent lines seem to approach zero from the right.

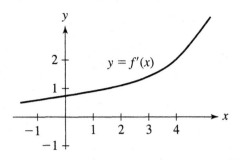

Classroom Discussion 3.2.5: Differentiability and Continuity

a. $\lim_{h \to 0} \dfrac{f(x+h) - f(x)}{h} = f'(x)$ and $\lim_{h \to 0} h = 0$.

b.

$$\lim_{h \to 0} \left[f(x+h) - f(x) \right] = \lim_{h \to 0} \frac{f(x+h) - f(x)}{h} \cdot h$$

$$= \lim_{h \to 0} \frac{f(x+h) - f(x)}{h} \cdot \lim_{h \to 0} h$$

$$= f'(x) \cdot 0 = 0.$$

Observe that the second equality holds since each term in the product has a limit as $h \to 0$.

c.

$$\lim_{h \to 0} f(x+h) = \lim_{h \to 0} \left\{ \left[f(x+h) - f(x) \right] + f(x) \right\}$$

$$= \lim_{h \to 0} [f(x+h) - f(x)] + \lim_{h \to 0} f(x)$$

$$= 0 + f(x) = f(x).$$

d. If f is differentiable at x, then $\lim_{h \to 0} f(x+h) = f(x)$, that is, f is continuous at x.

SECTION 3.3

Classroom Discussion 3.3.1: Velocity, Speed, and Acceleration

1. a. The graph of $h(t)$ is

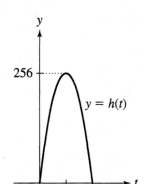

The graph does not represent the trajectory of the rocket; the rocket moves along a straight line.

b. $v(t) = (-16t^2 + 128t)' = -32t + 128$.

c. Here is the graph of $v(t)$.

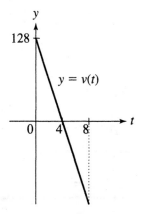

When launched, the rocket's velocity is positive; it decreases to zero—the value attained before descending; when falling, the rocket's velocity is negative since the rocket moves downward.

d. $v(2) = 64$, $v(3) = 32$, $v(5) = -32$. The rocket's velocity is positive when the rocket moves upward and negative when it moves downward.

e.

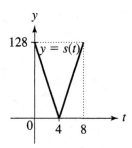

$s(2) = 64$, $s(3) = 32$, $s(5) = 32$.

f. The rocket starts falling after 4 seconds. This is the time when $v(t) = 0$. The maximum height attained by the rocket is $h(4) = 256$ feet.

g. From $h(t) = 0$, it follows that $t = 0$ or $t = 8$. The time $t = 0$ corresponds to the launching time, while $t = 8$ represents the moment when the rocket hits the ground.

h. $a(t) = (-32t + 128)' = -32$.

2. a. The graph of $h(t)$ is

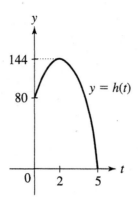

b. $v(t) = h'(t) = -32t + 64$. Here is the graph of $v(t)$.

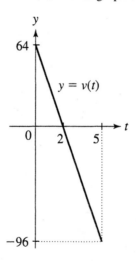

c. 64 represents the velocity at $t = 0$.

d. At the time of impact, the height is zero. The equation $h(t) = -16t^2 + 64t + 80 = 0$ has two solutions: $t = -1$ and $t = 5$. Since the rock's motion started at $t = 0$, we conclude that the rock hits the ground at $t = 5$. The velocity at the impact with the ground is $v(5) = -96$ feet per second.

e. Since $h(0) = 80$, 80 is the height from which the rock was thrown.

f. The rock has zero speed when $v(t) = 0$. This last equation implies $-32t + 64 = 0$, hence, $t = 2$. The speed is decreasing during the interval of time $[0, 2]$ and is increasing during the interval of time $[2, 5]$.

g. $a(t) = (-32t + 64)' = -32$.

SECTION 3.4

Classroom Discussion 3.4.1: Rules for Computing Derivatives

1. **Nonnegative integer powers of x.**

   a. $(c)' = 0$, $(x)' = 1$, $(x^2)' = 2x$

   b. $\displaystyle\lim_{h \to 0} \frac{(x+h)^3 - x^3}{h} = \lim_{h \to 0} \frac{(x+h-x)[(x+h)^2 + x(x+h) + x^2]}{h}$

   $$= \lim_{h \to 0} (3x^2 + 3xh + h^2) = 3x^2$$

   c. Using the hint, we can write

   $$(x+h)^4 - x^4 = [(x+h)^2 - x^2][(x+h)^2 + x^2]$$
   $$= (x+h-x)(x+h+x)[(x+h)^2 + x^2]$$
   $$= h(2x+h)(2x^2 + 2xh + h^2).$$

   Therefore,

   $$\lim_{h \to 0} \frac{(x+h)^4 - x^4}{h} = \lim_{h \to 0} (2x+h)(2x^2 + 2xh + h^2)$$

   $$= \lim_{h \to 0} (2x+h) \lim_{h \to 0} (2x^2 + 2xh + h^2) = 4x^3.$$

   d. In general, for any positive integer n, we have $(x^n)' = nx^{n-1}$.

   e.

   $$(y-x)(y^{n-1} + y^{n-2}x + \cdots + yx^{n-2} + x^{n-1})$$

   $$= y^n + y^{n-1}x + y^{n-2}x^2 + \cdots + y^2x^{n-2} + yx^{n-1}$$

   $$-y^{n-1}x - y^{n-2}x^2 - \cdots - y^2x^{n-2} - yx^{n-1} - x^n$$

   $$= y^n - x^n.$$

   Using formula (3) in Section 3.2 in combination with the previous, we have

   $$(x^n)' = \lim_{y \to x} \frac{y^n - x^n}{y - x} = \lim_{y \to x} (y^{n-1} + y^{n-2}x + \cdots + yx^{n-2} + x^{n-1})$$

   $$= ny^{n-1}.$$

2. **Constant Multiples of a Function**

   a. $(3x^2)' = 6x$ and $3(x^2)' = 3 \cdot 2x = 6x$, thus $(3x^2)' = 3(x^2)'$. Similarly, $(cx^2)' = c(x^2)'$ for any real number c.

   b. The calculations in a suggest that $(cf)' = cf'$.

   c. $\displaystyle\lim_{h \to 0} \frac{cf(x+h) - cf(x)}{h} = c \lim_{h \to 0} \frac{f(x+h) - f(x)}{h} = cf'(x)$.
   Hence, by Definition 3.2.1, the function cf is differentiable and $(cf)' = cf'$.

3. **Sums and Differences of Functions**

   a. $(x + x^2)' = 1 + 2x$ and $(x)' + (x^2)' = 1 + 2x$; thus $(x + x^2)' = (x)' + (x^2)'$.

b. $(x^2 + 3)' = 2x$ and $(x^2)' + (3)' = 2x + 0 = 2x$; thus $(x^2 + 3)' = (x^2)' + (3)'$.

c. The relationship suggested by these calculations is $(f + g)' = f' + g'$.

d. To prove that S is differentiable, we need to check that $\lim\limits_{h \to 0} \dfrac{S(x + h) - S(x)}{h}$ exists.

$$\lim_{h \to 0} \frac{S(x + h) - S(x)}{h} = \lim_{h \to 0} \frac{f(x + h) + g(x + h) - f(x) - g(x)}{h}.$$

Since f and g are both differentiable, $\lim\limits_{h \to 0} \dfrac{f(x + h) - f(x)}{h} = f'(x)$ and $\lim\limits_{h \to 0} \dfrac{g(x + h) - g(x)}{h} = g'(x)$. Hence,

$$S'(x) = \lim_{h \to 0} \frac{S(x + h) - S(x)}{h} = \lim_{h \to 0} \frac{f(x + h) + g(x + h) - f(x) - g(x)}{h}$$

$$= \lim_{h \to 0} \left[\frac{f(x + h) - f(x)}{h} + \frac{g(x + h) - g(x)}{h} \right]$$

$$= \lim_{h \to 0} \frac{f(x + h) - f(x)}{h} + \lim_{h \to 0} \frac{g(x + h) - g(x)}{h} = f'(x) + g'(x).$$

Thus, $(f + g)' = f' + g'$.

e. $(f - g)' = [f + (-1)g]' = f' + [(-1)g]' = f' + (-1)g' = f' - g'$.

f. $(2x^3 - 4x)' = 2(x^3)' - 4(x)' = 2 \cdot 3x^2 - 4 = 6x^2 - 4$
$(x^{99} + 9)' = (x^{99})' + (9)' = 99x^{98}$
$(5x^6 + 2x^4 - 4x^3)' = (5x^6)' + (2x^4 - 4x^3)' = 5(x^6)' + 2(x^4)' - 4(x^3)' = 30x^5 + 8x^3 - 12x^2$
$[(-x + 2)^2]' = (x^2 - 4x + 4)' = (x^2)' - 4(x)' + (4)' = 2x - 4$
$[(x - 1)(2x^2 + 3)]' = (2x^3 - 2x^2 + 3x - 3)' = 2(x^3)' - 2(x^2)' + 3(x)' - (3)' = 6x^2 - 4x + 3$

g. Using the rules obtained, we can write

$$f'(x) = (a_n x^n + a_{n-1} x^{n-1} + \cdots + a_1 x + a_0)'$$

$$= (a_n x^n)' + (a_{n-1} x^{n-1})' + \cdots + (a_1 x)' + (a_0)'$$

$$= a_n (x^n)' + a_{n-1} (x^{n-1})' + \cdots + a_1 (x)' + 0$$

$$= n a_n x^{n-1} + (n - 1) a_{n-1} x^{n-2} + \cdots + a_1.$$

For the second equality above we had to apply the result of Problem 3d $n - 1$ times.

4. Product of Two Functions

a. $f'(x) = x' = 1, g'(x) = (x^2)' = 2x$,
$(f \cdot g)'(x) = (x \cdot x^2)' = (x^3)' = 3x^2 \neq 1 \cdot 2x = f'(x) \cdot g'(x)$.
Hence, the formula $(f \cdot g)' = f' \cdot g'$ is not true for arbitrary f and g.

b. In order to prove that P is differentiable, we need to compute the limit
$$\lim_{h \to 0} \frac{P(x + h) - P(x)}{h} = \lim_{h \to 0} \frac{f(x + h)g(x + h) - f(x)g(x)}{h}.$$

d. If f and g are differentiable, then $\lim\limits_{h \to 0} \dfrac{f(x + h) - f(x)}{h} = f'(x)$ and $\lim\limits_{h \to 0} \dfrac{g(x + h) - g(x)}{h} = g'(x)$. In addition, g is continuous, thus $\lim\limits_{h \to 0} g(x + h) = g(x)$.

e. Combining what we have proved so far with the properties P1–P3 of limits, we have

$$\lim_{h \to 0} \frac{f(x + h)g(x + h) - f(x)g(x)}{h}$$

$$= \lim_{h \to 0} \frac{f(x + h) - f(x)}{h} g(x + h) + \lim_{h \to 0} f(x) \frac{g(x + h) - g(x)}{h}$$

$$= \lim_{h \to 0} \frac{f(x + h) - f(x)}{h} \lim_{h \to 0} g(x + h) + f(x) \lim_{h \to 0} \frac{g(x + h) - g(x)}{h}$$

$$= f'(x)g(x) + f(x)g'(x).$$

Thus, $P'(x)$ exists and is equal to $f'(x)g(x) + f(x)g'(x)$.

$(f \cdot g)' = f' \cdot g + f \cdot g'$

f. $(x^3)' = 3x^2$, $(x^6)' = 6x^5$, and $(x^3 \cdot x^6)' = (x^9)' = 9x^8$. Then, $(x^3)' \cdot x^6 + x^3 \cdot (x^6)' = 3x^2 \cdot x^6 + x^3 \cdot 6x^5 = 9x^8 = (x^3 \cdot x^6)'$.

5. The Function $\frac{1}{x}$

a. Since a tangent line to the graph $y = \frac{1}{x}$ exists at all points $(x, f(x))$ where $x \neq 0$, the derivative of f exists for all $x \neq 0$. All the slopes of the tangent lines to this graph are negative, thus the derivative is negative everywhere. If we denote by m_x the slope of the tangent line to the graph at a point $(x, f(x))$, $x \neq 0$, we observe that $m_x \to 0^-$ as $x \to -\infty$, $m_x \to -\infty$ as $x \to 0^-$, $m_x \to -\infty$ as $x \to 0^+$, and $m_x \to 0^-$ as $x \to \infty$.

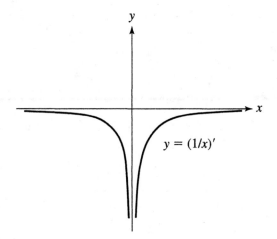

$y = (1/x)'$

b. $f(x + h) - f(x) = \frac{1}{x + h} - \frac{1}{x} = \frac{x - (x + h)}{(x + h)x} = \frac{-h}{(x + h)x}$

c. $f'(x) = \lim_{h \to 0} \frac{f(x + h) - f(x)}{h} = \lim_{h \to 0} \frac{\frac{-h}{(x + h)x}}{h} = \lim_{h \to 0} \frac{-1}{(x + h)x} = \frac{-1}{\lim_{h \to 0} (x + h)x} = -\frac{1}{x^2}$

$\left(\frac{1}{x}\right)' = -\frac{1}{x^2}$

d. The graph of $f'(x) = -\frac{1}{x^2}$ is similar to the graph we sketched in a.

6. More General Reciprocals

b. Since $f'(x)$ exists, we know that f is continuous; thus $\lim_{h \to 0} f(x + h) = f(x)$.

Consequently, $\lim_{h \to 0} \dfrac{1}{f(x + h)f(x)} = \dfrac{1}{\lim_{h \to 0} f(x + h)f(x)} = \dfrac{1}{[f(x)]^2}$.

c. By Definition 3.2.1, g is differentiable if $\lim_{h \to 0} \dfrac{g(x + h) - g(x)}{h}$ exists. Using a and the results from b, we have that

$$\lim_{h \to 0} \frac{g(x + h) - g(x)}{h} = \frac{-1}{\lim_{h \to 0} f(x + h)f(x)} \cdot \lim_{h \to 0} \frac{f(x + h) - f(x)}{h}$$

$$= \frac{-f'(x)}{[f(x)]^2},$$

thus g' exists and $g' = \dfrac{-f'}{f^2}$.

$$\left(\frac{1}{f}\right)' = \frac{-f'}{f^2}$$

d. $\left(\dfrac{1}{x^2}\right)' = -\dfrac{(x^2)'}{(x^2)^2} = -\dfrac{2x}{x^4} = -\dfrac{2}{x^3}$

$\left(\dfrac{1}{x^2 + 3}\right)' = -\dfrac{(x^2 + 3)'}{(x^2 + 3)^2} = -\dfrac{2x}{(x^2 + 3)^2}$

7. Negative Powers of x

If n is a negative integer, then $-n$ is a positive integer. According to Problem 1, $(x^{-n})'$ is differentiable and $(x^{-n})' = -nx^{-n-1}$. Using also Problem 6, we have that x^n is differentiable and that

$$(x^n)' = \left(\frac{1}{x^{-n}}\right)' = -\frac{(x^{-n})'}{(x^{-n})^2} = -\frac{-nx^{-n-1}}{x^{-2n}} = nx^{n-1}.$$

8. Quotient of Two Functions

a. Since g is differentiable, by Problem 6 we have that $\dfrac{1}{g}$ is also differentiable. According to Problem 4, the product between two differentiable functions is differentiable; thus $Q = f \cdot \dfrac{1}{g}$ is differentiable.

b. Based on Problem 6, we have that $\left(\dfrac{1}{g}\right)' = -\dfrac{g'}{g^2}$. Combining this with the formula obtained in Problem 4, we get that

$$Q' = \left(f \cdot \frac{1}{g}\right)' = f' \cdot \frac{1}{g} + f \cdot \left(\frac{1}{g}\right)' = \frac{f'}{g} - \frac{f \cdot g'}{g^2} = \frac{f' \cdot g - f \cdot g'}{g^2}.$$

$$\left(\frac{f}{g}\right)' = \frac{f'g - fg'}{g^2}$$

c. $\left(\dfrac{1}{x + 1}\right)' = \dfrac{(1)'(x + 1) - 1(x + 1)'}{(x + 1)^2} = \dfrac{0 - 1}{(x + 1)^2} = \dfrac{-1}{(x + 1)^2}$

$\left(\dfrac{x}{x^3 + 1}\right)' = \dfrac{(x)'(x^3 + 1) - x(x^3 + 1)'}{(x^3 + 1)^2} = \dfrac{x^3 + 1 - 3x^3}{(x^3 + 1)^2} = \dfrac{1 - 2x^3}{(x^3 + 1)^2}$

$\left(\dfrac{x^3 - 2x}{x^2 - 1}\right)' = \dfrac{(x^3 - 2x)'(x^2 - 1) - (x^3 - 2x)(x^2 - 1)'}{(x^2 - 1)^2}$

$\qquad = \dfrac{(3x^2 - 2)(x^2 - 1) - (x^3 - 2x)2x}{(x^2 - 1)^2} = \dfrac{3x^4 - 3x^2 - 2x^2 + 2 - 2x^4 + 4x^2}{(x^2 - 1)^2} = \dfrac{x^4 - x^2 + 2}{(x^2 - 1)^2}$

9. The Square Root of a Function

a.

$$\frac{g(x + h) - g(x)}{h} = \frac{\sqrt{f(x + h)} - \sqrt{f(x)}}{h}$$

$$= \frac{[\sqrt{f(x + h)} - \sqrt{f(x)}][\sqrt{f(x + h)} + \sqrt{f(x)}]}{h(\sqrt{f(x + h)} + \sqrt{f(x)})}$$

$$= \frac{f(x + h) - f(x)}{h} \cdot \frac{1}{\sqrt{f(x + h)} + \sqrt{f(x)}}$$

b. Being differentiable, f is also continuous, so $\lim_{h \to 0} f(x + h) = f(x)$. Consequently, using properties of limits, we have that $\lim_{h \to 0} \frac{1}{\sqrt{f(x + h)} + \sqrt{f(x)}} = \frac{1}{2\sqrt{f(x)}}$.

c. $\lim_{h \to 0} \frac{g(x + h) - g(x)}{h} = \lim_{h \to 0} \frac{f(x + h) - f(x)}{h} \cdot \lim_{h \to 0} \frac{1}{\sqrt{f(x + h)} + \sqrt{f(x)}} = f'(x) \cdot \frac{1}{2\sqrt{f(x)}}$, hence $(\sqrt{f})' = \frac{f'}{2\sqrt{f}}$.

d. $(\sqrt{x})' = \frac{(x)'}{2\sqrt{x}} = \frac{1}{2\sqrt{x}}$

$(\sqrt{x^4 - 3x^2 + 100})' = \frac{(x^4 - 3x^2 + 100)'}{2\sqrt{x^4 - 3x^2 + 100}} = \frac{4x^3 - 6x}{2\sqrt{x^4 - 3x^2 + 100}} = \frac{2x^3 - 3x}{\sqrt{x^4 - 3x^2 + 100}}$

10. Rational Powers of x

$$(x^{\frac{2}{3}})' = \frac{2}{3} x^{\frac{2}{3} - 1} = \frac{2}{3} x^{-\frac{1}{3}}$$

$$(x^{-\frac{1}{3}})' = -\frac{1}{3} x^{-\frac{1}{3} - 1} = -\frac{1}{3} x^{-\frac{4}{3}}$$

$$(x^{\frac{6}{5}})' = \frac{6}{5} x^{\frac{6}{5} - 1} = \frac{6}{5} x^{\frac{1}{5}}$$

SECTION 3.5

Classroom Discussion 3.5.1: Derivatives of Composite Functions

1. a. $k(x) = (2x + 1)^2 = 4x^2 + 4x + 1 \Rightarrow k'(x) = 8x + 4.$
 b. $f(y) = y^2.$
 c. $f'(y) = 2y, g'(x) = 2.$
 d. $f'(y) \cdot g'(x) = 2y \cdot 2 = 4y \ne k'(x)$ since $f'(y)$ depends on y, and the expression of $k'(x)$ does not contain the variable y.
 $k'(x) = f'(g(x)) \cdot g'(x)$ since $f'(g(x)) \cdot g'(x) = 2g(x) \cdot g'(x) = 2(2x + 1) \cdot 2 = 8x + 4 = k'(x).$

2. a. $(x^2 + x)^2 = x^4 + 2x^3 + x^2 \Rightarrow k'(x) = 4x^3 + 6x^2 + 2x.$
 b. $g(x) = x^2 + x, f(y) = y^2.$
 c. $f'(y) = 2y, g'(x) = 2x + 1.$
 d. $f'(g(x)) = 2g(x) = 2x^2 + 2x.$
 e. $f'(g(x)) \cdot g'(x) = (2x^2 + 2x)(2x + 1) = 4x^3 + 6x^2 + 2x.$
 f. $k'(x) = 4x^3 + 6x^2 + 2x = f'(g(x)) \cdot g'(x).$

Practice Problem

Let $f(y) = y^{100}$ and let $g(x) = x^3 - 2x + 7$. Then, $k(x) = f(g(x))$ and $k'(x) = f'(g(x)) \cdot g'(x)$. Since $f'(y) = 100y^{99}$ and $g'(x) = 3x^2 - 2$, it follows that $f'(g(x)) = 100(g(x))^{99} = 100(x^3 - 2x + 7)^{99}$, thus, $k'(x) = 100(x^3 - 2x + 7)^{99}(3x^2 - 2).$

To compute the derivative of $l(x)$, we define $f(y) = y^{\frac{2}{3}}$ and $g(x) = x^2 - 2x + 4$. Then $l(x) = f(g(x)), f'(y) = \frac{2}{3}y^{-\frac{1}{3}}, g'(x) = 2x - 2,$ and $f'(g(x)) = \frac{2}{3}(x^2 - 2x + 4)^{-\frac{1}{3}}$. Hence, by the chain rule, $l'(x) = \frac{2}{3}(x^2 - 2x + 4)^{-\frac{1}{3}}(2x - 2)$.

Classroom Discussion 3.5.2: General Power Rule

1. Let $f(x) = x^a$. Then $[g(x)]^a = f(g(x))$. Thus, by the chain rule, $\{[g(x)]^a\}' = f'(g(x)) \cdot g'(x) = a(g(x))^{a-1} \cdot g'(x)$.

2. $\sqrt{f(x)} = [f(x)]^{\frac{1}{2}} \Rightarrow [\sqrt{f(x)}]' = \frac{1}{2}[f(x)]^{\frac{1}{2}-1} \cdot f'(x) = \frac{f'(x)}{2\sqrt{f(x)}}$.

CHAPTER 4

SECTION 4.1

Classroom Discussion 4.1.1: Largest Area

1. The reasoning for answering question 5 was based on the table completed in task 4 in the Classroom Connection; it does not constitute a complete proof since not all possible rectangles with perimeter equal to 100 have been listed in the table. To show that no other rectangle with a perimeter of 100 meters will have an area larger than 625, one can reason algebraically as follows:

Take a rectangle with perimeter equal to 100, different from the square with side length equal to 25. Then, one side of this rectangle has length equal to $25 + a$ for some $a > 0$. Since the perimeter of the rectangle must be equal to 100, the length of the rectangle's other side must be $25 - a$. Then, the area of this rectangle is $(25 - a)(25 + a) = 625 - a^2$, which is a number smaller than 625 because $a^2 > 0$. It follows that among all rectangles with perimeter 100, the square with side length equal to 25 has the largest area.

2. a. $2(x + y) = 100$.

    b. $A = xy$ and $y = 50 - x \Rightarrow A(x) = x(50 - x) = 50x - x^2$. The domain of $A(x)$ is $(0, 50)$.

    c.

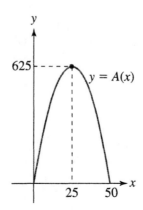

One possible value of x for which $A(x)$ seems to be the largest (obtained using a calculator's zoom and trace features) is $x = 24.876$. This is an approximate value, and it is different from the one obtained in Problem 1.

3. a. The slope of the tangent line to the graph of A at the highest point on the graph must be zero. This is equivalent to having $A'(x) = 0$ at that point.

b. $A'(x) = 50 - 2x$

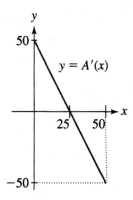

The condition $A'(x_0) = 0$ implies $x_0 = 25$.

c. $A'(x) > 0$ for $x \in (0, 25)$ and $A'(x) < 0$ for $x \in (25, 50)$. Thus, $A(x)$ is increasing on the interval $(0, 25)$ and decreasing on the interval $(25, 50)$. We can therefore conclude that $A(25) = 625$ is the largest value A attains on the interval $(0, 50)$. This is a complete proof of the fact that among all rectangles with perimeter 100, the one with the largest area is the square with side length equal to 25 with area 625.

Practice Problem

$f'(x) = (2x^2 - 3x + 1)' = 4x - 3$ and $f'(x) = 0$ implies $x = \frac{3}{4}$. Since $\frac{3}{4} \in \text{Domain}(f)$, $x = \frac{3}{4}$ is the critical point for f.

Classroom Discussion 4.1.2: Smallest Perimeter

1. The reasoning used in answering question 9 was based on a table containing examples of rectangles with area equal to 400. It does not constitute a complete proof since not all possible rectangles with area equal to 400 were listed in the table. To show that no other rectangle with area of 400 meters square will have a perimeter smaller than 80, one can reason by contradiction.

Suppose there exists a rectangle with area equal to 400, different from the square with side length equal to 20, whose perimeter is smaller than 80. Then, one side of this rectangle has length equal to $20 + a$ for some $a > 0$. Since there area of the rectangle must be equal to 400, it follows that the length of the rectangle's other side is $\frac{400}{20 + a}$. The perimeter of this rectangle is $2(20 + a + \frac{400}{20 + a})$. The assumption that the perimeter is smaller than 80 means that $2(20 + a + \frac{400}{20 + a}) < 80 \Rightarrow \frac{400}{20 + a} < 20 - a \Rightarrow 400 < 400 - a^2 \Rightarrow 0 < -a^2$. The latter is false. Hence, our assumption is false. It follows that, amongst all rectangles with area equal to 400, the square with side length equal to 20 has the smallest perimeter.

2. a. $xy = 400$.

b. $P = 2x + 2y$ and $y = \frac{400}{x} \Rightarrow P(x) = 2x + \frac{800}{x}$. The domain of $P(x)$ is $(0, \infty)$.

c.

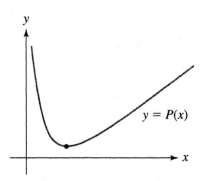

Using a calculator's zoom and trace features, we have obtained $x = 19.7418$ as the value for which $P(x)$ seems to be the smallest. This is an approximate value, and it is different from the one obtained in Problem 1.

3. a. $P'(x) = 2 - \frac{800}{x^2}$ and $P'(x) = 0 \Rightarrow x^2 = 400 \Rightarrow x = \pm 20$. Since -20 is not in the domain of P, it follows that $x_0 = 20$ is the only critical point of P.

b. $P'(x) < 0$ on $(0, 20)$ and $P'(x) > 0$ on $(20, \infty)$.

c. P is decreasing on the interval $(0, 20)$ and increasing on the interval $(20, \infty)$.

d. The smallest value for P is $P(20) = 80$. This is equal to the solution we obtained in Problem 1. The answer in Problem 2 is an approximate one.

Practice Problem

1. a. We follow the steps previously outlined.

Step 1. Let x and y be the length and width of an arbitrary rectangle with fixed perimeter P. We need to optimize the area function $A = xy$.

Step 2. The constraint is $2x + 2y = P$. Hence, $y = \frac{P}{2} - x$ and $A(x) = x\left(\frac{P}{2} - x\right)$.

Step 3. The domain of A is $(0, \frac{P}{2})$.

Step 4. $A'(x) = \frac{P}{2} - 2x$ and $A'(x) = 0 \Rightarrow x = \frac{P}{4}$.

Step 5. $A'(x) > 0$ for $x \in (0, \frac{P}{4})$ and $A'(x) < 0$ for $x \in (\frac{P}{4}, \frac{P}{2})$. Hence, A is increasing on the interval $(0, \frac{P}{4})$ and decreasing on the interval $(\frac{P}{4}, \frac{P}{2})$.

Step 6. Putting all the information in a table, we have

x	0		P/4		P/2
$A'(x)$		$+$	$A'(P/4) = 0$	$-$	
A		↗	$A(P/4)$ is the maximum value of A	↘	

Hence, A attains its maximum at $x = \frac{P}{4}$, which is equal to $\frac{P^2}{16}$.

Step 7. We have just proved that among all the rectangles with fixed perimeter P, the one with the largest area is the square with side length $\frac{P}{4}$.

b. Step 1. Let x and y be the length and width of an arbitrary rectangle with fixed area A. We need to optimize the perimeter $P = 2x + 2y$.

Step 2. The constraint is $xy = A$. Hence, $y = \frac{A}{x}$ and $P(x) = 2x + \frac{2A}{x}$.

Step 3. The domain of P is $(0, \infty)$.

Step 4. $P'(x) = 2 - \frac{2A}{x^2}$ and $P'(x) = 0 \Rightarrow x = \sqrt{A}$.

Step 5. $P'(x) < 0$ for $x \in (0, \sqrt{A})$ and $P'(x) > 0$ for $x \in (\sqrt{A}, \infty)$. Hence, P is decreasing on the interval $(0, \sqrt{A})$ and increasing on the interval $(\sqrt{A}, \infty)$.

Step 6. Putting all the information in a table, we have

x	0	$\sqrt{A}$	∞
$P'(x)$		$-$	$P'(\sqrt{A}) = 0$	$+$
P		$\searrow$	$P(\sqrt{A})$ is the minimum value of P	$\nearrow$

Hence, P attains its minimum at $x = \sqrt{A}$, which is equal to $4\sqrt{A}$.

Step 7. Among all the rectangles with fixed area A, the one with the smallest perimeter is the square with side length $\sqrt{A}$.

2. For an algebraic proof of the identity $\frac{P^2}{4} = (x - y)^2 + 4A$, we substitute $P = 2(x + y)$ and $A = xy$ in this identity to get $(x + y)^2 = (x - y)^2 + 4xy$, which is true for any x and y.

To prove Problem 1a, suppose x and y are the side lengths of a rectangle with fixed perimeter P. Then, using the identity we just proved, it follows that $(x - y)^2 + 4A$ has a fixed value. Under this condition, since $(x - y)^2 \geq 0$, the largest possible value for A is obtained when $(x - y)^2 = 0$. The latter is true if and only if $x = y$, when the rectangle is a square of side length $\frac{P}{4}$.

To prove Problem 1b, suppose x and y are the side lengths of a rectangle with fixed area A. Then, using the identity we proved, the rectangle's perimeter is equal to $(x - y)^2 + 4A$. Since A is fixed, the smallest value this last expression takes is $4A$ corresponding to the case when $x = y$. If $x = y$, the rectangle is a square of side length $\sqrt{A}$.

Classroom Discussion 4.1.3: The Behavior of Quadratic Functions

1. The case $f(x) = 2x^2 + 4x + 1$
   a. Algebraic approach

   i. The computation is called completing the square because all the terms involving x are used to write a complete square.

   ii. $2(x + 1)^2 \geq 0$ for all real values x; thus the smallest value $2(x + 1)^2$ takes is zero, which corresponds to $x = -1$.

   iii. Since the smallest value of $2(x + 1)^2$ is 0, we can conclude that $2(x + 1)^2 - 1 \geq -1$. Consequently, $f(x) \geq -1$ for all $x \in \mathbb{R}$ and $f(x) = -1$ when $x = -1$, the value corresponding to $2(x + 1)^2 = 0$.

   iv. $f(x)$ can be as large as we want since $2(x + 1)^2$ can take arbitrarily large values.

b. Calculus approach

x	$-\infty$	-1	∞
$f'(x)$	$-$	$f'(-1) = 0$	$+$
$f(x)$	↘	$f(-1) = -1$ is the minimum value of f	↗

2. The case $f(x) = -3x^2 + 6x - 2$

a. Algebraic approach

i. $-3x^2 + 6x - 2 = -3(x - 1)^2 + 1$.

ii. Since $-3(x - 1)^2 \leq 0$ for all $x \in \mathbb{R}$, it follows that $f(x) \leq 1$, and the value 1 is attained for $x = 1$.

b. Calculus approach

x	$-\infty$	1	∞
$f'(x)$	$+$	$f'(1) = 0$	$-$
$f(x)$	↗	$f(1) = 1$ is the minimum value of f	↘

3. The general case $f(x) = ax^2 + bx + c$

a. Algebraic approach

i. $ax^2 + bx + c = a\left(x^2 + 2 \cdot \dfrac{b}{2a}x + \dfrac{b^2}{4a^2}\right) - \dfrac{b^2}{2a} + \dfrac{c}{a}$

$= a\left(x + \dfrac{b}{2a}\right)^2 - \dfrac{b^2 - 4ac}{4a}$

ii. $\left(x + \dfrac{b}{2a}\right)^2 \geq 0$ for all $x \in \mathbb{R}$.

iii. Since $f(x) = a\left(x + \dfrac{b}{2a}\right)^2 - \dfrac{b^2 - 4ac}{4a}$, we conclude that $f(x) \geq -\dfrac{b^2 - 4ac}{4a}$ for $a > 0$ and that $f(x) \leq -\dfrac{b^2 - 4ac}{4a}$ for $a < 0$. As such, for any value of x on the real line, $f(x)$ has a minimum if $a > 0$ and a maximum if $a < 0$.

iv. The corresponding minimum or maximum is attained at $x = -\dfrac{b}{2a}$ and $f\left(-\dfrac{b}{2a}\right) = -\dfrac{b^2 - 4ac}{4a}$.

b. Calculus approach

The case a > 0:

x	$-\infty$	$-b/(2a)$	∞
$f'(x)$	$-$	$f'(-b/(2a)) = 0$	$+$
$f(x)$	↘	$f(-b/(2a))$ is the minimum value of f	↗

The case a < 0:

x	$-\infty$		$-b/(2a)$		∞
$f'(x)$		$+$	$f'(-b/(2a)) = 0$	$-$	
$f(x)$		$\nearrow$	$f(-b/(2a))$ is the maximum value of f	$\searrow$	

Classroom Discussion 4.1.4: Local Maximum/Minimum versus Absolute Maximum/Minimum

1. $f'(x) = 12x^3 - 12x^2 - 24x$
   $f'(x) = 0 \Rightarrow 12x(x^2 - x - 2) = 0 \Rightarrow 12x(x + 1)(x - 2) = 0 \Rightarrow x = 0, -1, 2$
   are the critical points of f (observe that they all belong to the domain of f). The subintervals determined by the critical points are $(-2, -1)$, $(-1, 0)$, $(0, 2)$, and $(2, 3)$.
2. $f' > 0$ on $(-1, 0) \cup (2, 3)$ and $f' < 0$ on $(-2, -1) \cup (0, 2)$.
   If the sign of f' would change within one of these subintervals then, based on the continuity of f', there would be some point within that interval where f' equals zero. This would yield an additional critical point, which is not possible.
3. f is increasing on $(-1, 0)$ and on $(2, 3)$, f is decreasing on $(-2, -1)$ and on $(0, 2)$.
4. Not all critical points yield local minima or local maxima. The function $f(x) = x^3$ has one critical point, $x_0 = 0$. The value of f at 0 is neither a local maximum nor a local minimum since, as seen from the following table, f is increasing on both $(-\infty, 0)$ and $(0, \infty)$.

x	$-\infty$		0		∞
$f'(x)$		$+$	0	$+$	
$f(x)$		$\nearrow$	0	$\nearrow$	

SECTION 4.2

Classroom Discussion 4.2.1: Concavity and the Second Derivative

1. f' is nondecreasing on (a, b) if and only if $f'' \geq 0$ on (a, b).
   f' is nonincreasing on (a, b) if and only if $f'' \leq 0$ on (a, b).
2.

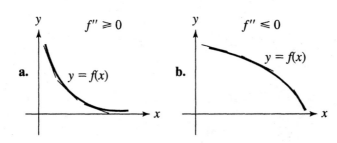

a. The slopes of the tangents lines to the graph $y = f(x)$ at points $(x, f(x))$ are nonpositive, and they increase as x increases. Thus, f' is nondecreasing.

b. The slopes of the tangents lines to the graph $y = f(x)$ at points $(x, f(x))$ are nonpositive, and they decrease as x increases. Thus, f' is nonincreasing.

c.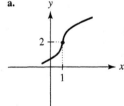

d.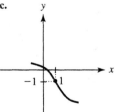

c. The slopes of the tangents lines to the graph $y = f(x)$ at points $(x, f(x))$ are nonnegative, and they decrease as x increases. Thus, f' is nonincreasing.

d. The slopes of the tangents lines to the graph $y = f(x)$ at points $(x, f(x))$ are nonnegative, and they increase as x increases. Thus, f' is nondecreasing.

3. $f'' \geq 0$ for the graphs in a and d since, for both graphs, f' is nondecreasing.

4. $f'' \leq 0$ for the graphs in b and c since, for both graphs, f' is nonincreasing.

Classroom Discussion 4.2.2: Inflection Points

a. The graph is concave up on $(-2, 0)$ and concave down on $(-\infty, -2) \cup (0, \infty)$.

b. The graph changes concavity at $x = -2$ and $x = 0$.

c. f'' must equal zero at an inflection point since f'' changes its sign at that point.

d. The equality $f''(x_0) = 0$ is not a sufficient condition for x_0 to be an inflection point. This can be seen by considering the function $f(x) = x^4$. It is immediate that $f''(x) = 12x^2$ and $f''(0) = 0$. In addition, since $f''(x) > 0$ for all $x \neq 0$, the function f is concave up to the left and right of 0. Hence, f does not change concavity at $x = 0$, so 0 is not an inflection point for f.

Classroom Discussion 4.2.3: Sketching Graphs

1. **a.**   **b.** 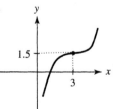 **c.**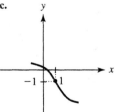

d. 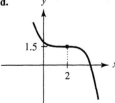 **e.** 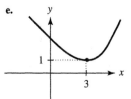 **f.**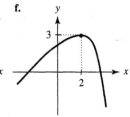

2. Limits at Infinity

a. $p(x) = 1$ for all real numbers x. As x gets larger and larger, the values of $p(x)$ stay equal to 1. This is why we can write $\lim\limits_{x \to \infty} p(x) = 1$. Similarly, $\lim\limits_{x \to -\infty} p(x) = 1$.

b. $\lim\limits_{x \to \infty} p(x) = c$, $\lim\limits_{x \to -\infty} p(x) = c$.

c. As x increases without bound, $p(x) = x$ increases without bound. The notation $\lim\limits_{x \to \infty} x = \infty$ is consistent with the notation we used in Definition 2.2.1 of limits, since informally we can view a function that increases without bound as "approaching" ∞. Moreover, this notation is consistent with the notation used earlier for sequences, where we saw that $\lim\limits_{n \to \infty} n = \infty$.

$$\lim\limits_{x \to -\infty} x = -\infty$$

d. $\lim\limits_{x \to \infty} x^2 = \infty$, $\lim\limits_{x \to -\infty} x^2 = \infty$. As x increases without bound, x^2 increases without bound. When x decreases without bound, x takes negative values that are larger and larger in absolute value. Consequently, x^2 takes larger and larger values, increasing without bound.

e. $\lim\limits_{x \to \infty} x^3 = \infty$, $\lim\limits_{x \to -\infty} x^3 = -\infty$. As x increases without bound, x^3 increases without bound. When x decreases without bound, x^3 decreases without bound.

f.

$$\lim\limits_{x \to \infty} x^n = \infty, \qquad \lim\limits_{x \to -\infty} x^n = \begin{cases} \infty & \text{if } n \text{ is even,} \\ -\infty & \text{if } n \text{ is odd.} \end{cases}$$

If $x \to \infty$, we also have $x^n \to \infty$.

If $x < 0$ and n is even, x^n is positive, so in this case, if $x \to -\infty$, then $x^n \to \infty$. On the other hand, if $x < 0$ and n is odd, then x^n is negative, and in this case, if $x \to -\infty$, then $x^n \to -\infty$.

g. As $x \to \infty$, the values of $\frac{1}{x}$ get closer and closer to zero. In fact, we can make $\frac{1}{x}$ as close to 0 as we want by selecting x large enough. This is why $\lim\limits_{x \to \infty} \frac{1}{x} = 0$.

As $x \to -\infty$, the values of $\frac{1}{x}$ that are negative get closer and closer to 0. In fact, we can make $\frac{1}{x}$ as close to 0 as we want by selecting $x < 0$, with $|x|$ large enough. This is why $\lim\limits_{x \to -\infty} \frac{1}{x} = 0$.

h. $\lim\limits_{x \to \infty} \frac{1}{x^2} = 0$, $\lim\limits_{x \to -\infty} \frac{1}{x^2} = 0$

i. $\lim\limits_{x \to \infty} \frac{1}{x^n} = 0$, $\lim\limits_{x \to -\infty} \frac{1}{x^n} = 0$

j. $\lim\limits_{x \to \infty} (x^2 - 2x - 3) = \infty$, $\lim\limits_{x \to -\infty} (x^2 - 2x - 3) = \infty$

As $x \to \infty$ or $x \to -\infty$, $\frac{1}{x} \to 0$ and $\frac{1}{x^2} \to 0$. Consequently, $1 - 2 \cdot \frac{1}{x} - 3 \cdot \frac{1}{x^2} \to 1$ as $x \to \infty$. On the other hand, $x^2 \to \infty$ as $x \to \infty$ or $x \to -\infty$. Since (for $x \neq 0$) $x^2 - 2x - 3 = x^2 \left(1 - 2 \cdot \frac{1}{x} - 3 \cdot \frac{1}{x^2} \right)$, we see that $x^2 - 2x - 3 \to \infty$ as $x \to \infty$ or $x \to -\infty$.

k. For $x \neq 0$ we have

$$5x^3 + 2x^2 - 4x + 12 = x^3 \left(5 + 2 \cdot \frac{1}{x} - 4 \cdot \frac{1}{x^2} + 12 \cdot \frac{1}{x^3} \right).$$

Using this identity together with

$$\lim_{x \to \infty} x^3 = \infty \quad \text{and} \quad \lim_{x \to \infty} \left(5 + 2 \cdot \frac{1}{x} - 4 \cdot \frac{1}{x^2} + 12 \cdot \frac{1}{x^3} \right) = 5,$$

we conclude that $\lim_{x \to \infty} (5x^3 + 2x^2 - 4x + 12) = \infty$.

Similarly, since

$$\lim_{x \to -\infty} x^3 = -\infty \quad \text{and} \quad \lim_{x \to -\infty} \left(5 + 2 \cdot \frac{1}{x} - 4 \cdot \frac{1}{x^2} + 12 \cdot \frac{1}{x^3} \right) = 5,$$

we see that $\lim_{x \to -\infty} (5x^3 + 2x^2 - 4x + 12) = -\infty$. The identity

$$-5x^3 + 2x^2 - 4x + 12 = x^3 \left(-5 + 2 \cdot \frac{1}{x} - 4 \cdot \frac{1}{x^2} + 12 \cdot \frac{1}{x^3} \right),$$

together with

$$\lim_{x \to \infty} x^3 = \infty \quad \text{and} \quad \lim_{x \to \infty} \left(-5 + 2 \cdot \frac{1}{x} - 4 \cdot \frac{1}{x^2} + 12 \cdot \frac{1}{x^3} \right) = -5,$$

yields $\lim_{x \to \infty} (-5x^3 + 2x^2 - 4x + 12) = -\infty$. Likewise,

$$\lim_{x \to -\infty} x^3 = -\infty \quad \text{and} \quad \lim_{x \to -\infty} \left(-5 + 2 \cdot \frac{1}{x} - 4 \cdot \frac{1}{x^2} + 12 \cdot \frac{1}{x^3} \right) = -5$$

give that $\lim_{x \to -\infty} (-5x^3 + 2x^2 - 4x + 12) = \infty$.

The next four limits are treated in the same way. For $x \neq 0$ we have that $3x^4 + x - 2 = x^4 \left(3 + \frac{1}{x^3} - \frac{2}{x^4} \right)$. Moreover,

$$\lim_{x \to \infty} \left(3 + \frac{1}{x^3} - \frac{2}{x^4} \right) = 3, \quad \lim_{x \to -\infty} \left(3 + \frac{1}{x^3} - \frac{2}{x^4} \right) = 3,$$

while $\lim_{x \to \infty} x^4 = \infty$ and $\lim_{x \to -\infty} x^4 = \infty$. Hence, $\lim_{x \to \infty} (3x^4 + x - 2) = \infty$ and $\lim_{x \to -\infty} (3x^4 + x - 2) = \infty$.

For $x \neq 0$ we have that $-3x^4 + x - 2 = x^4 \left(-3 + \frac{1}{x^3} - \frac{2}{x^4} \right)$, and

$$\lim_{x \to \infty} \left(-3 + \frac{1}{x^3} - \frac{2}{x^4} \right) = -3, \quad \lim_{x \to -\infty} \left(-3 + \frac{1}{x^3} - \frac{2}{x^4} \right) = -3.$$

Consequently, $\lim_{x \to \infty} (-3x^4 + x - 2) = -\infty$ and $\lim_{x \to -\infty} (-3x^4 + x - 2) = -\infty$.

l. First we write the given polynomial as ($x \neq 0$)

$$a_n x^n + a_{n-1} x^{n-1} + \cdots + a_1 x + a_0$$
$$= x^n \left(a_n + a_{n-1} \cdot \frac{1}{x} + \cdots + a_1 \cdot \frac{1}{x^{n-1}} + a_0 \cdot \frac{1}{x^n} \right).$$

Since

$$\lim_{x \to \infty} x^n = \infty, \quad \lim_{x \to -\infty} x^n = \begin{cases} \infty & \text{if } n \text{ is even,} \\ -\infty & \text{if } n \text{ is odd,} \end{cases}$$

$$\lim_{x \to \infty} \left(a_n + a_{n-1} \cdot \frac{1}{x} + \cdots + a_1 \cdot \frac{1}{x^{n-1}} + a_0 \cdot \frac{1}{x^n} \right) = a_n,$$

and

$$\lim_{x \to -\infty} \left(a_n + a_{n-1} \cdot \frac{1}{x} + \cdots + a_1 \cdot \frac{1}{x^{n-1}} + a_0 \cdot \frac{1}{x^n} \right) = a_n,$$

we see that

$$\lim_{x \to \infty} (a_n x^n + a_{n-1} x^{n-1} + \cdots + a_1 x + a_0) = \begin{cases} \infty, & \text{if } a_n > 0, \\ -\infty, & \text{if } a_n < 0, \end{cases}$$

while

$$\lim_{x \to -\infty} (a_n x^n + a_{n-1} x^{n-1} + \cdots + a_1 x + a_0)$$

$$= \begin{cases} \infty & \text{if } a_n > 0 \text{ and } n \text{ is even,} \\ -\infty & \text{if } a_n < 0 \text{ and } n \text{ is even,} \\ -\infty & \text{if } a_n > 0 \text{ and } n \text{ is odd,} \\ \infty & \text{if } a_n < 0 \text{ and } n \text{ is odd.} \end{cases}$$

3. **a.** $f'(x) = 5x^4 - 20x^3$ and the critical points are $x = 0$ and $x = 4$; $f' > 0$ on $(-\infty, 0)$ and on $(4, \infty)$, while $f' < 0$ on $(0, 4)$.
   b. f is increasing on $(-\infty, 0) \cup (4, \infty)$ and decreasing on $(0, 4)$.
   c. $f''(x) = 20x^3 - 60x^2, f''(x) = 0 \Rightarrow x = 0$ or $x = 3$.
   d. $f'' \le 0$ on $(-\infty, 3)$ and $f'' > 0$ on $(3, \infty)$. Consequently, the graph of f is concave down on $(-\infty, 3)$ and concave up on $(3, \infty)$; $x = 3$ is the only inflection point and $f(3) = -162$.
   e. $f(x) = 0 \Rightarrow x = 0$ or $x = 5$, thus the graph of f crosses the x-axis at the points $(0, 0)$ and $(5, 0)$.
   f. $\lim_{x \to \infty} f(x) = \lim_{x \to \infty} x^5 = \infty, \lim_{x \to -\infty} f(x) = \lim_{x \to -\infty} x^5 = -\infty.$
   g.

x	$-\infty$	0		3		4		∞
$f'(x)$		+	0	−		−	0	+
f	$-\infty$ ↗	0	↘	-162	↘	-256	↗	∞
$f''(x)$		−	0	−		0	+	+
f		⌢		⌢		⌣		⌣

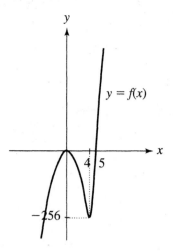

h. The function does not have an absolute maximum since we can make $f(x)$ as large as we want by choosing x large enough.

i. The function does not have an absolute minimum since we can make $f(x) < 0$ as large as we want in absolute value by choosing $x < 0$ large enough in absolute value.

j.

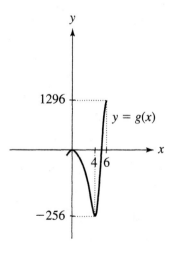

$g(4) = -256$ is the absolute minimum of g, and $g(6) = 1{,}296$ is the absolute maximum of g.

SECTION 4.3

Classroom Discussion 4.3.1: Irrational Exponents

1. $y_n < \pi < z_n$ for each n.

2. The sequence $\{y_n\}_n$ is nondecreasing and bounded; thus, by Weierstrass's theorem, $\{y_n\}_n$ is convergent. The sequence $\{z_n\}_n$ is nonincreasing and bounded; hence, by Weierstrass's theorem, it is convergent. In addition, $0 < \pi - y_n < \frac{1}{10^n}$, $0 < z_n - y_n < \frac{1}{10^n - 1}$, so applying the sandwich theorem for sequences (P6 in Section 1.2) we obtain that $\lim_{n\to\infty} y_n = \pi$ and $\lim_{n\to\infty} z_n = \pi$.

3. The values of 2^{y_n} increase as $n \to \infty$ but stay bounded; $|2^{y_n}| < 2^4$ for all n.

4. By Weierstrass's theorem, $\{2^{y_n}\}_n$ is convergent.

5. The values of 2^{z_n} decrease as $n \to \infty$ but stay bounded; $|2^{z_n}| < 2^4$ for all n.

6. By Weierstrass's theorem, $\{2^{z_n}\}_n$ is convergent.

7. Since $z_n = y_n + \frac{1}{10^n}$, it follows that $2^{z_n} - 2^{y_n} = 2^{y_n}(2^{\frac{1}{10^n}} - 1)$. If we take $a = 2$, then $\lim_{n\to\infty} \sqrt[n]{2} = 1$. Thus, the sequence $\{2^{\frac{1}{m}}\}_m$ is convergent to 1. If we consider the elements of this sequence corresponding to $m = 10^n$, we can also conclude that $\lim_{n\to\infty} 2^{\frac{1}{10^n}} = 1$; furthermore,

$$\lim_{n\to\infty} 2^{y_n}(2^{\frac{1}{10^n}} - 1) = \lim_{n\to\infty} 2^{y_n} \lim_{n\to\infty} (2^{\frac{1}{10^n}} - 1) = 0.$$

Hence, $\lim_{n\to\infty} (2^{z_n} - 2^{y_n}) = 0$.

8. It suggests the following definition: $2^\pi = \lim_{n\to\infty} 2^{y_n}$ or $2^\pi = \lim_{n\to\infty} 2^{z_n}$.

9. We can define $2^x = \lim_{n\to\infty} 2^{y_n}$. From the properties of $\{z_n\}_n$, we also have $2^x = \lim_{n\to\infty} 2^{z_n}$.

10. For an irrational number x, let $\{y_n\}_n$ be as in Problem 8. Then we can define $b^x = \lim_{n\to\infty} b^{y_n}$.

Classroom Discussion 4.3.2: Exponential versus Linear

1. $g(3) - g(2) = (3m + n) - (2m + n) = m$, $g(70) - g(69) = m$. We observe that the two expressions are equal.

2. $g(x + 1) - g(x) = [m(x + 1) + n] - (mx + n) = m$

3. $\frac{f(5)}{f(4)} = \frac{b^5}{b^4} = b$, $\frac{f(-5)}{f(-6)} = \frac{b^{-5}}{b^{-6}} = b$. We observe that the two expressions are equal.

4. $\frac{f(x + 1)}{f(x)} = \frac{b^{x+1}}{b^x} = b$

5. For a linear function f of slope m, increasing the input by 1 leads to a change of m units in the output $f(x + 1) = f(x) + m$. On the other hand, for an exponential function g with base b, increasing the input by 1 results in a multiplication by b of the output $g(x + 1) = g(x) \cdot b$.

Classroom Discussion 4.3.3: Graphs of Exponential Functions

1.

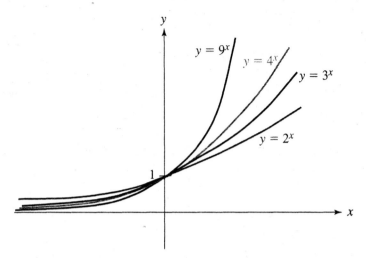

All the graphs have the same shape: they are graphs of increasing functions, they are concave up, they pass through the point $(0,1)$, they are getting closer and closer to the x-axis as $x \rightarrow -\infty$, and they increase without bound as $x \rightarrow \infty$. However, the rates at which the graphs approach the x-axis as $x \rightarrow -\infty$ and the rates at which they increase as $x \rightarrow \infty$ are different.

2.

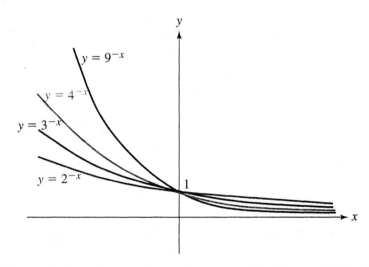

All the graphs have the same shape: they are graphs of decreasing functions, they are concave up, they pass through the point $(0,1)$, they are getting closer and closer to the x-axis as $x \rightarrow \infty$, and they increase without bound as $x \rightarrow -\infty$. However, the rates at which the graphs approach the x-axis as $x \rightarrow \infty$ and the rates at which they increase as $x \rightarrow -\infty$ are different.

3.

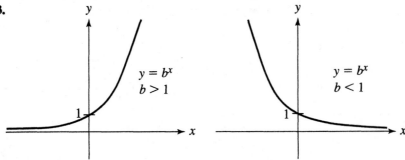

Classroom Discussion 4.3.4: Derivatives of Exponential Functions

1. a. $f'(x)$ is positive for all values of x.

  b. $f'(x) \to 0$ as $x \to -\infty$ and $f'(x) \to \infty$ as $x \to \infty$.

  c.

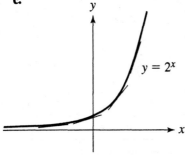 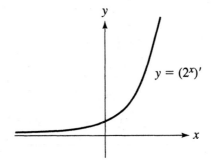

  d. They have similar shapes.

  e. $\dfrac{f(x + h) - f(x)}{h} = \dfrac{2^{x+h} - 2^{x}}{h}$

  g.

h	$\dfrac{2^{h} - 1}{h}$
-0.010	0.69075
-0.009	0.69099
-0.008	0.69123
-0.007	0.69147
-0.006	0.69171
-0.005	0.69195
-0.004	0.69219
-0.003	0.69243
-0.002	0.69267
-0.001	0.69291

h	$\dfrac{2^{h} - 1}{h}$
0.001	0.69339
0.002	0.69363
0.003	0.69387
0.004	0.69411
0.005	0.69435
0.006	0.69459
0.007	0.69483
0.008	0.69507
0.009	0.69531
0.010	0.69556

From the table it appears that $\lim\limits_{h \to 0} \frac{2^h - 1}{h}$ exists, and an approximate value for it is 0.69.

h. We can conclude that $f'(x) = c\, 2^x$ for $x \in \mathbb{R}$, where c is a real constant approximately equal to 0.69.

2. A sketch of the graph of g together with a few tangent lines and a sketch of the graph of g':

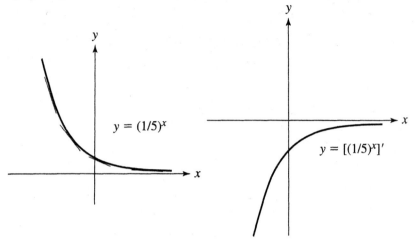

$g'(x)$ is negative for all values of x.

$g'(x) \to -\infty$ as $x \to -\infty$ and $g'(x) \to 0$ as $x \to \infty$.

The graphs of g and g' have the same shape.

$$\frac{g(x + h) - g(x)}{h} = \frac{(\frac{1}{5})^{x+h} - (\frac{1}{5})^x}{h} = (\frac{1}{5})^x \frac{(\frac{1}{5})^h - 1}{h}$$

h	$\dfrac{\left(\frac{1}{5}\right)^h - 1}{h}$
-0.010	-1.622
-0.009	-1.621
-0.008	-1.62
-0.007	-1.619
-0.006	-1.617
-0.005	-1.616
-0.004	-1.615
-0.003	-1.613
-0.002	-1.612
-0.001	-1.611

h	$\dfrac{\left(\frac{1}{5}\right)^h - 1}{h}$
0.001	-1.608
0.002	-1.607
0.003	-1.606
0.004	-1.604
0.005	-1.603
0.006	-1.602
0.007	-1.6
0.008	-1.599
0.009	-1.598
0.010	-1.597

From the table of values for $\frac{(\frac{1}{5})^h - 1}{h}$, with $h \neq 0$ varying from -0.01 to 0.01

and of step size $\Delta h = 0.001$, it appears that $\lim\limits_{h \to 0} \dfrac{\left(\frac{1}{5}\right)^h - 1}{h}$ exists; an approximate value for it is -1.6.

We can conclude that $g'(x) = c\left(\frac{1}{5}\right)^x$ for $x \in \mathbb{R}$, where c is a real constant approximately equal to -1.6.

3. a. $F'(x) = \lim\limits_{h \to 0} \dfrac{F(x + h) - F(x)}{h} = \lim\limits_{h \to 0} \dfrac{b^{x+h} - b^x}{h} = b^x \lim\limits_{h \to 0} \dfrac{b^h - 1}{h}$.

b. In order to estimate the values of k, we need to generate a table of values for $\frac{b^h - 1}{h}$ for each b as specified and for $h \neq 0$ varying from -0.01 to 0.01 with step size $\Delta h = 0.001$. If we want better approximations for $\lim\limits_{h \to 0} \dfrac{b^h - 1}{h}$, we need to decrease Δh. We observe that for $b > 1$, the value of k is positive, while for $0 < b < 1$, the value of k is negative. Since $\lim\limits_{h \to 0} \dfrac{1^h - 1}{h} = \lim\limits_{h \to 0} 0 = 0$, we have that $k = 0$ for $b = 1$. Based on the tables generated, we obtain that

$b = 0.2 \Rightarrow k \approx -1.61$,
$b = 0.6 \Rightarrow k \approx -0.51$,
$b = 1 \Rightarrow k = 0$,
$b = 1.5 \Rightarrow k \approx 0.40$,
$b = 2 \Rightarrow k \approx 0.69$,
$b = 5 \Rightarrow k \approx 1.60$.

As b increases, the values of k also increase.

base b	constant k
$0 < b < 1$	$k < 0$
$b = 1$	$k = 0$
$b > 1$	$k > 0$

c. Yes, it is: for $b > 1$ the slopes of the tangents to the graph $y = b^x$ are all positive; for $b < 1$, the slopes of the tangents to the graph $y = b^x$ are all negative; while for $b = 1$, we have $1^x = 1$ and $(1)' = 0$.

d. $(e^{-x})' = \left(\frac{1}{e^x}\right)' = -\dfrac{(e^x)'}{(e^x)^2} = -\dfrac{e^x}{e^{2x}} = -\dfrac{1}{e^x} = -e^{-x}$

e. Let $g(y) = e^y$. Then, $e^{f(x)} = g(f(x))$. By the chain rule we have $(g(f(x)))' = g'(f(x))f'(x)$. Hence, $f'(x) = e^{f(x)}f'(x)$.

CHAPTER 5

SECTION 5.1

Classroom Discussion 5.1.1: Antiderivatives

1. $F(x) = x^2$

2. Yes. For example, $x^2 + 1, x^2 - 2$, and $x^2 + \sqrt{7}$ since $(x^2 + 1)' = (x^2 - 2)' = (x^2 + \sqrt{7})' = 2x$.

3. Any two antiderivatives differ by a constant. Indeed, $(x^2 + 1) - x^2 = 1$, $(x^2 - 2) - x^2 = -2$, and $(x^2 + \sqrt{7}) - x^2 = \sqrt{7}$.

4. the derivative, differentiation

5. an antiderivative, antidifferentiation

6. Let $G(x)$ be an arbitrary function satisfying $G'(x) = 2x$ for all real numbers. Then, $G'(x) = (x^2)'$ and thus $G'(x) - (x^2)' = 0$. Using the difference rule for differentiation, you see that $[G(x) - x^2]' = 0$ for all real numbers. This means that the function $G(x) - x^2$ is constant on the real line; that is, $G(x) - x^2 = C$ for some numerical constant C independent of x. Therefore, $G(x) = x^2 + C$ for all real numbers.

7. Since $G'(x) = 2x$, the function G must be of the form $G(x) = x^2 + C$ for some constant C. Since $G(1) = 20$, this constant C must satisfy $1 + C = 20$. Thus, $C = 19$. Therefore, the function $G(x) = x^2 + 19$ is the only function satisfying the two requirements.

Classroom Discussion 5.1.2: Graphs of Antiderivatives

1. Since $F' = f$, the function f takes nonnegative values on the intervals where F is increasing and nonpositive values on the intervals where F is decreasing.

2. Since $F'' = f'$, the function f' takes nonnegative values on the intervals where F is concave up and nonpositive values on the intervals where F is concave down. Therefore, the function f is increasing on the intervals where F is concave up and decreasing on the intervals where F is concave down.

3. Every antiderivative of f is of the form $F + C$ for some constant C, and the graph of $F + C$ can be obtained by shifting the graph of F vertically with constant C.

4.

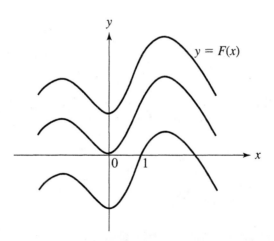

5. Let G be such that $G'(x) = g(x)$. An analysis of the graph of g leads to the information summarized in the following table.

x		-3		-2		-1		0		1	
$G' = g$	$-$	0	$+$		0		$-$		0	$+$	
G	↘		↗		↗		↘		↗		
$G'' = g'$		$+$		0		$-$		0		$+$	
Concavity		⌣				⌢				⌣	

The graph of G has roughly the following shape. The graph of any antiderivative of g can be obtained from the graph of G by a vertical shift.

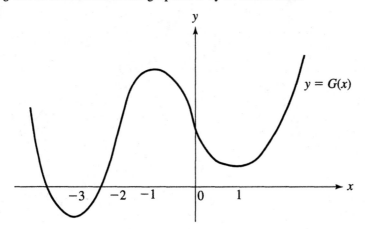

Exercise.

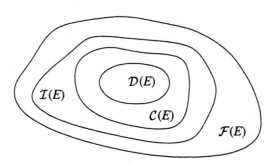

Classroom Discussion 5.1.3: Integrating Nonnegative Powers of x

1. $\int 1 \, dx = x + C$
2. $\int 5x^7 \, dx = \frac{5}{8} x^8 + C$
3. $\int \sqrt{x} \, dx = \int x^{\frac{1}{2}} \, dx = \frac{2}{3} x^{\frac{3}{2}} + C$
4. $\int x^n \, dx = \frac{1}{n+1} x^{n+1} + C$

Classroom Discussion 5.1.4: Integrating Negative Powers of x

1. $\int \frac{1}{x^2} \, dx = \int x^{-2} \, dx = -x^{-1} + C = -\frac{1}{x} + C$. The domain of the integrand $\frac{1}{x^2}$ is $(-\infty, 0) \cup (0, \infty)$.
2. For $x < 0$, $F'(x) = \left(-\frac{1}{x} + 1\right)' = \frac{1}{x^2}$; for $x > 0$, $F'(x) = \left(-\frac{1}{x} + 3\right)' = \frac{1}{x^2}$.
   Therefore, the function F is an antiderivative of the function $\frac{1}{x^2}$.
3. For $x < 0$, $F(x) - \left(-\frac{1}{x}\right) = 1$; for $x > 0$, $F(x) - \left(-\frac{1}{x}\right) = 3$.
4. According to Theorem 5.1.1, the antiderivatives of a function differ by a constant on *a given interval*. In this example, the integrand's domain is composed of two intervals $(-\infty, 0)$ and $(0, \infty)$. Theorem 5.1.1 is satisfied on each interval.

5. The answer to Problem 1 should be understood as follows:

$$\int \frac{1}{x^2}\,dx = \begin{cases} -\dfrac{1}{x} + C_1 \text{ if } x < 0, \\[2mm] -\dfrac{1}{x} + C_2 \text{ if } x > 0, \end{cases}$$

where C_1 and C_2 are two arbitrary constants.

6. $\int x^n\,dx = \dfrac{1}{n+1}x^{n+1} + C$. As the domain of the function x^n is $(-\infty, 0) \cup (0, \infty)$ for $n < 0$, this answer should be understood as

$$\int x^n\,dx = \begin{cases} \dfrac{1}{n+1}x^{n+1} + C_1 \text{ if } x < 0, \\[3mm] \dfrac{1}{n+1}x^{n+1} + C_2 \text{ if } x > 0. \end{cases}$$

7. The case $n = -1$ was excluded since the reciprocal of $n + 1 = 0$ is not defined.

Classroom Discussion 5.1.5: Integrating Exponential Functions

1. $\int e^x\,dx = e^x + C$ since $(e^x)' = e^x$. Let $F(x) = e^x + C$ be such that $F(0) = 3$; that is, $e^0 + C = 1 + C = 3$. Thus, $C = 2$ and $F(x) = e^x + 2$.

2. For $k \neq 0$, $\int e^{kx}\,dx = \frac{1}{k}e^{kx} + C$ since $(e^{kx})' = k\,e^{kx}$. Let $F(x) = \frac{1}{k}e^{kx} + C$ be such that $F(0) = 3$; that is, $\frac{1}{k}e^0 + C = \frac{1}{k} + C = 3$. Thus, $C = 3 - \frac{1}{k}$ and $F(x) = \frac{1}{k}e^{kx} + (3 - \frac{1}{k})$.

3. When $k = 0$, $e^{kx} = e^0 = 1$. In this case, $\int e^{kx}\,dx = \int 1\,dx = x + C$.

Classroom Discussion 5.1.6: Integrating Sums and Constant Multiples of Functions

1. Since $[kF(x)]' = kF'(x) = kf(x)$, the function $kF(x)$ is an antiderivative of the function $kf(x)$.

2. Since $[F(x) + G(x)]' = F'(x) + G'(x) = f(x) + g(x)$, the function $F(x) + G(x)$ is an antiderivative of the function $f(x) + g(x)$.

3. **a.** $k\int f(x)\,dx$
   b. $\int f(x)\,dx + \int g(x)\,dx$
   c. $\int f(x)\,dx - \int g(x)\,dx$

4. $\int (3x^2 + 6x)\,dx = \int 3x^2\,dx + 3\int 2x\,dx = x^3 + 3x^2 + C$.

Classroom Discussion 5.1.7: Integrating Products of Functions

1. $\int x^2(x - 1)\,dx = \int (x^3 - x^2)\,dx = \int x^3\,dx - \int x^2\,dx = \frac{1}{4}x^4 - \frac{1}{3}x^3 + C$.

2. $\int x^2\,dx = \frac{1}{3}x^3 + C$.

3. $\int (x - 1)\,dx = \int x\,dx - \int 1\,dx = \frac{1}{2}x^2 - x + C$.

4. Clearly, $\int x^2(x - 1)\,dx \neq \int x^2\,dx \int (x - 1)\,dx$.

5. $\int f(x)\,dx \int g(x)\,dx$.

Classroom Discussion 5.1.8: Motion and Antiderivatives

1. Since $h''(t) = -32$, $h'(t) = -32t + C$ for some constant C. Since $h'(0) = 10$, $C = 10$; thus, $h'(t) = -32t + 10$.
2. Since $h'(t) = -32t + 10$, $h(t) = -16t^2 + 10t + C$ for some constant C. Since $h(0) = 6$, $C = 6$; thus, $h(t) = -16t^2 + 10t + 6$.
3. The penny reaches the maximal height at the instant t when its velocity vanishes; that is, $h'(t) = -32t + 10 = 0$. The solution of this equation is $t = 5/16$ second.
4. The penny hits the ground at the instant t when $h(t) = 0$; that is, $-16t^2 + 10t + 6 = 0$. The solutions of this equation are $-3/8$ and 1. Since this event occurs after the penny was thrown, $t > 0$. Thus, the penny hits the ground at $t = 1$ second.
5. $h'(1) = -32 + 10 = -22$ feet per second.
6. If the penny is thrown upward, $v_0 \geq 0$; if it is thrown downward, $v_0 \leq 0$.
7. Using the same arguments as in Problems 1 and 2, you get $h'(t) = -32t + v_0$ and $h(t) = -16t^2 + v_0 t + h_0$.

SECTION 5.2

Classroom Discussion 5.2.1: Variations of Antiderivatives over Bounded Intervals

1. $F(x) = x^4$ is an antiderivative of $f(x)$. $F(2) - F(0) = 2^4 - 0 = 16$.
2. $G(x) = x^4 + C$ for some constant C. $G(2) - G(0) = (2^4 + C) - (0 + C) = 16$.
3. The antiderivatives of $f(x) = 4x^3$ all have the same change over the interval $[0,2]$.
4. Let $G(x) = x^4 + C$ for some constant C. $G(b) - G(a) = (b^4 + C) - (a^4 + C) = b^4 - a^4$. This change is independent of C, and thus all of the antiderivatives of $f(x)$ have the same change over the interval $[a, b]$.
5. Let $F(x)$ be an antiderivative of $f(x)$. Then, any other antiderivative of $f(x)$ is of the form $G(x) = F(x) + C$ for some constant C. The change of $G(x)$ over the interval $[a,b]$ is $G(b) - G(a) = [F(b) + C] - [F(a) + C] = F(b) - F(a)$. Thus, all of the antiderivatives of $f(x)$ have the same change over the interval $[a,b]$.

Classroom Discussion 5.2.2: Functions Defined by Means of Definite Integrals

1. **a.** $g(x) = \int_1^x t^2 \, dt = \frac{1}{3}t^3 \big|_1^x = \frac{1}{3}(x^3 - 1)$.
   b. $g'(x) = \frac{1}{3}(x^3 - 1)' = \frac{1}{3}3x^2 = x^2$.
   c. $g(1) = \frac{1}{3}(1^3 - 1) = 0$.
   d. antiderivative, $x^2, 0, 1$
2. **a.** Since $F(x)$ is an antiderivative of $f(x)$,

$$g(x) = F(t)\big|_a^x = F(x) - F(a) = F(x) + C,$$

   where $C = -F(a)$. Thus, $g(x)$ is also an antiderivative of $f(x)$.
   b. $g(a) = F(a) - F(a) = 0$. Thus, $g(x)$ is the antiderivative of $f(x)$ that takes the value 0 at $x = a$.
   c. antiderivative, $f_1(x) = \frac{1}{x}, 0, 1$
   antiderivative, $f_2(x) = \dfrac{1}{3x^2 + e^x}, 0, 0$

Classroom Discussion 5.2.3: Properties of Definite Integrals

Let $F(x)$ and $G(x)$ be antiderivatives of $f(x)$ and $g(x)$, respectively.

1. $\int_a^a f(x)\,dx = F(x)\big|_a^a = F(a) - F(a) = 0.$

2. $\int_b^a f(x)\,dx = F(x)\big|_b^a = F(a) - F(b) = -[F(b) - F(a)] = -\int_a^b f(x)\,dx.$

3.
$$\int_a^b f(x)\,dx + \int_b^c f(x)\,dx = F(x)\big|_a^b + F(x)\big|_b^c = F(b) - F(a) + F(c) - F(b)$$

$$= F(c) - F(a) = \int_a^c f(x)\,dx.$$

4. Since $kF(x)$ is an antiderivative of the function $kf(x)$,

$$\int_a^b kf(x)\,dx = \big[kF(x)\big]\big|_a^b = \big[kF(b) - kF(a)\big] = k\big[F(b) - F(a)\big] = k\int_a^b f(x)\,dx.$$

5. Since $F(x)$ and $G(x)$ are antiderivatives of $f(x)$ and $g(x)$, respectively, $F(x) + G(x)$ is an antiderivative of $f(x) + g(x)$. Thus,

$$\int_a^b \big[f(x) + g(x)\big]\,dx = \big[F(x) + G(x)\big]\Big|_a^b = \big[F(b) + G(b)\big] - \big[F(a) + G(a)\big]$$

$$= \big[F(b) - F(a)\big] + \big[G(b) - G(a)\big] = \int_a^b f(x)\,dx + \int_a^b g(x)\,dx.$$

6. This follows directly from Problems 4 and 5 with $k = -1$.

Practice Problem

In order to integrate the function $f(x)$ over the interval $[0,2]$, we need to distinguish between $x \le 1$ and $x \ge 1$. Thus, we will integrate this function over the subintervals $[0,1]$ and $[1,2]$, then add the results.

$$\int_0^1 f(x)\,dx = \int_0^1 x^2\,dx = \frac{1}{3}x^3\Big|_0^1 = \frac{1}{3}.$$

$$\int_1^2 f(x)\,dx = \int_1^2 x^3\,dx = \frac{1}{4}x^4\Big|_1^2 = \frac{1}{4}2^4 - \frac{1}{4} = \frac{15}{4}.$$

Thus,

$$\int_0^2 f(x)\,dx = \int_0^1 f(x)\,dx + \int_1^2 f(x)\,dx = \frac{1}{3} + \frac{15}{4} = \frac{49}{12}.$$

SECTION 5.3

Classroom Discussion 5.3.1: General Power Rule of Integration

1. $f(x) = 3x^2(x^3 + 1)^2 = 3x^2(x^6 + 2x^3 + 1) = 3x^8 + 6x^5 + 3x^2.$ Therefore, $\int f(x)\,dx = \frac{1}{3}x^9 + x^6 + x^3 + C.$

2. Expand into a polynomial form the binomial $(x^3 + 1)^n$, then $3x^2(x^3 + 1)^n$, and then integrate the latter term by term.

3. $f(x) = 3x^2(x^3 + 1)^2 = (x^3 + 1)'(x^3 + 1)^2$; thus $\int f(x)\,dx = \frac{1}{3}(x^3 + 1)^3 + C$. In order to compare this solution with the one obtained in Problem 1, expand it into a polynomial form: $\frac{1}{3}(x^3 + 1)^3 + C = \frac{1}{3}(x^9 + 3x^6 + 3x^3 + 1) + C = \frac{1}{3}x^9 + x^6 + x^3 + \frac{1}{3} + C = \frac{1}{3}x^9 + x^6 + x^3 + D'$, where C and D' are arbitrary constants. The two solutions are identical.

4. $f(x) = 3x^2(x^3 + 1)^n = (x^3 + 1)'(x^3 + 1)^n$; thus $\int f(x)\,dx = \frac{1}{n+1}(x^3 + 1)^{n+1} + C$.

5. The answer to Problem 4 is valid for any real number $n \neq -1$ as the reasoning therein still works.

6. $3x^4\sqrt{x^5 + 6} = 3x^4(x^5 + 6)^{1/2} = \frac{3}{5}(x^5 + 6)'(x^5 + 6)^{1/2}$; thus $\int 3x^4\sqrt{x^5 + 6}\,dx = \frac{3}{5}\left(\frac{2}{3}(x^5 + 6)^{3/2}\right) + C = \frac{2}{5}(x^5 + 6)^{3/2} + C.$

7. $\frac{5x}{(x^2+1)^2} = 5x(x^2 + 1)^{-2} = \frac{5}{2}(x^2 + 1)'(x^2 + 1)^{-2}$. It follows that $\int \frac{5x}{(x^2 + 1)^2}\,dx = \frac{5}{2}\left(\frac{1}{-1}(x^2 + 1)^{-1}\right) + C = \frac{-5}{2(x^2 + 1)} + C.$ Thus, $\int_0^1 \frac{5x}{(x^2 + 1)^2}\,dx = \frac{-5}{2(x^2 + 1)}\Big|_0^1 = \frac{-5}{4} - \frac{-5}{2} = \frac{5}{4}.$

Classroom Discussion 5.3.2: General Exponential Rule of Integration

1. $u(x) = x^2$

2. $\int 2x\,e^{x^2}\,dx = \int (x^2)'\,e^{x^2}\,dx = e^{x^2} + C$

3. Let $u(x) = \sqrt{x} = x^{1/2}$. Then, $u'(x) = \frac{1}{2}x^{-1/2} = \frac{1}{2\sqrt{x}}$. It follows that $\int \frac{e^{\sqrt{x}}}{\sqrt{x}}\,dx = \int 2u'(x)\,e^{u(x)}\,dx = 2e^{u(x)} + C = 2e^{\sqrt{x}} + C.$

Classroom Discussion 5.3.3: The Method of Substitution

1. **a.** $\int u'(x)\,dx = u(x) + C.$
   b. $\int 1\,du = u + C.$
   c. When expressed in terms of x, $u + C$ is identical to $u(x) + C.$
   d. It follows from your findings in 1a–1c that $1\,du = u'(x)\,dx$; that is $du = u'(x)\,dx.$

2. **a.** $u'(x) = (x^2 - 3)' = 2x.$
   b. $du = u'(x)\,dx = 2x\,dx.$
   c. $I = \int 2x\sqrt{x^2 - 3}\,dx = \int \sqrt{u}\,du.$
   d. $I = \int \sqrt{u}\,du = \int u^{1/2}\,du = \frac{2}{3}u^{3/2} + C.$
   e. $I = \frac{2}{3}u^{3/2} + C = \frac{2}{3}(x^2 - 3)^{3/2} + C.$
   f. $\int 2x\sqrt{x^2 - 3}\,dx = \int (x^2 - 3)'(x^2 - 3)^{1/2}\,dx = \frac{2}{3}(x^2 - 3)^{3/2} + C.$
   g. The two findings are identical.

3. **a.** $u(x) = x^3 + x + 2.$
   b. $u'(x) = 3x^2 + 1$; thus $du = (3x^2 + 1)\,dx.$
   c. $J = \int e^u\,du.$
   d. $J = e^u + C.$

e. $J = e^{x^3+x+2} + C.$

f. $\int (3x^2 + 1) e^{x^3+x+2} \, dx = \int (x^3 + x + 2)' \, e^{x^3+x+2} \, dx = e^{x^3+x+2} + C.$

g. The two findings are identical.

4. Let $u(x) = x^3 + x^2 + x + 1$. Then $u'(x) = 3x^2 + 2x + 1$ and $du = (3x^2 + 2x + 1) \, dx$. Thus, $K = \int \frac{1}{u^4} \, du = \int u^{-4} \, du = \frac{-1}{3} u^{-3} + C = \frac{-1}{3(x^3 + x^2 + x + 1)^3} + C.$

CHAPTER 6

SECTION 6.1

Classroom Discussion 6.1.1: Area of an Arbitrary Polygonal Region

1. Denote by $\mathcal{A}$ the area of the rectangle of interest. Consider the square whose sides each have length $a + b$. Its area is $(a + b)^2$.

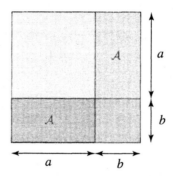

This square can be divided as in the preceding figure, and thus its area equals $2\mathcal{A} + a^2 + b^2$. This means that $(a + b)^2 = 2\mathcal{A} + a^2 + b^2$. Solving this equation for $\mathcal{A}$, you get $\mathcal{A} = ab$.

2. Consider the rectangle $ACDF$ obtained from $\triangle BDF$ as in the following figure.

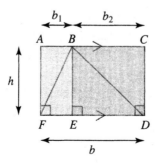

It is easy to check that $\triangle ABF$ and $\triangle EFB$ are congruent and thus have the same area $\mathcal{A}_1$. Since the rectangle $ABEF$ is formed of these two triangles, its area $b_1 h$ is twice the area $\mathcal{A}_1$; that is, $\mathcal{A}_1 = \frac{1}{2}b_1 h$. Similarly, it is easy to check that $\triangle BCD$ and $\triangle DEB$ are congruent and thus have the same area $\mathcal{A}_2$. Since the rectangle

BCDE is composed of these two triangles, its area b_2h is twice the area A_2; that is, $A_2 = \frac{1}{2}b_2h$. Since $\triangle BDF$ is composed of $\triangle EFB$ and $\triangle DEB$, its area A is the sum of their areas; that is,

$$A = A_1 + A_2 = \frac{1}{2}b_1h + \frac{1}{2}b_2h = \frac{1}{2}(b_1 + b_2)h = \frac{1}{2}bh.$$

3. The trapezoid can be divided into two triangles along one of its diagonals as in the following figure.

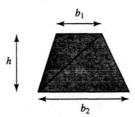

Using Problem 2, you see that one of the triangles has area $\frac{1}{2}b_1h$ and the other has area $\frac{1}{2}b_2h$. Therefore, the area A of the trapezoid is given by $A = \frac{1}{2}b_1h + \frac{1}{2}b_2h = \frac{1}{2}(b_1 + b_2)h$.

4. A region that has a polygonal boundary can always be decomposed into triangles. The following are examples of such decompositions:

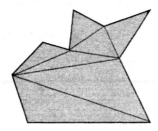 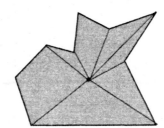

The area of a polygonal region is the sum of the areas of the triangles used in a given decomposition, and the latter can be computed using the formula obtained in Problem 2.

SECTION 6.2

Classroom Connection 6.2.2: Area of a Foot

1. A lower bound for the printed foot's area can be obtained by counting the unit squares that lie inside the foot. By doing so, you obtain the lower bound $A_1 = 120$ square units. If you wish to have better lower bounds, you may count fractions of unit squares as in Classroom Connection 6.2.1.

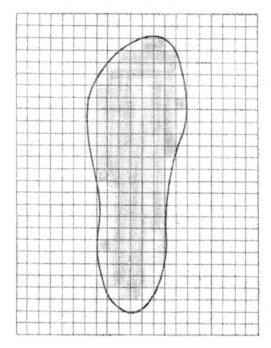

2. An upper bound for the printed foot's area can be obtained by counting the unit squares in the grid that touch the foot. By doing so, you obtain the upper bound $A_2 = 170$ square units. Again, if you wish to have better upper bounds, you may count fractions of unit squares instead of whole squares.

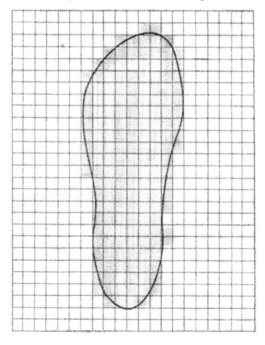

3. Denote by $\mathcal{A}$ the (unknown) exact value of the area of the printed foot. Using our answers to Problems 1 and 2, we see that $120 \leq \mathcal{A} \leq 170$. It follows that the error generated by Alice's approximation of $\mathcal{A}$ with 131 square units is at most 39 units.

4. Bob approximates the area $\mathcal{A}$ by $(120 + 170)/2 = 145$ square units. The magnitude of the error generated by this approximation is 25 square units.

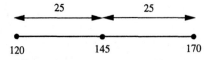

5. It would be better to work with Bob's approximation as the magnitude of the error generated by his approximation is smaller than the one generated by Alice's approximation. However, keep in mind that Alice's approximation might be better if the area $\mathcal{A}$ turns out to be much closer to 131 square units than to 145 square units. The point is that since such information is unknown, in general, it is better to use the average value between the lower and upper bounds as an approximate value.

6. Using the given grid, we saw that the area $\mathcal{A}$ ranges in the interval $[120, 170]$. The use of a finer grid will lead to a smaller interval and thus to more accurate approximations for the area $\mathcal{A}$.

7. If you continue using finer and finer grids, you will reduce increasingly more the lengths of the intervals in which the area $\mathcal{A}$ ranges, and thus you will be able to progressively locate $\mathcal{A}$.

Classroom Discussion 6.2.1: The Trapezoidal Method

1. The error generated when approximating the area $\mathcal{A}$ of the irregular shape with T_n can be reduced if you let n take large values. Indeed, with a large number of points selected throughout the curved path, the segment lines joining each pair of consecutive points will approximate more closely the portion of the curved path that is in between. Thus, the approximating polygonal region will better fit the region of interest leading to a better approximation for the area $\mathcal{A}$.

2. I do agree with Louise since, as the number n of points selected *throughout* the curved path goes to ∞, the approximating polygonal regions increasingly better fit the region of interest; thus, their areas T_n tend to its area $\mathcal{A}$. Here, by *throughout* we mean that the width δ_n of the tile with the greatest width approaches 0.

3. The areas T_n are necessarily lower bounds for the areas of the regions bounded by the graphs of functions that are concave down.

4. The areas T_n are necessarily upper bounds for the areas of the regions bounded by the graphs of functions that are concave up.

5. In this case, we have

$$\text{Area of Tile } 1 = \frac{1}{2} \delta \left[f(a) + f(x_1) \right]$$

$$\text{Area of Tile } 2 = \frac{1}{2}\delta\big[f(x_1) + f(x_2)\big]$$

$$\vdots \qquad\qquad\qquad \vdots$$

$$\text{Area of Tile } i = \frac{1}{2}\delta\big[f(x_{i-1}) + f(x_i)\big]$$

$$\vdots \qquad\qquad\qquad \vdots$$

$$\text{Area of Tile } n = \frac{1}{2}\delta\big[f(x_{n-1}) + f(b)\big].$$

Thus,

$$T_n = \frac{\delta}{2}\Big[f(a) + f(x_1) + f(x_1) + \ldots + f(x_{n-1}) + f(x_{n-1}) + f(b)\Big]$$

$$= \frac{\delta}{2}\Big[f(a) + 2\sum_{i=1}^{n-1} f(x_i) + f(b)\Big].$$

6. $\delta = \dfrac{b - a}{n}$ and $T_n = \dfrac{b - a}{2n}\Big[f(a) + 2\sum_{i=1}^{n-1} f(x_i) + f(b)\Big]$.

7. $x_0 = a, x_1 = a + \delta, x_2 = x_1 + \delta = a + 2\delta, x_3 = x_2 + \delta = a + 3\delta$, etc. More generally,

$$x_i = a + i\delta = a + i\frac{b - a}{n} \text{ for all integers } 0 \le i \le n.$$

8. The following is a program for a TI-89 calculator.

Step 1: Tell the calculator how to select the points on the curved path by entering

$$a + i * (b - a)/n \to x(i, a, b, n)$$

then press ENTER. The calculator should say "Done." To get the $\to$ symbol, press STO$\to$ (it is located above the ON button). To store the function x, go to VAR-LINK, press the key F1 followed by the key 8 for "Archive Variable." The function will stay in your calculator's memory even if you clear it from the home screen; you will not have to redefine $x(i, a, b, n)$ for the calculator again, but you will not be allowed to store anything else under the name "x" without unarchiving the latter.

Step 2: Tell the calculator how to compute the Riemann sums by entering

$$(b - a)/(2 * n) * \Big(f(a) + f(b) + 2 * \sum_{i=1}^{n-1} f(x(i, a, b, n))\Big) \to t(a, b, n)$$

then, press ENTER. The calculator should say "Done." Go to VAR-LINK and archive $t(a, b, n)$. To get the Σ symbol, press the keys F3 and then 4. To enter the sum $\sum_{i=1}^{n-1} f(x(i, a, b, n))$, type: $\Sigma\,(f(x(i, a, b, n)), i, 1, n - 1)$

Step 3: Tell the calculator what function to consider. For example, if your function is $f(x) = 1/x^3$, then enter

$$1/x^3 \to f(x)$$

Do not archive this function; you will want to change it later.

Step 4: Evaluate the desired Riemann sums. For example, to evaluate the Riemann sum corresponding to the function in Step 3 defined on the interval $[1,6]$ with 70 tiles, type "$t(1,6,70)$" and then, press ENTER. The calculator should return the value of this Riemann sum. To work with a different function, return to Step 3 and enter your new function as described.

Classroom Discussion 6.2.2: The Rectangular Methods

1. **The Left Rectangular Method**

   a.

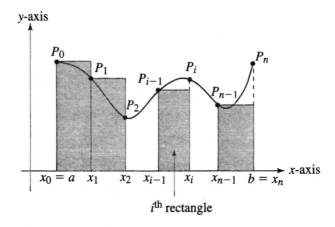

   i^{th} rectangle

   b. In this case, we have

$$\text{Area of each tile} = \text{width} \cdot \text{length}$$

$$\text{Area of Tile 1} = (x_1 - x_0)f(x_0)$$

$$\text{Area of Tile 2} = (x_2 - x_1)f(x_1)$$

$$\vdots \qquad\qquad \vdots$$

$$\text{Area of Tile } i = (x_i - x_{i-1})f(x_{i-1})$$

$$\vdots \qquad\qquad \vdots$$

$$\text{Area of Tile } n = (x_n - x_{n-1})f(x_{n-1})$$

   c. The Riemann sum corresponding to these rectangular tiles is given by the formula $\sum_{i=1}^{n}(x_i - x_{i-1})f(x_{i-1})$.

d. Divide the interval $[a, b]$ into n congruent segments. The common length of these segments is $\delta = (b - a)/n$. Thus, for each $1 \le i \le n, x_i - x_{i-1} = \delta = (b - a)/n$. The resulting Riemann sums are given by

$$L_n = \sum_{i=1}^{n} (x_i - x_{i-1}) f(x_{i-1}) = \sum_{i=1}^{n} \frac{b - a}{n} f(x_{i-1}) = \frac{b - a}{n} \sum_{i=1}^{n} f(x_{i-1})$$

$$= \frac{b - a}{n} \sum_{i=0}^{n-1} f(x_i).$$

As in the trapezoidal method, for each $0 \le i \le n, x_i = a + i\frac{b - a}{n}$.

e. As n gets larger and larger, the curved path becomes increasingly more covered by the points $P_0, P_1, \ldots, P_n$; thus, the corresponding polygonal approximating region increasingly better fits the region of interest. Consequently, the Riemann sums L_n become closer and closer to the area $\mathcal{A}$. Taking the limit as $n \to \infty$, we get

$$\mathcal{A} = \lim_{n \to \infty} L_n = \lim_{n \to \infty} \frac{b - a}{n} \sum_{i=1}^{n} f(x_{i-1}).$$

f. The Riemann sums L_n always provide lower bounds for the areas of the regions that are bounded by the graphs of increasing functions.

g. The areas L_n always provide upper bounds for the areas of the regions that are bounded by the graphs of decreasing functions.

h. Follow Steps 1–4 in the calculator program described in the case of the trapezoidal method, with the following change in Step 2:

$$(b - a)/n * \sum_{i=1}^{n} f(x(i - 1, a, b, n)) \to l(a, b, n)$$

2. The Right Rectangular Method

a.   i.

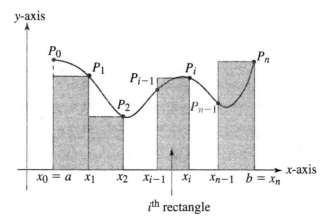

i^{th} rectangle

ii. In this case, we have

$$\text{Area of each tile} = \text{width} \cdot \text{length}$$

$$\text{Area of Tile 1} = (x_1 - x_0)f(x_1)$$

$$\text{Area of Tile 2} = (x_2 - x_1)f(x_2)$$

$$\vdots \qquad\qquad \vdots$$

$$\text{Area of Tile } i = (x_i - x_{i-1})f(x_i)$$

$$\vdots \qquad\qquad \vdots$$

$$\text{Area of Tile } n = (x_n - x_{n-1})f(x_n)$$

iii. The Riemann sums corresponding to these rectangular tiles are $\sum_{i=1}^{n}(x_i - x_{i-1})f(x_i)$.

iv. Divide the interval $[a,b]$ into n congruent segments. The common length of these segments is $\delta = (b - a)/n$. Thus, $x_i - x_{i-1} = \delta = (b - a)/n$ for each $1 \le i \le n$. The resulting Riemann sums are given by

$$R_n = \sum_{i=1}^{n}(x_i - x_{i-1})f(x_i) = \sum_{i=1}^{n}\frac{b - a}{n}f(x_i) = \frac{b - a}{n}\sum_{i=1}^{n}f(x_i).$$

As in the trapezoidal and left rectangular methods, for each $0 \le i \le n$, $x_i = a + i\frac{b - a}{n}$.

v. As n gets larger and larger, the curved path becomes increasingly more covered by the points $P_0, P_1, \ldots, P_n$; thus, the obtained polygonal approximating region increasingly better fits the region of interest. Consequently, the Riemann sums R_n become closer and closer to the area $\mathcal{A}$. Taking the limit as $n \to \infty$, we get

$$\mathcal{A} = \lim_{n\to\infty} R_n = \lim_{n\to\infty}\frac{b - a}{n}\sum_{i=1}^{n}f(x_i).$$

vi. The Riemann sums R_n always provide lower bounds for the areas of the regions that are bounded by the graphs of decreasing functions.

vii. The areas R_n always provide upper bounds for the areas of the regions that are bounded by the graphs of increasing functions.

viii. Follow Steps 1–4 in the calculator program described in the case of the trapezoidal method, with the following change in Step 2:

$$(b - a)/n * \sum_{i=1}^{n} f(x(i,a,b,n)) \to r(a,b,n)$$

b. If you average between the Riemann sums corresponding to the left and right rectangular methods, you obtain the Riemann sum corresponding to the trapezoidal method. Indeed,

$$
\frac{L_n + R_n}{2} = \frac{1}{2}\left[\frac{b-a}{n}\sum_{i=1}^{n}f(x_{i-1}) + \frac{b-a}{n}\sum_{i=1}^{n}f(x_i)\right]
$$

$$
= \frac{b-a}{2n}\left[f(x_0) + f(x_1) + \dots + f(x_{n-1}) + f(x_1)\right.
$$

$$
\left. + \dots + f(x_{n-1}) + f(x_n)\right]
$$

$$
= \frac{b-a}{2n}\left[f(a) + 2[f(x_1) + \dots + f(x_{n-1})] + f(b)\right]
$$

$$
= \frac{b-a}{2n}\left[f(a) + 2\sum_{i=1}^{n-1}f(x_i) + f(b)\right]
$$

$$
= T_n.
$$

This can be seen geometrically as well; the ith trapezoidal tile is obtained by "averaging" between the two i^{th} tiles from the left and right rectangular methods.

c. For n large enough, the Riemann sum T_n is bound to give a better approximation for the area A than the Riemann sums L_n and R_n. Indeed, the approximating polygonal region from the trapezoidal method fits the region of interest better than the approximating polygonal regions from the left and right rectangular methods. To convince yourself, consider two consecutive points P_{i-1} and P_i on the curved path. The two points are very close to each other when n is large, and thus the portion of the curved path between these two points looks very much like the line segment joining them.

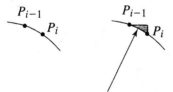

Error generated at the level of the ith tile using the left rectangular method

Error generated at the level of the ith tile using the right rectangular method

3. The Midpoint Rectangular Method

a. For each $1 \le i \le n$, denote by c_i the midpoint of the interval $[x_{i-1}, x_i]$ and draw the rectangular tile with height $f(c_i)$ as shown in the following figure.

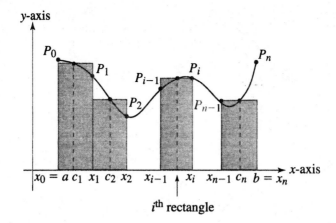

i^{th} rectangle

b. In this case, we have

$$\text{Area of Tile } 1 \;=\; (x_1 - x_0)f(c_1)$$

$$\text{Area of Tile } 2 \;=\; (x_2 - x_1)f(c_2)$$

$$\vdots \qquad\qquad \vdots$$

$$\text{Area of Tile } i \;=\; (x_i - x_{i-1})f(c_i)$$

$$\vdots \qquad\qquad \vdots$$

$$\text{Area of Tile } n \;=\; (x_n - x_{n-1})f(c_n)$$

c. The Riemann sums corresponding to these rectangular tiles are given by
$$\sum_{i=1}^{n}(x_i - x_{i-1})f(c_i).$$

d. Divide the interval $[a,b]$ into n congruent segments. The common length of these segments is $\delta = (b - a)/n$. Thus, $x_i - x_{i-1} = \delta = (b - a)/n$ for each $1 \le i \le n$. The resulting Riemann sums M_n are given by

$$M_n = \sum_{i=1}^{n}(x_i - x_{i-1})f(c_i) = \frac{b - a}{n}\sum_{i=1}^{n}f(c_i),$$

where

$$c_i = x_{i-1} + \frac{\delta}{2} = a + (i - 1)\delta + \frac{\delta}{2} = a + \left(i - \frac{1}{2}\right)\delta$$

$$= a + \left(i - \frac{1}{2}\right)\frac{b - a}{n} = a + \frac{(2i - 1)(b - a)}{2n}.$$

e. For the same reasons as in Problems 1e and 2a.vi, we have

$$A = \lim_{n \to \infty} M_n = \lim_{n \to \infty} \frac{b - a}{n} \sum_{i=1}^{n} f(c_i).$$

f. Unless the curved path is a horizontal line segment, the Riemann sums M_n will not provide lower bounds for the area A of the region of interest for each value of n. To be convinced, analyze Figure 1.

g. Similarly, unless the curved path is a horizontal line segment, the Riemann sums M_n will not provide upper bounds for the area A of the region of interest for each value of n. To be convinced, analyze Figure 2.

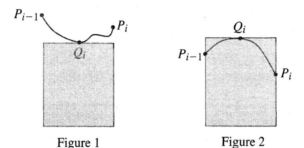

Figure 1   Figure 2

h. Follow Steps 1–4 in the calculator program described in the case of the trapezoidal method, with the following changes.

In Step 1:

$$a + (2 * i - 1) * (b - a)/(2 * n) \to c(i, a, b, n)$$

In Step 2:

$$(b - a)/n * \sum_{i=1}^{n} f(c(i, a, b, n)) \to m(a, b, n)$$

4. Generalization

a. Pick $n - 1$ arbitrary points $P_1, P_2, \ldots, P_{n-1}$ on the curved path between the endpoints $(a, f(a))$ and $(b, f(b))$ and denote by $x_1, x_2, \ldots x_{n-1}$ their x-coordinates, respectively. For convenience, let $x_0 = a$ and $x_n = b$. For each $1 \le i \le n$, pick randomly a point Q_i on the curved path between points P_{i-1} and P_i and denote by d_i its x-coordinate. Then, draw the rectangular tile with height $f(d_i)$ as shown in the following figure. The Riemann sums corresponding to these rectangular tiles are given by the formula

$$G_n = \sum_{i=1}^{n} (x_i - x_{i-1}) f(d_i).$$

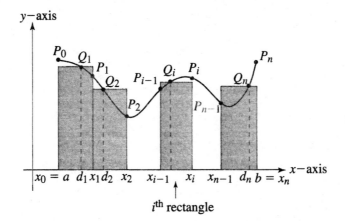

Picking the Q_i's such that $Q_1 = P_0, Q_2 = P_1, \ldots, Q_n = P_{n-1}$ leads to the left rectangular method. Picking the Q_i's such that $Q_1 = P_1, Q_2 = P_2, \ldots, Q_n = P_n$ leads to the right rectangular method. Picking the Q_i's such that $d_1 = c_1, d_2 = c_2, \ldots, d_n = c_n$ leads to the midpoint rectangular method.

b. In order for the Riemann sum G_n to be necessarily a lower bound for the area A of the region above the x-axis and below the graph of f, pick the Q_i's as follows: for each $1 \leq i \leq n, f(d_i)$ is the minimum value of f on the interval $[x_{i-1}, x_i]$. In order for G_n to be necessarily an upper bound for the area A, pick the Q_i's as follows: for each $1 \leq i \leq n, f(d_i)$ is the maximum value of f on the interval $[x_{i-1}, x_i]$.

Classroom Discussion 6.2.3: Riemann Sums for Functions Taking Negative Values

1. Consider the function $g(x) = -f(x)$ defined on the interval $[a, b]$. Note that since f takes only nonpositive values, g must take only nonnegative values. Denote by A the area of the region that is above the graph of f and below the x-axis; denote by B the area of the region that is below the graph of g and above the x-axis.

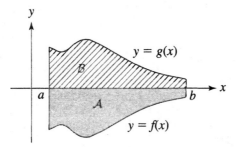

For symmetry reasons, it is clear that $A = B$. Now, since g takes only nonnegative values, we have seen that the area B can be approximated by the Riemann sums for g. For example, using the right rectangular method with n tiles, we have

$$B \approx \frac{b - a}{n} \sum_{i=1}^{n} g(x_i),$$

where the x_i's are defined by $x_i = a + \frac{b-a}{n}i$ for $1 \leq i \leq n$. Thus,

$$\mathcal{A} = \mathcal{B} \approx \frac{b-a}{n}\sum_{i=1}^{n}g(x_i) = \frac{b-a}{n}\sum_{i=1}^{n}-f(x_i) = -\frac{b-a}{n}\sum_{i=1}^{n}f(x_i).$$

Therefore,

$$-\mathcal{A} \approx \frac{b-a}{n}\sum_{i=1}^{n}f(x_i).$$

In other words, the Riemann sums for f provide approximate values for the negative of the area of the region that is above the graph of f and below the x-axis. This reasoning clearly works also for the Riemann sums obtained using the other methods.

2. Consider the functions f^+ and f^- defined by $f^+(x) = f(x)$ if $f(x) > 0$ and 0 otherwise, and $f^-(x) = f(x)$ if $f(x) < 0$ and 0 otherwise. The graphs of f^+ and f^- are shown in the following figure.

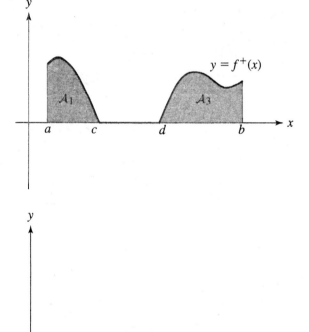

We have $f = f^+ + f^-$. Indeed, for $f(x) > 0, f^+(x) + f^-(x) = f(x) + 0 = f(x)$; for $f(x) < 0, f^+(x) + f^-(x) = 0 + f(x) = f(x)$; and for $f(x) = 0, f^+(x) + f^-(x) = 0 = f(x)$. It follows that the Riemann sums for f are the sum of the Riemann sums for f^+ and those for f^-. Since f^+ is a nonnegative function and f^-

is a nonpositive function, the Riemann sums for f^+ provide approximate values for the area of the region bounded by the graph $y = f^+(x)$ and by the x-axis, that is, $A_1 + A_3$; the Riemann sums for f^- provide approximate values for the negative of the area of the region bounded by the graph $y = f^-(x)$ and by the x-axis, that is, $-A_2$. Putting it all together, you see that the Riemann sums for f provide approximate values for $A_1 - A_2 + A_3$.

SECTION 6.3

Classroom Discussion 6.3.1: The Fundamental Theorem of Calculus

1. $A(a) = 0$. $A(\frac{a+b}{2})$ is the area of the region bounded by the x-axis, the graph of f, and the vertical lines through points $(a, 0)$ and $(\frac{a+b}{2}, 0)$. $A(b) = A$.

2. As x increases, the region bounded by the x-axis, the graph of f, and the vertical lines through points $(a, 0)$ and $(x, 0)$ becomes larger; its area $A(x)$ becomes larger. Thus, the function A is increasing on the interval $[a, b]$.

3. If Δx is very small, the portion of the graph of f lying between points $(x, f(x))$ and $(x + \Delta x, f(x + \Delta x))$ looks very much like the line segment joining them; thus, the trapezoid with vertices $(x, 0), (x + \Delta x, 0), (x, f(x))$, and $(x + \Delta x, f(x + \Delta x))$ is a good candidate for approximating the additional region.

4. The area ΔA of the additional region can be approximated with the area of the trapezoid described in Problem 3. The latter has width Δx and bases $f(x)$ and $f(x + \Delta x)$. Thus,

$$\Delta A \approx \frac{1}{2} \Delta x \left[f(x) + f(x + \Delta x) \right].$$

5. Using the approximation obtained in Problem 4, we see that

$$\frac{\Delta A}{\Delta x} \approx \frac{1}{2} \left[f(x) + f(x + \Delta x) \right].$$

6. Since f is continuous on $[a, b]$, $\lim_{\Delta x \to 0} f(x + \Delta x) = f(x)$ for each x.

7.

$$\lim_{\Delta x \to 0} \frac{\Delta A}{\Delta x} = \lim_{\Delta x \to 0} \frac{1}{2} \left[f(x) + f(x + \Delta x) \right]$$

$$= \frac{1}{2} \left[f(x) + \lim_{\Delta x \to 0} f(x + \Delta x) \right]$$

$$= \frac{1}{2} \left[f(x) + f(x) \right]$$

$$= f(x).$$

8. Using the definition of the derivative and our findings from Problem 7, we get

$$A'(x) = \lim_{\Delta x \to 0} \frac{A(x + \Delta x) - A(x)}{(x + \Delta x) - x} = \lim_{\Delta x \to 0} \frac{\Delta A}{\Delta x} = f(x).$$

9. Using Problems 1 and 8, you see that the function $A(x)$ is the antiderivative of f that takes the value 0 at a. It follows from Classroom Discussion 5.2.2 that

$$A(x) = \int_a^x f(t)\,dt.$$

10. $A = A(b) = \int_a^b f(t)\,dt.$

Practice Problems

1. Using the Fundamental Theorem of Calculus, the area of the shaded region is

$$\int_0^2 2x\,dx = x^2\big|_0^2 = (4 - 0) = 4.$$

On the other hand, the shaded region is a triangle whose base has length 2 and whose height is 4. Using the formula for the area of a triangle derived in Section 6.1, you see that its area is $2 \cdot 4/2 = 4$.

2. Using the Fundamental Theorem of Calculus, the area of the shaded region is

$$\int_2^4 (x + 1)\,dx = \left(\frac{1}{2}x^2 + x\right)\Big|_2^4 = \left(\frac{16}{2} + 4\right) - \left(\frac{4}{2} + 2\right) = 8.$$

On the other hand, the shaded region is a trapezoid with width 2 and whose bases have lengths 3 and 5. Using the formula for the area of a trapezoid derived in Section 6.1, you see that its area is $2(3 + 5)/2 = 8$.

3. Using the Fundamental Theorem of Calculus, you see that the area of the shaded region is

$$\int_0^2 x^2\,dx = \frac{1}{3}x^3\Big|_0^2 = \frac{1}{3}(8 - 0) = \frac{8}{3}.$$

4. Using the Fundamental Theorem of Calculus, you see that the area of the shaded region is

$$\int_1^5 \frac{1}{x^2}\,dx = \int_1^5 x^{-2}\,dx = -x^{-1}\big|_1^5 = -\frac{1}{5} + 1 = \frac{4}{5} = 0.8.$$

This confirms our conjecture in the Example following Classroom Discussion 6.2.1.

Classroom Discussion 6.3.2: Area of a Region above a Curve and below the x-axis

1. a.

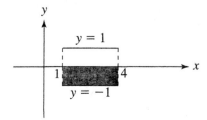

b. $\int_1^4 f(x)\,dx = \int_1^4 -1\,dx = -x|_1^4 = (-4) - (-1) = -3.$

c. Since the area of the shaded rectangle is 3, $\int_1^4 f(x)\,dx$ can be geometrically interpreted as negative the area of this rectangle. To explain this, consider the graph $y = 1$ for $1 \le x \le 4$. Using the Fundamental Theorem of Calculus, you see that the area of the rectangle above the x-axis and below the graph $y = 1$ for $1 \le x \le 4$ is $\int_1^4 1\,dx$. Since this rectangle is congruent to the shaded one, the two rectangles have the same area. Thus, the area $\mathcal{A}$ of the shaded rectangle is

$$\mathcal{A} = \int_1^4 1\,dx = -\int_1^4 -1\,dx = -\int_1^4 f(x)\,dx.$$

2. Consider the function $-f(x)$. Since $f(x) \le 0$, $-f(x) \ge 0$. Using the Fundamental Theorem of Calculus, you see that the region below the graph of $-f$ and above the x-axis has area $\int_a^b -f(x)\,dx$. Since this region is congruent to the shaded region, the two regions have the same area. Thus, the area $\mathcal{A}$ of the shaded region is given by

$$\mathcal{A} = \int_a^b -f(x)\,dx = -\int_a^b f(x)\,dx,$$

where, for the last equality, we used the constant multiple property of definite integrals.

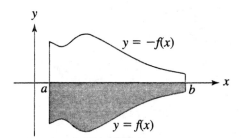

3. a.

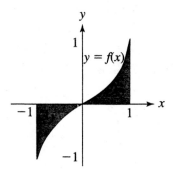

b. $\int_{-1}^0 x^3\,dx = \frac{1}{4}x^4|_{-1}^0 = \frac{1}{4}[0 - (-1)^4] = -\frac{1}{4}.$

  $\int_0^1 x^3\,dx = \frac{1}{4}x^4|_0^1 = \frac{1}{4}(1^4 - 0) = \frac{1}{4}.$

c. $\int_{-1}^1 x^3\,dx = \int_{-1}^0 x^3\,dx + \int_0^1 x^3\,dx = -\frac{1}{4} + \frac{1}{4} = 0.$

d. The area A of the region between the graph of f and the x-axis is equal to $A_1 + A_2$, where A_1 is the area of the region below the x-axis and above the portion of the graph of f corresponding to $-1 \le x \le 0$; A_2 is the area of the region above the x-axis and below the portion of the graph of f corresponding to $0 \le x \le 1$. Using Problem 2 together with the Fundamental Theorem of Calculus, you see that

$$A = A_1 + A_2 = -\int_{-1}^{0} x^3\, dx + \int_{0}^{1} x^3\, dx = -\left(-\frac{1}{4}\right) + \frac{1}{4} = \frac{1}{2}.$$

e. $\int_{-1}^{1} f(x)\, dx = \int_{-1}^{0} f(x)\, dx + \int_{0}^{1} f(x)\, dx = -A_1 + A_2$.

4. a. $A = A_1 + A_2 + A_3 = \int_{a}^{c} f(x)\, dx - \int_{c}^{d} f(x)\, dx + \int_{d}^{b} f(x)\, dx$.

b. $\int_{a}^{b} f(x)\, dx = \int_{a}^{c} f(x)\, dx + \int_{c}^{d} f(x)\, dx + \int_{d}^{b} f(x)\, d = A_1 - A_2 + A_3$.

c. As we have seen in Classroom Discussion 6.2.3, the Riemann sums for the function f over the interval $[a, b]$ provide approximate values for $A_1 - A_2 + A_3$; that is, for $\int_{a}^{b} f(x)\, dx$.

SECTION 6.4

Classroom Discussion 6.4.1: Approximating the Circle by Inscribed Regular *n*-gons

1. Each n-gon has an area $A_n = \dfrac{1}{2}\, a_n p_n$. This formula can be obtained by dividing the region inside the regular n-gon into n congruent triangular regions (each with a vertex at the center of the n-gon), and then applying the formula for the area of a triangle obtained in Section 6.1.

2. As n increases, the n-gons approach the circle. In particular, the a_n's approach the radius of the circle, the p_n's approach the perimeter of the circle, and the A_n's approach the area A of the disc. Thus, $\lim\limits_{n \to \infty} a_n = r$, $\lim\limits_{n \to \infty} p_n = 2\pi r$, and $\lim\limits_{n \to \infty} A_n = A$.

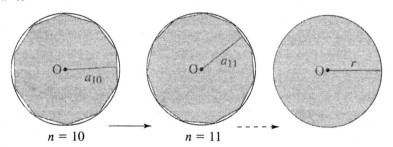

$n = 10$ $n = 11$

3. Letting n tend to infinity in the formula $A_n = \dfrac{1}{2}\, a_n p_n$ leads to $A = \frac{1}{2}(r)(2\pi r) = \pi r^2$.

4. The method described in this Classroom Discussion is called the *method of exhaustion*, as, when $n \to \infty$, the n-gons exhaust the disc.

Classroom Discussion 6.4.2: Area of a Disc Using the Fundamental Theorem of Calculus

1. Point (x, y) lies on the circle centered at the origin with radius r if and only if x and y satisfy $x^2 + y^2 = r^2$.

2. A point (x, y) is on the upper semicircle if and only if $x^2 + y^2 = r^2$ and $y \geq 0$.

3. Since $y \geq 0$, solving for y in the equation from Problem 1 leads to $y = \sqrt{r^2 - x^2}$.

4. The Fundamental Theorem of Calculus says that the area of the upper semidisc is $\int_{-r}^{r} \sqrt{r^2 - x^2} \, dx$; the area of the entire disc is obtained by doubling the result.

5. Using your calculator, you see that $A = 2\int_{-r}^{r} \sqrt{r^2 - x^2} \, dx = \pi |r| r = \pi r^2$.

SECTION 6.5

Classroom Discussion 6.5.1: Area of a Region Bounded by Two Graphs

1. **a.** Using the Fundamental Theorem of Calculus, you get $A_f = \int_a^b f(x) \, dx$.

   b. Using the Fundamental Theorem of Calculus, you get $A_g = \int_a^b g(x) \, dx$.

   c. The area A is the difference between the areas A_f and A_g; that is, $A = A_f - A_g$.

   d. Using Problems 1a–1c and the properties of definite integrals, we see that

   $$A = A_f - A_g = \int_a^b f(x) \, dx - \int_a^b g(x) \, dx = \int_a^b [f(x) - g(x)] \, dx.$$

   e. Since $0 \leq x^2 \leq 4$, you can apply the formula obtained in Problem 1d to compute the area A of the region bounded by the parabola $y = x^2$ and the line $y = 4$. In order to determine the limits of integration, look for the points where the two graphs intersect, which is done by solving the equation $x^2 = 4$. The solutions are $x = 2, x = -2$. Hence,

   $$A = \int_{-2}^{2} (4 - x^2) \, dx = \left(4x - \frac{x^3}{3} \right)\Big|_{-2}^{2}.$$

   Consequently,

   $$A = \left(8 - \frac{8}{3} \right) - \left(-8 - \frac{-8}{3} \right) = \frac{32}{3}.$$

2. **a.** Any vertical translation by C, where C is any real number satisfying $C \geq |g(x)|$ for all $a \leq x \leq b$, will shift the shaded region above the x-axis without changing its shape.

   b. $F(x) = f(x) + C$ and $G(x) = g(x) + C$ for all $a \leq x \leq b$.

   c. Area $= \int_a^b [F(x) - G(x)] \, dx$.

   d. The new and original regions have the same area. Thus,

   $$A = \int_a^b [F(x) - G(x)] \, dx = \int_a^b [(f(x) + C) - (g(x) + C)] \, dx.$$

Consequently,

$$A = \int_a^b \left[f(x) - g(x) \right] dx.$$

e. Since $1 - x \le x^2$ for all $1 \le x \le 3$, you can apply the formula obtained in Problem 2d to evaluate the area A of the shaded region.

$$A = \int_1^3 \left[x^2 - (1 - x) \right] dx = \int_1^3 (x^2 + x - 1) \, dx.$$

Therefore,

$$A = \left(\frac{x^3}{3} + \frac{x^2}{2} - x \right) \Big|_1^3 = \left(\frac{27}{3} + \frac{9}{2} - 3 \right) - \left(\frac{1}{3} + \frac{1}{2} - 1 \right) = \frac{32}{3}.$$

3. a. The area A of the shaded region is equal to $A_1 + A_2$. A_1 is the area of the region between the portions of the graphs of f and g corresponding to $a \le x \le c$; A_2 is the area of the region between the portions of the graphs of f and g corresponding to $c \le x \le b$.

b. Using Problem 2d, you see that

$$A = A_1 + A_2 = \int_a^c \left[f(x) - g(x) \right] dx + \int_c^b \left[g(x) - f(x) \right] dx.$$

c. On the interval $[a, c]$, $f(x) - g(x) \ge 0$ and thus $f(x) - g(x) = |f(x) - g(x)|$. Similarly, on the interval $[c, b]$, $g(x) - f(x) \ge 0$ and thus $g(x) - f(x) = |f(x) - g(x)|$. It follows that

$$A = \int_a^c |f(x) - g(x)| \, dx + \int_c^b |f(x) - g(x)| \, dx = \int_a^b |f(x) - g(x)| \, dx.$$

d. The two graphs intersect three times. To find the intersection points, you need to solve the equation $x^3 - 1 = x - 1$; that is, $x^3 - x = 0$. Equivalently, $x(x^2 - 1) = 0$. The solutions of this equation are $-1, 0,$ and 1. The region enclosed by the two graphs can be partitioned into two subregions:

• The left subregion in which $x - 1 \le x^3 - 1$; its area is given by

$$A_1 = \int_{-1}^0 \left[(x^3 - 1) - (x - 1) \right] dx = \int_{-1}^0 (x^3 - x) \, dx = \frac{x^4}{4} - \frac{x^2}{2} \Big|_{-1}^0$$

$$= \frac{1}{4}.$$

• The right subregion in which $x^3 - 1 \le x - 1$; its area is given by

$$A_2 = \int_0^1 \left[(x - 1) - (x^3 - 1) \right] dx = \int_0^1 (x - x^3) \, dx = \frac{x^2}{2} - \frac{x^4}{4} \Big|_0^1 = \frac{1}{4}.$$

Therefore, the area A of the region that is enclosed by the two graphs is $A = A_1 + A_2 = 1/2$.

CHAPTER 7

SECTION 7.1

Classroom Discussion 7.1.1: Setting the Stage

1. Yes, I agree with Katie.
2. Katie's solution is not practical in general. For example, it cannot be used to find the moon's circumference.
3. Yes, Katie's answer defines unambiguously the length of a curve.
4. One can approximate a given curve by polygonal paths and then use the ruler to measure the lengths of these paths to obtain approximate values for the length of the curve. The better the polygonal paths fit the curve, the more accurate these approximate values become.

Classroom Discussion 7.1.2: Approximate Values for the Length of an Arbitrary Plane Curve

4. As n becomes larger and larger, the polygonal path formed by joining consecutive points on the given curve increasingly better fits the curve. Consequently, the length $\mathcal{L}_n$ of the polygonal path becomes closer and closer to the length $\mathcal{L}$ of the curve. The exact value of $\mathcal{L}$ can be obtained by taking the limit of $\mathcal{L}_n$ as $n \to \infty$.

Classroom Discussion 7.1.3: Approximate Values for the Length of a Graph in the Plane

1.

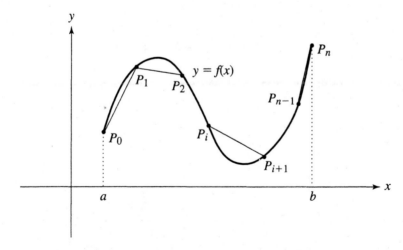

2. Consider the triangle with vertices P_i, P_{i+1}, and Q_i, where Q_i is the point $(x_i, f(x_{i+1}))$. This triangle has a right angle at vertex Q_i.

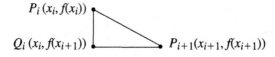

Using the Pythagorean Theorem, we have $(Q_iP_i)^2 + (Q_iP_{i+1})^2 = (P_iP_{i+1})^2$. Since $Q_iP_i = |f(x_{i+1}) - f(x_i)|$ and $Q_iP_{i+1} = x_{i+1} - x_i$, it follows that

$$P_iP_{i+1} = \sqrt{(x_{i+1} - x_i)^2 + [f(x_{i+1}) - f(x_i)]^2}.$$

3. The polygonal path formed by joining the n line segments has length $\mathcal{L}_n$ given by

$$\mathcal{L}_n = \sum_{i=0}^{n-1} P_iP_{i+1} = \sum_{i=0}^{n-1} \sqrt{(x_{i+1} - x_i)^2 + [f(x_{i+1}) - f(x_i)]^2}.$$

4. Since the shortest path between two points is the line segment joining them, $\mathcal{L}_n \leq \mathcal{L}$ for each value of n.

5. As n becomes larger and larger, $\mathcal{L}_n$ becomes closer and closer to $\mathcal{L}$. Here, the P_i's are to be selected throughout the curve in order to "cover" increasingly more of the latter as n increases. We have $\mathcal{L} = \lim_{n \to \infty} \mathcal{L}_n$.

6. In order to simplify the expression of $\mathcal{L}_n$, choose the P_i's such that their x-coordinates are equally spaced; that is, $x_1 - x_0 = x_2 - x_1 = \ldots = x_n - x_{n-1} = (b - a)/n$. It follows that $x_i = a + i(b - a)/n$ for each $0 \leq i \leq n$.

7. By choosing the P_i's as in Problem 6, we get

$$\mathcal{L}_n = \sum_{i=0}^{n-1} \sqrt{\frac{(b - a)^2}{n^2} + [f(x_{i+1}) - f(x_i)]^2},$$

where $x_i = a + i\frac{b-a}{n}$ and $x_{i+1} = a + (i + 1)\frac{b-a}{n}$.

8. The following is a program for a TI-89 calculator.

Step 1: Tell the calculator how to select the points on the curved path by entering

$$a + i * (b - a)/n \to x(i, a, b, n)$$

then press ENTER. The calculator should say "Done." To get the $\to$ symbol, press STO$\to$ (it is located above the ON button). To store the function x, go to VAR-LINK, press the keys F1 and then 8 for "Archive Variable." The function will stay in your calculator's memory even if you clear it from the home screen; you will not have to redefine $x(i, a, b, n)$ for the calculator again, but you will not be allowed to store anything else under the name "x" without unarchiving the latter.

Step 2: Tell the calculator how to compute $\mathcal{L}_n$ by entering

$$\sum_{i=0}^{n-1} \sqrt{(b - a)^2/n^2 + (f(x_{i+1}) - f(x_i))^2} \to L(a, b, n)$$

then, press ENTER. The calculator should say "Done." Go to VAR-LINK and archive $L(a, b, n)$. To get the Σ symbol, press the keys F3 and then 4. To enter the sum

$$\sum_{i=0}^{n-1} \sqrt{(b - a)^2/n^2 + (f(x_{i+1}) - f(x_i))^2},$$

type: $\Sigma\left(\sqrt{(b - a)^2/n^2 + (f(x_{i+1}) - f(x_i))^2}, i, 0, n - 1\right)$

Step 3: Tell the calculator what function to consider. For example, if your function is $f(x) = 1/x^3$, then enter

$$1/x^3 \to f(x)$$

Do not archive this function; you will want to change it later.

Step 4: Evaluate the desired length $\mathcal{L}_n$. For example, to evaluate the length $\mathcal{L}_n$ of the graph of the function in Step 3 that is defined on the interval $[1,6]$ with $n = 70$, type "$L(1,6,70)$", and then press ENTER. The calculator should return the value of this length.

To work with a different function, return to Step 3 and enter your new function as described.

9.

n	$\mathcal{L}_n$
1	$\sqrt{2}$
2	1.4604
3	1.47065
4	1.47428
5	1.47596
6	1.47687
7	1.47742
8	1.47778
9	1.47802
10	1.4782

Table 2

10. $\mathcal{L} \approx 1.4782$.

Classroom Discussion 7.1.4: Exact Value for the Length of a Graph in the Plane

1. $\mathcal{L}(a) = 0$. $\mathcal{L}(\frac{a+b}{2})$ is the length of the portion of the graph lying between points $(a, f(a))$ and $(c, f(c))$ where $c = \frac{a+b}{2}$ is the midpoint of the interval $[a, b]$. $\mathcal{L}(b) = \mathcal{L}$.

2. As x increases, the portion of the graph lying between points $(a, f(a))$ and $(x, f(x))$ increases; its length becomes larger. Thus, the function $\mathcal{L}(x)$ is increasing on the interval $[a, b]$.

3. If Δx is very small, the additional portion of the graph that is between points $(x, f(x))$ and $(x + \Delta x, f(x + \Delta x))$ might be viewed as the line segment joining them.

4. Using the Pythagorean Theorem, the length of the line segment joining points $(x, f(x))$ and $(x + \Delta x, f(x + \Delta x))$ is $\sqrt{(\Delta x)^2 + (\Delta y)^2}$. Thus,

$$\Delta \mathcal{L} \approx \sqrt{(\Delta x)^2 + (\Delta y)^2}.$$

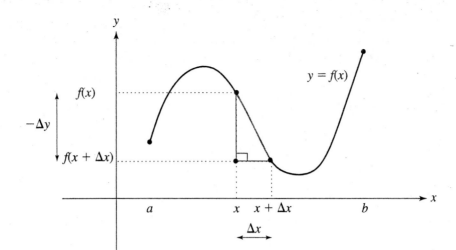

5. The approximation obtained in Problem 4 leads to the following approximation:

$$\frac{\Delta \mathcal{L}}{\Delta x} \approx \frac{\sqrt{(\Delta x)^2 + (\Delta y)^2}}{\Delta x} = \sqrt{\frac{(\Delta x)^2 + (\Delta y)^2}{(\Delta x)^2}} = \sqrt{1 + \left(\frac{\Delta y}{\Delta x}\right)^2}.$$

6. Since the function f is differentiable,

$$\lim_{\Delta x \to 0} \frac{\Delta y}{\Delta x} = \lim_{\Delta x \to 0} \frac{f(x + \Delta x) - f(x)}{\Delta x} = f'(x).$$

It follows that

$$\mathcal{L}'(x) = \lim_{\Delta x \to 0} \frac{\Delta \mathcal{L}}{\Delta x} = \lim_{\Delta x \to 0} \sqrt{1 + \left(\frac{\Delta y}{\Delta x}\right)^2} = \sqrt{1 + \lim_{\Delta x \to 0} \left(\frac{\Delta y}{\Delta x}\right)^2}$$

$$= \sqrt{1 + \left(\lim_{\Delta x \to 0} \frac{\Delta y}{\Delta x}\right)^2} = \sqrt{1 + [f'(x)]^2}.$$

7. Since the function $f'(x)$ is continuous, the function $[f'(x)]^2$ is continuous, as is the function $1 + [f'(x)]^2$. Therefore, the function $\sqrt{1 + [f'(x)]^2}$ is also continuous.

The function $\mathcal{L}(x)$ is the antiderivative of $\sqrt{1 + [f'(x)]^2}$ that takes the value 0 at $x = a$. It follows from Classroom Discussion 5.2.2 that

$$\mathcal{L}(x) = \int_a^x \sqrt{1 + [f'(t)]^2}\, dt.$$

8. $\mathcal{L} = \mathcal{L}(b) = \displaystyle\int_a^b \sqrt{1 + [f'(t)]^2}\, dt.$

Practice Problems

1. Since $y'(x) = 2x$, $\sqrt{1 + [y'(x)]^2} = \sqrt{1 + 4x^2}$. Using your calculator, you see that $\mathcal{L} \approx 1.4789$, which is correct to four decimal places. This confirms our conjecture in Classroom Discussion 7.1.3.

2. **a.**

$$y'(x) = 10\left(\frac{1}{20}e^{x/20} - \frac{1}{20}e^{-x/20}\right) = \frac{1}{2}\left(e^{x/20} - e^{-x/20}\right).$$

   b.

$$1 + [y'(x)]^2 = 1 + \frac{1}{4}\left(e^{x/20} - e^{-x/20}\right)^2$$

$$= 1 + \frac{1}{4}\left[\left(e^{x/20}\right)^2 - 2e^{x/20}e^{-x/20} + \left(e^{-x/20}\right)^2\right].$$

Using the properties of the exponential function, you get

$$1 + [y'(x)]^2 = 1 + \frac{1}{4}\left(e^{2x/20} - 2e^0 + e^{-2x/20}\right)$$

$$= 1 + \frac{1}{4}\left(e^{x/10} - 2 + e^{-x/10}\right).$$

Therefore,

$$1 + [y'(x)]^2 = \frac{1}{4}\left(e^{x/10} + 2 + e^{-x/10}\right).$$

   c. Using again the properties of the exponential function, you see that

$$1 + [y'(x)]^2 = \frac{1}{4}\left(e^{x/20} + e^{-x/20}\right)^2.$$

   d. It follows that

$$\sqrt{1 + [y'(x)]^2} = \frac{1}{2}\left(e^{x/20} + e^{-x/20}\right).$$

Therefore,

$$\mathcal{L} = \int_{-20}^{20} \frac{1}{2}\left(e^{x/20} + e^{-x/20}\right)dx$$

$$= \frac{1}{2}\left(20e^{x/20} - 20e^{-x/20}\right)\Big|_{-20}^{20}$$

$$= 10\left[(e - e^{-1}) - (e^{-1} - e)\right]$$

$$= 20(e - e^{-1}) \approx 47.008.$$

Classroom Discussion 7.1.5: Circumference of a Circle from Different Points of View

2. Circumference of a Circle Using Calculus

a. A point (x, y) lies on the upper semicircle if and only if $x^2 + y^2 = r^2$ and $y \geq 0$.

b. It follows that $y = \sqrt{r^2 - x^2}$.

c. **i.** Rewrite $y(x)$ as $(r^2 - x^2)^{1/2}$, then use the square-root rule to evaluate its derivative.

$$y'(x) = \frac{1}{2}\left(r^2 - x^2\right)^{-1/2}\left(r^2 - x^2\right)'$$

$$= \frac{1}{2}\left(r^2 - x^2\right)^{-1/2}(-2x)$$

$$= \frac{-x}{\sqrt{r^2 - x^2}}.$$

ii. We have

$$1 + \left[y'(x)\right]^2 = 1 + \frac{x^2}{r^2 - x^2} = \frac{r^2}{r^2 - x^2}.$$

Therefore,

$$\sqrt{1 + \left[y'(x)\right]^2} = \frac{r}{\sqrt{r^2 - x^2}}.$$

iii. Using Theorem 7.1.1, we obtain

$$\mathcal{L} = 2\int_{-r}^{r}\sqrt{1 + \left[y'(x)\right]^2}\,dx = 2\int_{-r}^{r}\frac{r}{\sqrt{r^2 - x^2}}\,dx = 2r\int_{-r}^{r}\frac{1}{\sqrt{r^2 - x^2}}\,dx.$$

Using your calculator to evaluate the integral, you see that $\mathcal{L} = 2\pi r$.

SECTION 7.2

Classroom Discussion 7.2.1: Generalities about Surfaces

1.

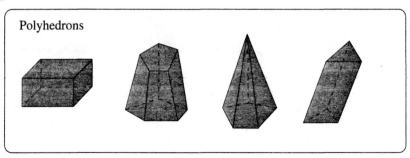

Polyhedrons

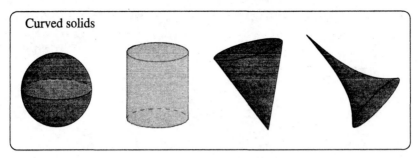

Curved solids

2. The surface area of a polyhedron is the sum of the areas of all of its faces. The faces are polygons, and thus we know how to compute their areas.
3. I agree with Brian since, in a sense, this is similar to flattening the surface and then evaluating the area of the resulting plane region.

Classroom Discussion 7.2.2: Area of the Lateral Surface of a Cylinder

2. **Approximating the Cylinder with Appropriate Prisms**
   a. The lateral surface of each n-hedron has an area $S_n = p_n h$.
   b. As n increases, these n-hedrons approach the cylinder. In particular, the perimeters p_n approach the circumference of its base, and the areas S_n approach the area of its lateral surface.
   c. Letting n tend to infinity in the formula $S_n = p_n h$ leads to

$$S = \lim_{n \to \infty} S_n = \lim_{n \to \infty} (p_n h) = \left(\lim_{n \to \infty} p_n \right) h = 2\pi r h.$$

Classroom Discussion 7.2.3: Area of the Lateral Surface of a Cone

1. **Cutting the Lateral Surface of a Cone into a Sector**
   a. Using the Pythagorean Theorem, we see that $r^2 + h^2 = R^2$; therefore $R = \sqrt{h^2 + r^2}$.
   b. The circumference of the cone's base and the arc length of the sector are equal, thus $2\pi r = \frac{\alpha}{360} 2\pi R$. It follows that $r/R = \alpha/360$.
   c. Since the area S of the lateral surface of the cone is equal to the area of the sector, you see that

$$S = \frac{\alpha}{360} \pi R^2 = \frac{r}{R} \pi R^2 = \pi r R = \pi r \sqrt{h^2 + r^2}.$$

Classroom Discussion 7.2.4: Area of the Lateral Surface of a Truncated Cone

1. Using the formula obtained in Classroom Discussion 7.2.3, you see that $S_1 = \pi R \sqrt{R^2 + (h + a)^2}$.
2. Using the same formula, you get $S_2 = \pi r \sqrt{r^2 + a^2}$.
3. It follows that $S = S_1 - S_2 = \pi R \sqrt{R^2 + (h + a)^2} - \pi r \sqrt{r^2 + a^2}$.
4. Using similarity, $r/R = a/(a + h)$. Solving for a, you get $a = rh/(R - r)$.

5. In the formula for the area S obtained in Problem 3, replace a by its expression in terms of r, R, and h; then, simplify as follows:

$$S = \pi R\sqrt{R^2 + (h + a)^2} - \pi r\sqrt{r^2 + a^2}$$

$$= \pi R\sqrt{R^2 + \left(h + \frac{rh}{R - r}\right)^2} - \pi r\sqrt{r^2 + \left(\frac{rh}{R - r}\right)^2}$$

$$= \pi R\sqrt{R^2 + \frac{R^2 h^2}{(R - r)^2}} - \pi r\sqrt{r^2 + \frac{r^2 h^2}{(R - r)^2}}$$

$$= \pi R\sqrt{\frac{R^2}{(R - r)^2}\left[(R - r)^2 + h^2\right]} - \pi r\sqrt{\frac{r^2}{(R - r)^2}\left[(R - r)^2 + h^2\right]}$$

$$= \pi \frac{R^2}{R - r}\sqrt{(R - r)^2 + h^2} - \pi\frac{r^2}{R - r}\sqrt{(R - r)^2 + h^2}$$

$$= \pi \frac{R^2 - r^2}{R - r}\sqrt{(R - r)^2 + h^2}$$

$$= \pi (R + r)\sqrt{h^2 + (R - r)^2}.$$

Classroom Discussion 7.2.5: Area of an Arbitrary Surface of Revolution

1. $S(a) = 0$. $S\left(\frac{2a + b}{3}\right)$ is the area of the surface swept out by revolving about the x-axis the portion of the graph lying between the points $(a, f(a))$ and $(c, f(c))$, where $c = \frac{2a + b}{3} = a + \frac{b - a}{3}$. $S\left(\frac{a + b}{2}\right)$ is the area of the surface swept out by revolving about the x-axis the portion of the graph lying between the points $(a, f(a))$ and $(m, f(m))$, where $m = \frac{a + b}{2}$ is the midpoint of the interval $[a, b]$. $S(b) = S$.

2. As x increases, the portion of the graph lying between the points $(a, f(a))$ and $(x, f(x))$ becomes larger. In turn, the surface swept out by revolving about the x-axis this portion of the graph becomes larger; its area becomes larger. Thus, the function $S(x)$ is increasing on the interval $[a, b]$.

3. If Δx is very small, the additional region can be approximated by the truncated cone obtained by revolving about the x-axis the line segment joining points $(x, f(x))$ and $(x + \Delta x, f(x + \Delta x))$.

4. Recall that the surface area of a truncated cone of height h and radii r and R is $\pi(R + r)\sqrt{h^2 + (R - r)^2}$; see Classroom Discussion 7.2.4. The truncated cone described in Problem 3 has height Δx and radii $f(x)$ and $f(x + \Delta x)$. Thus, its lateral surface has area $\pi[f(x) + f(x + \Delta x)]\sqrt{(\Delta x)^2 + [f(x + \Delta x) - f(x)]^2}$. Thus,

$$\Delta S \approx \pi[f(x) + f(x + \Delta x)]\sqrt{(\Delta x)^2 + [f(x + \Delta x) - f(x)]^2}.$$

5.

$$\frac{\Delta S}{\Delta x} \approx \pi[f(x) + f(x + \Delta x)] \frac{\sqrt{(\Delta x)^2 + [f(x + \Delta x) - f(x)]^2}}{\Delta x}$$

$$= \pi[f(x) + f(x + \Delta x)] \sqrt{\frac{(\Delta x)^2 + [f(x + \Delta x) - f(x)]^2}{(\Delta x)^2}}$$

$$= \pi[f(x) + f(x + \Delta x)] \sqrt{1 + \left[\frac{f(x + \Delta x) - f(x)}{\Delta x}\right]^2}.$$

6. Since f is differentiable, it is continuous, and thus $\lim_{\Delta x \to 0} f(x + \Delta x) = f(x)$. By definition of the derivative, $\lim_{\Delta x \to 0} \frac{f(x + \Delta x) - f(x)}{\Delta x} = f'(x)$.

7. Using Problem 6, you see that $\lim_{\Delta x \to 0} [f(x) + f(x + \Delta x)] = 2f(x)$. Using Problem 5 and the properties of limits, you get

$$\lim_{\Delta x \to 0} \frac{\Delta S}{\Delta x} = 2\pi f(x)\sqrt{1 + [f'(x)]^2}.$$

8.

$$S'(x) = \lim_{\Delta x \to 0} \frac{\Delta S}{\Delta x} = 2\pi f(x)\sqrt{1 + [f'(x)]^2}.$$

9. The function $S(x)$ is the antiderivative of the function $2\pi f(x)\sqrt{1 + [f'(x)]^2}$ that takes the value 0 at $x = a$. It follows from Classroom Discussion 5.2.2 that

$$S(x) = \int_a^x 2\pi f(t)\sqrt{1 + [f'(t)]^2}\, dt.$$

10. $S = S(b) = \int_a^b 2\pi f(t)\sqrt{1 + [f'(t)]^2}\, dt.$

Practice Problems

1. The lateral surface of the cone is generated by revolving $\overline{OA}$ about the x-axis, where A is the point of coordinates (h, r) and O is the origin. $\overline{OA}$ is the graph of the function $f(x) = ax$ where $a = r/h$, defined on the interval $[0, h]$. The area of the cone's lateral surface is given by Theorem 7.2.1. For $0 \le x \le h, f'(x) = a$ and $\sqrt{1 + [f'(x)]^2} = \sqrt{1 + a^2}$. Therefore,

$$S = \int_0^h 2\pi a x \sqrt{1 + a^2}\, dx = \pi a \sqrt{1 + a^2} \int_0^h 2x\, dx = \pi a \sqrt{1 + a^2}\, x^2 \Big|_0^h$$

$$= \pi a h^2 \sqrt{1 + a^2} = \pi \frac{r}{h} h^2 \sqrt{1 + \frac{r^2}{h^2}} = \pi r \sqrt{h^2 + r^2}.$$

This is exactly the formula obtained earlier for the area of the lateral surface of a right cone of radius $r > 0$ and height $h > 0$.

2. You have $f'(x) = 3x^2$ on $[0, 1]$. Applying Theorem 7.2.1, you get

$$S = \int_0^1 2\pi x^3 \sqrt{1 + (3x^2)^2} \, dx = 2\pi \int_0^1 x^3 \sqrt{1 + 9x^4} \, dx.$$

This integral can be computed using the chain rule or a substitution. Let $u = 1 + 9x^4$; then $du = 36x^3 \, dx$. When $x = 0$, $u = 1$; when $x = 1$, $u = 10$. Therefore,

$$S = 2\pi \int_1^{10} u^{1/2} \frac{du}{36} = \frac{\pi}{18} \int_1^{10} u^{1/2} \, du = \frac{\pi}{18} \frac{2}{3} u^{3/2} \Big|_1^{10} = \frac{\pi}{27} (10^{3/2} - 1).$$

Classroom Discussion 7.2.6: Revolving about the x-axis the Graphs of Functions Taking Negative Values

1.

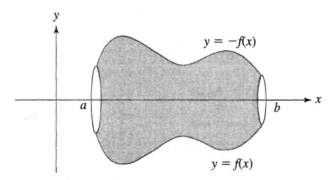

The surface generated by revolving the graph $y = f(x)$ about the x-axis is the same as the one generated by revolving the graph $y = -f(x)$ about the x-axis. Thus, using Theorem 7.2.1, we have

$$S = \int_a^b 2\pi [-f(x)] \sqrt{1 + [(-f)'(x)]^2} \, dx.$$

Using the constant multiple rules for derivatives and integrals, we get

$$S = \int_a^b 2\pi [-f(x)] \sqrt{1 + [-f'(x)]^2} \, dx = - \int_a^b 2\pi f(x) \sqrt{1 + [f'(x)]^2} \, dx.$$

2.

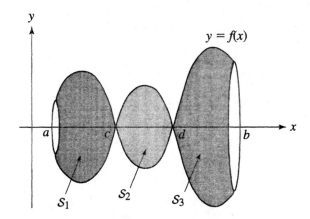

Let S_1, S_2, and S_3 be the areas of the surfaces swept out by revolving about the x-axis the portions of the graph $y = f(x)$ whose endpoints are $(a, f(a))$ and $(c, f(c))$, $(c, f(c))$ and $(d, f(d))$, and $(d, f(d))$ and $(b, f(b))$, respectively. Clearly, $S = S_1 + S_2 + S_3$. Using Theorem 7.2.1 together with Problem 1, we have

$$S_1 = \int_a^c 2\pi f(x)\sqrt{1 + [f'(x)]^2}\, dx$$

$$S_2 = -\int_c^d 2\pi f(x)\sqrt{1 + [f'(x)]^2}\, dx$$

$$S_3 = \int_d^b 2\pi f(x)\sqrt{1 + [f'(x)]^2}\, dx.$$

Classroom Discussion 7.2.7: Area of a Sphere from Different Points of View

2. Areas of Spheres Using Calculus

a. Points (x, y) on the upper semicircle are those satisfying the equation $x^2 + y^2 = r^2$, where $y \geq 0$. Thus, this semicircle is the graph of $f(x) = \sqrt{r^2 - x^2}$ defined on $[-r, r]$.

b. Using the square-root rule,

$$f'(x) = \left[(r^2 - x^2)^{\frac{1}{2}}\right]' = \frac{1}{2}(r^2 - x^2)^{-\frac{1}{2}}(-2x) = -\frac{x}{\sqrt{r^2 - x^2}}.$$

c. The expression can be simplified as follows.

$$\sqrt{1 + [f'(x)]^2} = \sqrt{1 + \left(\frac{-x}{\sqrt{r^2 - x^2}}\right)^2} = \sqrt{1 + \frac{x^2}{r^2 - x^2}}$$

$$= \sqrt{\frac{r^2}{r^2 - x^2}} = \frac{r}{\sqrt{r^2 - x^2}}.$$

d. The surface area S can be computed as follows.

$$S = \int_{-r}^{r} 2\pi f(x) \sqrt{1 + \left[f(x)'\right]^2}\, dx$$

$$= \int_{-r}^{r} 2\pi \sqrt{r^2 - x^2}\, \frac{r}{\sqrt{r^2 - x^2}}\, dx$$

$$= 2\pi r \int_{-r}^{r} dx.$$

Therefore, $S = 2\pi r x \Big|_{-r}^{r} = 2\pi r \left[r - (-r)\right] = 4\pi r^2$.

SECTION 7.3

Classroom Discussion 7.3.1: Volumes of Polyhedrons and Cylinders

1. Since ℓ, w, and h are whole numbers, the rectangular right prism can be divided into $\ell w h$ unit cubes; its volume is $\ell w h$ cubic units.
2. Since ℓ, w, and h are rational numbers, $\ell = \frac{\ell_1}{\ell_2}$, $w = \frac{w_1}{w_2}$, and $h = \frac{h_1}{h_2}$, where ℓ_1, ℓ_2, w_1, w_2, h_1, and h_2 are whole numbers with ℓ_2, w_2, and h_2 being nonzero. By duplicating appropriately each edge of the prism of length ℓ, w, or h a number of times equal to ℓ_2, w_2, or h_2, respectively, you obtain a larger cube of length ℓ_1, width w_1, and height h_1. Since these dimensions are all whole numbers, using Problem 1, you see that its volume is $\ell_1 w_1 h_1$. The resulting prism is composed of $\ell_2 w_2 h_2$ copies of the original prism. Thus, the volume of the latter is $\frac{\ell_1 w_1 h_1}{\ell_2 w_2 h_2}$; that is, $\ell w h$.
3. Since ℓ, w, and h are irrational numbers, there exist sequences of rational numbers (ℓ_n), (w_n), and (h_n) satisfying $\lim_{n\to\infty} \ell_n = \ell$, $\lim_{n\to\infty} w_n = w$, and $\lim_{n\to\infty} h_n = h$. For each $n \geq 1$, consider the prism of length ℓ_n, width w_n, and height h_n. Its volume is equal to $\ell_n w_n h_n$ by Problem 2. Since these prisms approach increasingly more the original prism as n becomes larger, the volume of the latter is $\lim_{n\to\infty} \ell_n w_n h_n = \ell w h$.

   Therefore, the volume of a rectangular right prism is the product of its length, width, and height. Equivalently, it is the product of its height and the area of its base.
4. Complete the triangular bases of the given prism into rectangles and then the right triangular prism into a right rectangular one as shown in the following figure. From the construction, you can see that the right triangular prism occupies half of the space that is occupied by the right rectangular one. The right rectangular prism has volume $2\mathcal{A}h$ by what we have done earlier. Thus, the right triangular one has volume $\mathcal{A}h$.

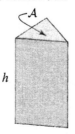

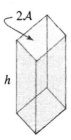

base

Now consider an arbitrary right prism, then divide one of its polygonal bases into triangles and denote by $A_1, A_2, \ldots, A_n$ the areas of the resulting triangles. Using this decomposition, divide the right prism into right triangular prisms as shown in the following figure. The resulting prisms have volumes $A_1h, A_2h, \ldots,$ A_nh. Thus, the volume of the original prism is

$$\sum_{i=1}^{n} A_i h = \left(\sum_{i=1}^{n} A_i \right) h = Ah.$$

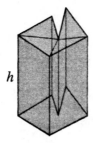

h

5. Approximate one of the bases of the generalized right cylinder by polygonal regions of areas $A_1, \ldots, A_n, \ldots$ so that the area A of the generalized right cylinder's base is $A = \lim_{n \to \infty} A_n$. Using these polygonal regions, approximate the generalized right cylinder by right prisms. These approximating prisms have volumes $A_1h, \ldots, A_nh, \ldots$ using Problem 4. As n increases, the approximating right prisms approach increasingly more the original prism, and thus its volume is

$$\lim_{n \to \infty} A_n h = \left(\lim_{n \to \infty} A_n \right) h = Ah.$$

For a cylinder, $A = \pi r^2$, where r is the radius of the cylinder's base; thus, its volume is $\pi r^2 h$.

6. The formulas obtained in Problems 4 and 5 are still valid.

7. No, the techniques used so far do not extend to this type of solids, as their shapes are too irregular.

8. The volume of the cylindrical cup is $\pi \cdot 4^2 \cdot 12 = 192\pi$ cubic centimeters; thus the volume of the container is 192π cubic centimeters. Clearly, this procedure is not practical for measuring the volumes of arbitrary solids.

Classroom Discussion 7.3.2: Formula for the Volumes of Solids

2. $V(a) = 0$. $V\left(\frac{2a + b}{3}\right)$ is the volume of the portion of the solid that lies between the cross-sections corresponding to a and $c = \frac{2a + b}{3} = a + \frac{b - a}{3}$. $V\left(\frac{a + b}{2}\right)$ is the volume of the portion of the solid that lies between the cross-sections corresponding to a and the midpoint $m = \frac{a + b}{2}$ of the interval $[a, b]$. $V(b) = V$.

3. As x increases, the portion of the solid that lies between the cross-sections corresponding to a and x becomes larger and so does its volume. Thus, the function $V(x)$ is increasing on the interval $[a, b]$.

4. If Δx is very small, the additional portion of the solid can be approximated by the generalized cylinder whose base is the cross-section corresponding to x and whose width is Δx.

5. $\Delta V \approx A(x)\Delta x$.

6. $\dfrac{\Delta V}{\Delta x} \approx A(x)$.

7.
$$V'(x) = \lim_{\Delta x \to 0} \frac{\Delta V}{\Delta x} = \lim_{\Delta x \to 0} A(x) = A(x).$$

8. Using Problems 2 and 7, you see that the function $V(x)$ is the antiderivative of $A(x)$ that takes the value 0 at a. It follows from Classroom Discussion 5.2.2 that

$$V(x) = \int_a^x A(t)\,dt.$$

9. $V = V(b) = \displaystyle\int_a^b A(t)\,dt.$

Practice Problem

The cross-sections of the given cylinder are all congruent to its base, so for each real number $0 \le x \le h$, the cross-section corresponding to x has area $A(x) = A$. The volume of the cylinder is given by:

$$V = \int_0^h A(x)\,dx = \int_0^h A\,dx = A \int_0^h 1\,dx = A\,x\Big|_0^h = A(h - 0) = Ah.$$

This formula is the same as the one obtained in Classroom Discussion 7.3.1.

Classroom Discussion 7.3.3: Volumes of Cones

2. **Volumes of Cones Using Calculus**

   a. If an oblique cone and a right cone have the same height and congruent bases, then they have the same volume. To see this, imagine again that your oblique cone is a stack of circular sheets of paper.

   b. For each $0 \le x \le h$, the cross-section of the cone corresponding to x is a disc of radius $f(x) = ax$, where $a = r/h$; its area is

   $$A(x) = \pi[f(x)]^2 = \pi(ax)^2 = \pi a^2 x^2.$$

   c. Using Theorem 7.3.1, you see that the volume of the cone is

   $$V = \int_0^h A(x)\,dx = \int_0^h \pi a^2 x^2\,dx = \pi a^2 \int_0^h x^2\,dx = \pi a^2 \frac{x^3}{3}\Big|_0^h$$

   $$= \pi \frac{r^2}{h^2}\left(\frac{h^3}{3} - 0\right) = \frac{1}{3}\pi r^2 h.$$

   So, the volume of a cone is $1/3$ the volume of a cylinder that has the same radius and height as the cone.

Classroom Discussion 7.3.4: Volumes of Pyramids

3. Volumes of Pyramids Using Calculus

a. Use Cavalieri's idea: imagine that the pyramid is made up of a stack of cross-sections that are sheets of paper, then rectify it into a right pyramid. The original pyramid and the resulting one have the same height, the same base, and the same volume (they occupy the same amount of space).

b. Using similarity, $A(x)/A_0 = x^2/h^2$. Therefore, $A(x) = A_0 x^2/h^2$.

c. Using Theorem 7.3.1, you see that the volume V of the pyramid is

$$V = \int_0^h A(x)\, dx = \int_0^h \frac{A_0}{h^2} x^2\, dx = \frac{A_0}{h^2} \int_0^h x^2\, dx = \frac{A_0}{h^2} \frac{x^3}{3}\Big|_0^h = \frac{A_0}{h^2}\frac{h^3}{3}$$

$$= \frac{1}{3} A_0 h.$$

So, the volume of a pyramid is $1/3$ the volume of a prism that has the same height and the same base area as the pyramid.

Classroom Discussion 7.3.5: Volumes of Spheres

3. Volumes of Spheres Using Calculus

a. For each $-r \le x \le r$, the cross-section of the sphere corresponding to x is the disc of radius $f(x) = \sqrt{r^2 - x^2}$ centered at point $(x, 0)$. Thus, its area is

$$A(x) = \pi[f(x)]^2 = \pi(r^2 - x^2).$$

b. Using Theorem 7.3.1, you see that the sphere's volume is

$$V = \int_{-r}^r A(x)\, dx = \int_{-r}^r \pi(r^2 - x^2)\, dx = \pi\left(r^2 x - \frac{x^3}{3}\right)\Big|_{-r}^r$$

$$= \pi\left[\left(r^3 - \frac{r^3}{3}\right) - \left(-r^3 + \frac{r^3}{3}\right)\right] = \pi\left(r^3 - \frac{r^3}{3} + r^3 - \frac{r^3}{3}\right)$$

$$= \frac{4}{3}\pi r^3.$$

Answers to Selected Exercises

CHAPTER 1

SECTION 1.1

1. $x_1 = 5, x_{n+1} = 3x_n$ if $n \geq 1$.

3. $y_1 = 4, y_{n+1} = y_n + 6$ if $n \geq 1$.

5. $x_1 = 3, x_2 = 6, x_{n+1} = x_n + x_{n-1}$ if $n \geq 2$.

7. $x_1 = 5, x_{n+1} = x_n - 3$ if $n \geq 1$.

9. $y_1 = 16, y_{n+1} = y_n\left(-\frac{1}{4}\right)$ if $n \geq 1$.

11. We have two possibilities:

$$y_1 = \frac{3}{2}, y_{n+1} = y_n \cdot 2 \text{ if } n \geq 1,$$

$$y_1 = \frac{3}{2}, y_{n+1} = y_n(-2) \text{ if } n \geq 1.$$

13. $x_n = 100 - 2(n - 1), n \geq 1$.

15. $y_n = 1{,}024\left(\frac{1}{2}\right)^{n-1}, n \geq 1$.

17. $x_n = (n + 2)^2, n \geq 1$.

19. $x_1 = 27, x_n = 27 - 2(n - 1), n \geq 1$.

21. $y_n = \frac{1}{3}(3)^{n-1} = 3^{n-2}, n \geq 1$.

23. $y_n = 10(-1)^{n-1}, n \geq 1$.

25. $S_9 = \frac{365}{4}$.

27. Divergent; oscillates between 1 and 3.

29. Convergent; $\lim_{n\to\infty} x_n = 0$.

31. Divergent; $\lim_{n\to\infty} x_n = \infty$.

33. Convergent; $\lim_{n\to\infty} a_n = 0$.

35. Convergent; $\lim_{n\to\infty} a_n = 0$.

37. Convergent; $\lim_{n\to\infty} a_n = \frac{5}{2}$.

39. Convergent; $\lim_{n\to\infty} x_n = 0$.

41. Convergent; $\lim_{n\to\infty} \frac{n!}{n^n} = 0$.

43. If a^2, b^2, c^2 are consecutive terms of an arithmetic sequence, then $c^2 - b^2 = b^2 - a^2$. Use this identity to show that $\left(\frac{1}{a+b} - \frac{1}{a+c}\right) - \left(\frac{1}{a+c} - \frac{1}{b+c}\right) = 0$.

45. $\sum_{i=4}^{1{,}000} 2i$

47. $\sum_{i=3}^{100} i^2$

49. $\sum\limits_{i=0}^{333} (11 + 3i)$

51. $\sum\limits_{i=0}^{8} 10(2)^i$

53. $\sum\limits_{i=1}^{27} 5i = 5 + 10 + 15 + \cdots + 135$

$\sum\limits_{j=0}^{26} (5 + 5j) = 5 + 10 + 15 + \cdots + 135$

The two sums are equal, thus there is more than one way of writing a sum using the summation notation.

55. $8 + 11 + 14 + \cdots + 59$

57. $\dfrac{3}{2}$

59. a. $x_n = 11 + (n - 1) \cdot 3, n \geq 1$.

   b. S has 334 terms; $x_{12} = 44$.

   c. $S = 170{,}507$

61. a. $x_n = 9\left(\dfrac{1}{3}\right)^{n-1}, n \geq 1$.

   b. S has 9 terms; $x_5 = \dfrac{1}{9}$.

   c. $S = \dfrac{3{,}280}{243}$

63. a. $x_n = \sqrt{2} \cdot n^2, n \geq 1$.

   b. S has 60 terms.

   c. $S = 73{,}810\sqrt{2}$

65. Let $(-a, a)$ be an arbitrary open interval centered at 0, where $a > 0$. If $a > \dfrac{1}{5}$, then all the terms of the sequence belong to $(-a, a)$. If $0 < a \leq \dfrac{1}{5}$, then for $n \geq n_0 + 1$ we have $0 < \dfrac{1}{n + 4} \leq \dfrac{1}{n_0 + 5} < a$. Hence, all but n_0 terms of the sequence are contained in the interval $(-a, a)$.

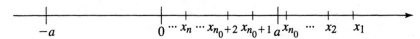

67. True. Let a be an arbitrary positive number. Whenever $n > \dfrac{1}{a}$, x_n will be contained in the interval $(-a, a)$. Therefore, an arbitrary open interval containing zero contains all but finitely many terms of the sequence.

   A similar problem is as follows. Consider a sequence $\{x_n\}_n$ generated by randomly picking a number x_n in the interval $\left(\dfrac{1}{2^{n+1}}, \dfrac{1}{2^n}\right)$ for each integer $n \geq 1$. Is it true that any such sequence will converge to 0?

SECTION 1.2

1. $0.\overline{53} = \dfrac{\frac{53}{100}}{1 - \frac{1}{100}} = \dfrac{53}{99}$ by Theorem 1.2.2. Alternatively, if $x = 0.\overline{53}$, then

$$\begin{array}{r} 100x = 53.\overline{53} \\ -x = -0.\overline{53} \\ \hline 99x = 53 \end{array},$$

so $x = \dfrac{53}{99}$.

3. $0.0\overline{25} = \dfrac{\frac{25}{1,000}}{1 - \frac{1}{1,000}} = \dfrac{25}{999}$ by Theorem 1.2.2. Alternatively, if $x = 0.0\overline{25}$, then

$$
\begin{aligned}
1,000x &= 25.\overline{025} \\
-x &= -0.\overline{025} \\
\hline
999x &= 25
\end{aligned},
$$

so $x = \dfrac{25}{999}$.

5. $1.12\overline{34} = \dfrac{112}{100} + \dfrac{34}{9,900} = \dfrac{5,561}{4,950}$ by Theorem 1.2.2. Alternatively, if $x = 1.12\overline{34}$, then

$$
\begin{aligned}
100x &= 112.\overline{34} \\
-x &= -1.12\overline{34} \\
\hline
99x &= 111.22
\end{aligned},
$$

so $x = \dfrac{5,561}{4,950}$.

7. Converges. Its sum is $\dfrac{\frac{2}{3}}{1 - \frac{1}{3}} = 1$.   **9.** Diverges.

11. Converges. Its sum is $\dfrac{1}{1 - \left(-\frac{1}{2}\right)} = \dfrac{2}{3}$.   **13.** Diverges.

15. Converges. Its sum is $\dfrac{\frac{3}{4}}{1 - \left(-\frac{3}{4}\right)} = \dfrac{3}{7}$.   **17.** Diverges.

19. Diverges.   **21.** The total vertical distance the ball travelled is 70 feet.

23. $S = \dfrac{1}{(1-x)^2}$

CHAPTER 1 REVIEW

1. $x_n = \dfrac{n}{n+1}, n \geq 1.$

3. $x_n = \dfrac{1}{n^2}, n \geq 1.$

5. $a_{2n+2} = \dfrac{(2n)^2 \cdot [2(n-1)]^2 \cdots 2^2}{(2n+2)!} = \dfrac{2^{2n}(n!)^2}{(2n+2)!}, n \geq 1, a_2 = \dfrac{1}{2}.$

7. $y_{n+1} = \sqrt{3}\,y_n, n \geq 1,$ and $y_1 = \dfrac{1}{6}.$

9. $x_{n+1} = x_n + \dfrac{27}{23}, n \geq 1,$ and $x_1 = 20.$

11. Since a, b, c are consecutive terms of an arithmetic sequence, it follows that $b - a = c - b$. Use this identity to show that $c^2 + ca + a^2 - (b^2 + bc + c^2) = a^2 + ab + b^2 - (c^2 + ca + a^2)$ by factoring out $a + b + c$ on both sides of this equality.

13. Divergent (see Example 2c in Section 1.1).

15. Convergent; $\lim\limits_{n \to \infty} |x_n| = 0.$ **17.** Convergent; $\lim\limits_{n \to \infty} a_n = 2.$

19. Convergent; $\lim\limits_{n \to \infty} a_n = 0.$ **21.** $107 \cdot 320 = 34,240$

23. $10\sqrt{5}\,\dfrac{(\sqrt{5})^{16} - 1}{\sqrt{5} - 1}$

25. Denote by S this sum. If $a = 1$ or $a = -1$, then $S = 4n$. If $a \neq \pm 1$, $S = \frac{a^{2n} - 1}{a^2 - 1}\left(a^2 + \frac{1}{a^{2n}}\right) + 2n$.

27. $S = y_1^3 \cdot r^3 \frac{r^{3n} - 1}{r^3 - 1}$

29. Rationalize the denominators; $\frac{\sqrt{x_1 + (n-1)d} - \sqrt{x_1}}{d}$.

31. Converges to 5.

33. Divergent.

35. Divergent.

37. Converges to 10.

39. $99 \cdot 2^{100} + 1$

41. $\lim\limits_{n \to \infty} \frac{4}{3} \cdot \frac{\frac{1}{2^{n+1}} - 1}{\frac{1}{3^{n+1}} - 1} = \frac{4}{3}$

CHAPTER 2

SECTION 2.1

1. Function: Image $= \{0, 1, 4, 9, 16, 25, 36, \ldots\}$.

3. Not a function

5. **a.** $[0, \frac{5\sqrt{10}}{2}]$

b. $h(0) = 1{,}000$, $h(1) = 984$, $h(3) = 856$, $h(t + 1) = -16(t + 1)^2 + 1{,}000$.

c. $h(t + 1) - h(t)$ represents the distance covered by the falling object between t and $t + 1$ seconds.

d. False. Corresponds to $h(t + 1) = h(t) - h(1)$, while $h(t + 1) = h(t) - 32t - 16 \neq h(t) - h(1)$.

7. $y = 4x - 2$ is a function of x.

9. $y = \pm\sqrt{x^2 + 1}$ is not a function of x.

11. $y = \sqrt[5]{\frac{100 + 2x}{3}}$ is a function of x.

13. $y = -3$ is a function of x.

15. Image $= (-\infty, 1] \cup (5, \infty)$.

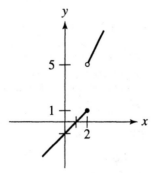

17. $\mathbb{R}, [0, \infty)$

19. $\mathbb{R}\setminus\{0\}, \mathbb{R}\setminus\{0\}$

21. $\mathbb{R}, (0, 3]$

23. $\mathbb{R}\setminus\{-1\}, \mathbb{R}\setminus\{0\}$

25. $\mathbb{R}, [0,\infty)$

27. $[1,\infty), [0,\infty)$

29. $\mathbb{R}, \mathbb{R}$

31. $(-2,\infty), (0,\infty)$

33. $(f \circ g)(x) = 3 + 2(-2x + 1) = 5 - 4x$, Domain $= \mathbb{R}$

35. $(f \circ g)(x) = | - x| = |x|$, Domain $= \mathbb{R}$

37. $(g \circ f)(x) = \dfrac{1}{x + 3}$, Domain $= \mathbb{R} \setminus \{-3\}$

39. $(g \circ f)(x) = (x^4 + 1)^{\frac{1}{3}}$, Domain $= \mathbb{R}$

41. $f(g(-1)) = f(1) = 2$

43. $g(f(1)) = g(2) = 5$

45. $f(f(-1)) = f(3) = 0$

47. $f(x) \geq 0$ for all $x \in$ Domain(f).

49. The domain of f contains $[0,\infty)$.

51. $f(x) > -1$ for all $x \in$ Domain(f).

53. $f(x + 1) = (x + 2)^2$

55. Suppose such functions exist. If we first set $x = 0$ and then set $y = 0$ in $f(x) = g(y) = xy$, we obtain $f(0) + g(y) = 0$ for all $y \in \mathbb{R}$ and $f(x) + g(0) = 0$ for all $x \in \mathbb{R}$, respectively. Now we can conclude that $f(x) = f(0)$ for all $x \in \mathbb{R}$ and $g(y) = g(0)$ for all $y \in \mathbb{R}$. These facts plus setting $y = 1$ in the identity $f(x) + g(y) = xy$ give $f(x) = x - g(1) = x - g(0)$ for all $x \in \mathbb{R}$, leading to a contradiction.

57. $g_1(x) = \frac{1}{2}x + \frac{9}{2}$ and $g_2(x) = \frac{1}{2}x + \frac{15}{2}$ for all $x \in \mathbb{R}$.

SECTION 2.2

1. $0, 1.5, \lim\limits_{x \to 2} f(x)$ does not exist.

3. $2, 2, 2$

5. $-1, 0, 0, 0$

7. $2, +\infty, 2$, does not exist.

9. $0, 0, 0, 0$

11. -9

13. $\sqrt{3}$

15. $\dfrac{1}{4}$

17. 0

19. -1

21. $\sqrt[3]{-3}$

23. 0

25. 2

27. -7

29. **a.** $\mathbb{R} \setminus \{1\}$. We cannot compute $\lim\limits_{x \to 1} f(x)$ using P4 since the denominator of $f(x)$ is zero at $x = 1$.

c. $\lim\limits_{x \to 1} \dfrac{x^{n+1} - 1}{x - 1} = \lim\limits_{x \to 1} (x^n + x^{n-1} + \cdots + x + 1) = n + 1$

31. Let $f(x) = 3x^3 - 1$. Fix an open interval I centered at 2; and let $2l$ be the length of the interval I; that is, $I = (2 - l, 2 + l)$. The condition $f(x) \in I$ is equivalent to $1 - \dfrac{l}{3} < x^3 < 1 + \dfrac{l}{3}$. This suggests that a good candidate for the interval J is $J = (\sqrt[3]{1 - \frac{l}{3}}, \sqrt[3]{1 + \frac{l}{3}})$. Indeed, $1 \in J, J \setminus \{1\}$ is contained in the domain of f, and $x \in J \setminus \{1\} \Rightarrow -\dfrac{l}{3} < x^3 - 1 < \dfrac{l}{3} \Rightarrow f(x) \in I$.

SECTION 2.3

1. Continuous on $\mathbb{R}\backslash\{-1\}$; discontinuous at $x = 1$.

3. Continuous on $\mathbb{R}$.   **5.** $a = 4$

7. $a = 2$ **9.** $f(x) = x^2 - x$

11. $f(x) = \begin{cases} 1 & \text{if } x \neq 4 \\ 0 & \text{if } x = 4 \end{cases}$ **13.** $f(x) = \begin{cases} 0 & \text{if } x < 4 \\ 1 & \text{if } x \geq 4 \end{cases}$

15. We have four possibilities: $f_1(x) = x$, $f_2(x) = -x$, $f_3(x) = |x|$, and $f_4(x) = -|x|$.

CHAPTER 2 REVIEW

1. $y = \pm\sqrt{x^3 - 2}$ is not a function of x.

3. $y = \frac{5}{2}x + \frac{1}{2}$ is a function of x.

5. $f(x) = \begin{cases} x^2 + x & \text{if } x \leq 0 \\ x^2 - x & \text{if } x > 0 \end{cases}$ Image $(f) = [-\frac{1}{4}, \infty)$.

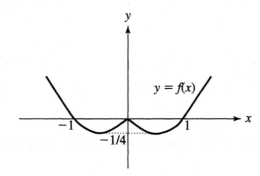

7. Domain $(f) = \mathbb{R}$; Image $(f) = [\frac{7}{4}, \infty)$

9. Domain $(f) = [0, \infty)$; Image $(f) = [0, \infty)$

11. Domain $(f) = \mathbb{R}$; Image $(f) = \mathbb{R}$

13. Domain $(f) = \mathbb{R}\backslash\{-2\}$; Image $(f) = \{-1, 1\}$

15. Domain $(f) = \mathbb{R}\backslash\{-3, 3\}$; Image $(f) = \mathbb{R}\backslash\{0, -\frac{1}{6}\}$

17. $f(g(x)) = \frac{x^2 + 2x}{x^2 + 2x - 3}$; Domain $(f \circ g) = \mathbb{R}\backslash\{-3, 1\}$

19. $g(f(x)) = \sqrt{x - 3}$; Domain $(g \circ f) = [3, \infty)$

21. $f(g(x)) = \begin{cases} 2x + 5 & \text{if } x < -1 \\ x^2 + 2x + 4 & \text{if } x \geq -1 \end{cases}$ $g(f(x)) = \begin{cases} 2x + 4 & \text{if } x < 0 \\ x^2 + 4 & \text{if } x \geq 0 \end{cases}$

23. f is not defined at $-2, \frac{5}{6}, \frac{5}{6}, \frac{5}{6}$.

25. $1, -\infty, 1$, does not exist.   **27.** $0, 0, 0, 0$

29. 56

31. -1

33. 0

35. -3

37. $f(x) = \begin{cases} x + 1 & \text{if } x \le 0 \\ -x & \text{if } x > 0 \end{cases}$

39. $f(x) = \begin{cases} 2x - 1 & \text{if } x \ne 1 \\ 0 & \text{if } x = 1 \end{cases}$

41. $f(x) = x^2 + 3x - 1$

43. f is continuous on $\mathbb{R} \backslash \{0\}$ and is discontinuous at 0.

45. f is continuous on $\mathbb{R}$.

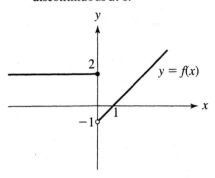

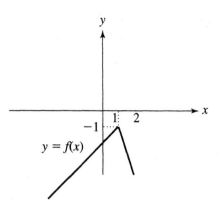

47. $a = -1$

49. $f(x) = x$ or $f(x) = -x + b$ for some $b \in \mathbb{R}$.

CHAPTER 3

SECTION 3.1

1. The team won 40% of its games, or 4 out of 10 games. The number of total games played is $x = 20$.

3. **i.** 33.1667 miles per gallon
   ii. 31.0256 miles per gallon
   iii. 29.1463 miles per gallon
   iv. 30.9266 miles per gallon

5. **a.** $f(x) = 0.50x$ where $x \in \{0, 1, 2, 3, \ldots\}$.

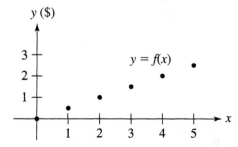

   b. f is a linear function. The rate of change of f over any interval $[a, b]$ is constant and equal to 0.5.

SECTION 3.2

1. $f'(x) = 2x + 2$ **3.** $A'(s) = 4s + 1$

5. $y = -5x + 8$

7. a. A is decreasing. $A(t)$ will never be zero, but as $t \to \infty$, $A(t) \to 0$.

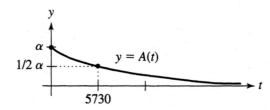

     b. A' is negative because A is decreasing. $A'(t)$ will never be zero, since the graph of A never has a horizontal tangent line. A' is increasing and $A'(t) \to 0$ as $t \to \infty$.

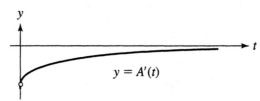

9. a. iii
     b. i
     c. ii

11.

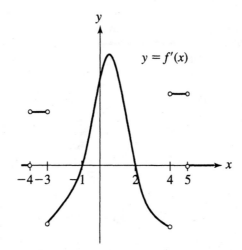

13. a. If we set $x_1 = f(x), x_2 = g(x)$, then $x_1 \le x_2$; since f is nondecreasing, it follows that $f(f(x)) = f(x_1) \le f(x_2) = f(g(x))$.
     b. If we set $y = g(x)$, since $f(y) \le g(y)$ for all $y \in \mathbb{R}$, it follows that $f(g(x)) = f(y) \le g(y) = g(g(x))$.

c. $f(f(x)) \leq f(g(x)) \leq g(g(x))$.

d. Since $g(x) \leq h(x)$ and g is nondecreasing, it follows that $g(g(x)) \leq g(h(x))$. Since $g(y) \leq h(y)$ for all $y \in \mathbb{R}$, if we let $y = g(x)$, then $g(g(x)) \leq h(g(x))$. Hence, $g(g(x)) \leq g(h(x)) \leq h(h(x))$.

e. From c and d, it follows that $f(f(x)) \leq g(g(x)) \leq h(h(x))$, and the answer is true.

15. a. Since $|f(x)| \leq |x|^\alpha$ for all $x \in \mathbb{R}$, it follows that $f(0) = 0$. Hence, $f'(0) = \lim\limits_{h \to 0} \frac{f(h)}{h}$. Based on the assumptions on f, we have $\left| \frac{f(h)}{h} \right| \leq |h|^{\alpha-1}$ for any $h \neq 0$, so $\lim\limits_{h \to 0} \frac{f(h)}{h} = 0$. Consequently, f is differentiable at 0 and $f'(0) = 0$.

b. Under the new assumptions, we can conclude that $\left| \frac{f(h)}{h} \right| \geq |h|^{\alpha-1}$ for all $h \neq 0$. Reasoning as before, it follows that $\lim\limits_{h \to 0} \frac{f(h) - f(0)}{h}$ does not exist.

c. We can conclude that $f(0) = 0$ but cannot conclude anything about the differentiability of f at 0. For example, if $f(x) = x$, then f is differentiable at $x = 0$ and $f'(0) = 1$, while the function $g(x) = |x|$ fails to be differentiable at $x = 0$.

SECTION 3.3

1. a. $v(t) = d'(t) = 110$ miles/hour

b. $d(t) = -110t + \frac{880}{3}$, where $\frac{3}{2} \leq t \leq \frac{8}{3}$.

c. $d(t) = -110t + \frac{1{,}045}{3}$, where $\frac{19}{6} \leq t \leq \frac{13}{3}$.

d. $d(t) = 110t - 715$, where $\frac{16}{3} \leq t \leq \frac{13}{2}$.

e. The graph of the displacement $d(t)$ and the velocity $v(t)$ are sketched here.

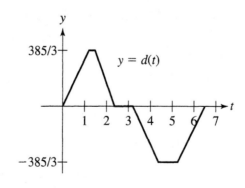

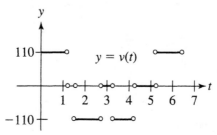

3. **a.** $v(t) = -32t + 100$. The object was fired from the ground with the initial velocity $v(0) = 100$ feet per second.
   b. $v(1.2) = 61.6$ feet per second
   c. $h(3.125) = 156.25$ feet is the maximum height.
   d. $[0, 3.125)$
   e. $(3.125, 6.25]$
   f. $a(1) = a(2) = -32$

5. **a.** $v(t) = \begin{cases} 0.51\,t^2, & 0 < t < 1 \\ -0.34(t - 2), & 1 < t < 2 \end{cases}$ The velocity at $t = 1$ cannot be determined $[\lim\limits_{t \to 1^-} v(t) = 0.51$ and $\lim\limits_{t \to 1^+} v(t) = 0.34]$.

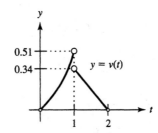

b. $a(t) = \begin{cases} 1.02\,t, & 0 < t < 1 \\ -0.34, & 1 < t < 2 \end{cases}$ The acceleration at $t = 1$ cannot be determined.

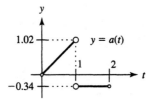

Since the car is always traveling in the positive direction, the acceleration is positive when the car is speeding up and negative when it is slowing down.

SECTION 3.4

1. $f'(x) = (3x^2)' + (-4x)' + (2)'$ sum rule
   $= 3(x^2)' - 4(x)' + (2)'$ constant multiple rule
   $= 6x - 4$ power rule

3. $g'(t) = -6 + 20{,}040t^{2{,}003}$ sum, constant multiple, power rules

5. $f'(s) = 4s - 1$ product, sum, constant multiple, power rules

7. $g'(t) = -5 + 2t$ sum, constant multiple, power rules

9. $h'(t) = \frac{2}{t^2}$ constant multiple, reciprocal rules

11. $h'(t) = \frac{3}{t^4}$ constant multiple, power rules (or constant multiple, reciprocal, and power rules)

13. $f'(x) = \frac{-2x - 3}{(x^2 + 3x)^2}$ reciprocal, sum, constant multiple, power rules

15. $f'(x) = \frac{-1}{(x + 1)^2}$ quotient, sum, power rules

17. $f'(x) = \dfrac{2x + 3}{2\sqrt{x^2 + 3x}}$ square root, sum, constant multiple, power rules

19. $h'(x) = 0$ power rule

21. $f'(s) = \dfrac{2s - 1}{2\sqrt{s^2 - s}}$

23. $A'(x) = \sqrt{4 - x^2} - \dfrac{x^2}{\sqrt{4 - x^2}}$

25. $f'(s) = \dfrac{s + 2}{2(s + 1)^{\frac{3}{2}}}$

27. $f'(x) = 8x^3 + 9x^2 - 10x - 3$

29. $g'(x) = -5x^{-\frac{8}{3}}$

31. $f'(x) = \dfrac{3x + 3}{4\sqrt{x}}$

33. $y = \dfrac{4}{5}x + \dfrac{9}{5}$

35. $f'(x) = -\dfrac{1}{(x + 2)^2}$, negative for all $x \neq -2$; $f(x)$ is decreasing on $(-\infty, -2)$ and $(-2, \infty)$.

37. $g'(x) = 0$ when $x = 2^{\frac{1}{3}}$, $g'(x) < 0$ when $0 < x < 2^{\frac{1}{3}}$, and $g'(x) > 0$ when $x > 2^{\frac{1}{3}}$ and when $x < 0$. $g(x)$ is decreasing on $(0, 2^{\frac{1}{3}})$ and increasing on $(2^{\frac{1}{3}}, \infty)$ and $(-\infty, 0)$.

39. $f'(x) = \dfrac{2}{(x + 1)^2} > 0$ for all $x \neq -1$. $f(x)$ is increasing on $(-\infty, -1)$ and on $(-1, \infty)$.

41. $f'(x) = \dfrac{-x^2 - 2x + 2}{(x^2 + 2)^2}$ is negative for $x \in (-\infty, -1 - \sqrt{3}) \cup (-1 + \sqrt{3}, \infty)$ and positive for $x \in (-1 - \sqrt{3}, -1 + \sqrt{3})$. $f(x)$ is decreasing on $(-\infty, -1 - \sqrt{3})$ and on $(-1 + \sqrt{3}, \infty)$ and is increasing on $(-1 - \sqrt{3}, -1 + \sqrt{3})$.

43. Taking the derivative of $f(x) = g(x)Q(x)$ and using the product rule in the right-hand side, we obtain $f'(x) = g'(x)Q(x) + g(x)Q'(x)$. Hence, $Q'(x) = \dfrac{f'(x)g(x) - f(x)g'(x)}{[g(x)]^2}$.

SECTION 3.5

1. $f'(g(x))g'(x) = [153(2 - 3x + x^3)^{50} - 156(2 - 3x + x^3)^{38}](-3 + 3x^2)$

3. $f'(g(x))g'(x) = \dfrac{4x}{(-x^2 + 3)^3}$

5. $f'(g(x))g'(x) = \dfrac{4(10 - 4x^3 + 5x)^3}{3[(10 - 4x^3 + 5x)^4 + 6]^{\frac{2}{3}}}(-12x^2 + 5)$

7. $2{,}004(x^3 - 4x + 7)^{2{,}003}(3x^2 - 4)$

9. $\dfrac{2x - 1}{5(x^2 - x - 1)^{\frac{4}{5}}}$

11. $\dfrac{-4{,}004}{(2x - 1)^{1{,}002}}$

13. $\dfrac{1}{5}(x^4 - 3x^2 + 4x)^{-\frac{4}{5}}(4x^3 - 6x + 4)$

15. $\dfrac{15(2t - 3t^2)}{(t^2 - t^3)^4}$

17. $-\dfrac{4}{3}\left(\dfrac{x + 2}{x - 2}\right)^{-\frac{2}{3}}\dfrac{1}{(x - 2)^2}$

19. $y = 4x + 3$ **21.** $(f \circ g)'(9) = 1$

23. $(f \circ g)'(2) = \frac{7}{18}$

25. $\{D(s(t))\}'\big|_{t=-2} = \frac{-2}{\sqrt{2,500 + (-2)^2}} \cdot 1 \approx 0.04$ meters per second

$\{D(s(t))\}'\big|_{t=3} \approx 0.06$ meters per second

$\{D(s(t))\}'\big|_{t=600} \approx 1$ meter per second

27. $\{D(h(t))\}'\big|_{t=2} = \frac{192}{\sqrt{400 + (192)^2}}[-32(2) + 128] \approx 63.6556$ feet per second.

$\{D(h(t))\}'\big|_{t=5} = \frac{240}{\sqrt{400 + (240)^2}}[-32(5) + 128] \approx -31.8895$ feet per second.

29. $(f \circ g)'(1) = 12 + \frac{2}{\sqrt{5}}, (h \circ f)'(0) = 0$

CHAPTER 3 REVIEW

1. $y = \frac{1}{5}(x - 2) + 2 = \frac{1}{5}x + \frac{8}{5}$

3.

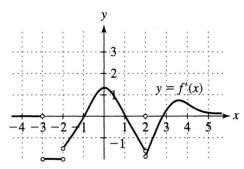

5. a. $s(0) = |v(0)| = 16$ feet/second

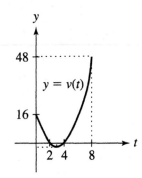

b. $a(t) = 4t - 12, a(0) = -12$ feet per second squared

c. The skater reverses his motion at $t = 2$ and $t = 4$.

d. When $0 \leq t < 2$, $v(t) > 0$; thus $h(t)$ is increasing, and the skater is going forward. When $2 < t < 4$, $v(t) < 0$; thus $h(t)$ is decreasing, and the skater is going backward. During these 2 seconds, his speed increased and then decreased. When $4 < t < 8$, $v(t) > 0$; thus $h(t)$ is increasing, and the skater is going forward again. At $t = 2$ and $t = 4$, the skater reverses his motion. His velocity is 0 at these times.

e. The skater's velocity is the smallest at $t = 3$ when $v(3) = -2$ feet per second.

7. a. $v(t) = -32t$ feet per second, $s(t) = 32t$ feet per second, $a(t) = -32$ feet per second squared

 b. $\frac{25\sqrt{2}}{4}$ seconds

 c. $-200\sqrt{2}$ feet per second

9. a. 81 feet

 b. 4.5 seconds

 c. -72 feet per second

11. $g'(x) = 12x^3 + 2 + 12x^{-3}$

13. $f'(x) = -x^4 - 6x^2$

15. $14x^6 - 10x^4 - 4x^3 + 6x^2 + 2x - 2$

17. $\dfrac{-x^4 - 4x^2 - 2x - 4}{(x^3 + 2x - 1)^2}$

19. $\dfrac{t^2 - 2t - 2}{(t-1)^2}$

21. $10(t^2 - 3t)^9(2t - 3)$

23. $\dfrac{3}{2}\dfrac{1}{\sqrt{t-1}} + \dfrac{3}{2}\dfrac{1}{t\sqrt{t}}$

25. $\dfrac{5}{2}\dfrac{(x^3 + x - 2)^4(3x^2 + 1)}{\sqrt{1 + (x^3 + x - 2)^5}}$

27. $-\dfrac{1}{\sqrt{x}(\sqrt{x} - 1)^2}$

29. $-\dfrac{1}{2\sqrt{x}(\sqrt{x} + 2)^2}$

31. $5\left(\dfrac{x^3}{3} - x^2 + \dfrac{2}{x}\right)^4(x^2 - 2x - \dfrac{2}{x^2})$

33. $\dfrac{29}{4}$

35. $-\dfrac{3}{4}$

37. $\dfrac{10}{9\pi}$ feet per minute

CHAPTER 4

SECTION 4.1

1. $f'(x) = 3$; no critical points

 $f'(x) > 0$ on the entire domain $[-4, 12]$.

 $f(x)$ is increasing on $[-4, 12]$.

3. $f'(x) = x^2 - 9$; critical points: $-3, 3$

 On $(-\infty, -3)$, $f'(x) > 0$, and $f(x)$ is increasing.

 On $(-3, 3)$, $f'(x) < 0$, and $f(x)$ is decreasing.

 On $(3, \infty)$, $f'(x) > 0$, and $f(x)$ is increasing.

5. **1.** absolute minimum: $f(-4) = -14$; absolute maximum: $f(12) = 34$

  2. absolute minimum: $f(1.5) = -0.25$; absolute maximum: $f(8) = 42$

  3. local minimum: $f(3) = -18$; local maximum: $f(-3) = 18$; no absolute minimum or maximum

  4. local and absolute maximum: $f(2) = 16$; no local or absolute minimum

7. The volume is maximized when squares of side length $\frac{5}{3}$ inches are cut out.

9. The area of the playground will be maximized when its dimensions are $2r = 200$ meters by $\ell = 100\pi$ meters. In this case, there will be $\pi r^2 = 10,000\pi$ square meters available for planting flowers.

11. To earn the maximum revenue of \$98, the gardener should sell his crop after 2 weeks.

13. The cost for the cable will be minimized if the cable is run along the river for $10 - \frac{\sqrt{3}}{3} \approx 9.42$ miles, then angled across the river to the other house.

15. In order to minimize cost, the store should order 20 refrigerators at a time and place 25 orders per year.

SECTION 4.2

1. concave up

3. concave down

5. If $a > 0$, f is concave up; if $a < 0$, f is concave down.

7.

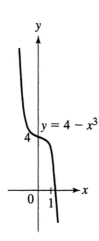

9.

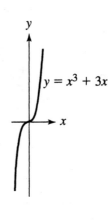

11.

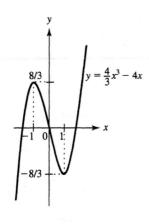

13.

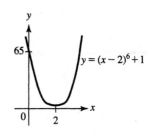

15.

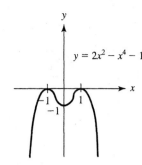

$y = 2x^2 - x^4 - 1$

17.

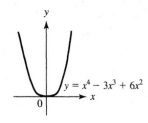

$y = x^4 - 3x^3 + 6x^2$

19. a. Increasing
  b. Concave up
  c. Concave down
  d. No
  e. No

SECTION 4.3

1. $f'(x) = \frac{4}{5}e^x - 15x^4 + 6$

3. $f'(x) = e^{x^2 - x + 6}(2x - 1)$

5. $f'(x) = \sqrt{e^{2x} - 1}$

7. $f'(x) = e^x + xe^x$

9. $f'(x) = e^{\frac{2}{3}x}\left(\frac{1}{3}x^{-\frac{2}{3}} + \frac{2}{3}x^{\frac{1}{3}}\right)$

11. $f'(x) = e^{-5x+1}\dfrac{-5x^4 - 4x^3 - 5x^2 - 2x - 20}{(x^4 + x^2 + 4)^2}$

13. $f'(x) = \dfrac{1 - x^2e^x}{(1 + xe^x)^2}$

15.

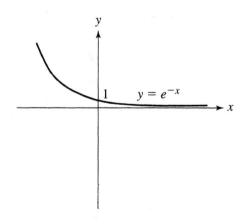

$y = e^{-x}$

17.

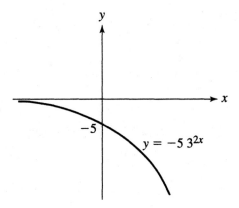

$y = -5 \, 3^{2x}$

19.

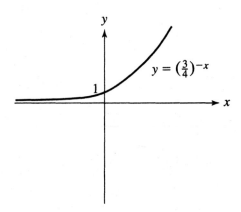

$y = \left(\frac{3}{4}\right)^{-x}$

21.

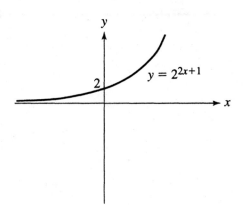

$y = 2^{2x+1}$

23. $[1, 2]$

25. $[0, \infty)$

27. $x \in \left(\frac{5}{3}, \infty\right)$

CHAPTER 4 REVIEW

1. $x = 2$

3.

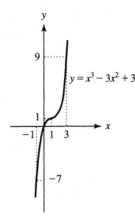

5.

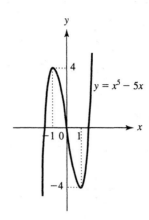

7. $6x^5 + \dfrac{1}{x^2} + 3e^x$

9. $e^{x^2-3x+1}(2x - 3)$

11. $e^{\sqrt{x}}\dfrac{1}{2\sqrt{x}}$

13. $e^{-x}(5 - 2x)$

15. $\dfrac{2e^x}{(2+e^x)^2}$

17. $e^{\frac{x+1}{x^2+1}}\left[\dfrac{1-2x-x^2}{(x^2+1)^2}\right]$

19.

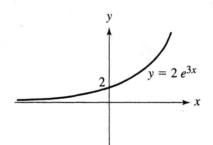

21.

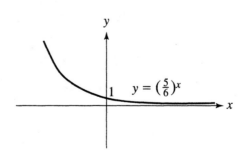

23. $[1, e]$

25. $\left(0, \dfrac{1}{3}\right]$

27. $(1, 2)$

CHAPTER 5

SECTION 5.1

1. $\mathbb{R}, \frac{1}{4}x^4 + C$

3. $\mathbb{R}, 2x + C$

5. $\mathbb{R}, C$

7. $\mathbb{R}, \frac{1}{3}t^3 + t + C$

9. $\mathbb{R}, \frac{5}{7}t^{\frac{7}{5}} + C$

11. $[0,\infty), 2t^{\frac{3}{2}} + C$

13. $[0,\infty), \frac{2}{5}x^{\frac{5}{2}} + C$

15. $(-\infty,0) \cup (0,\infty), \begin{cases} -\frac{1}{2x^2} + C \text{ if } x < 0 \\ -\frac{1}{2x^2} + D \text{ if } x > 0 \end{cases}$

17. $(-\infty,0) \cup (0,\infty), \begin{cases} 2x - \frac{1}{x} + C \text{ if } x < 0 \\ 2x - \frac{1}{x} + D \text{ if } x > 0 \end{cases}$

19. $\mathbb{R}, e^t + C$

21. $\mathbb{R}, \frac{3}{7}t^{\frac{7}{3}} + e^t - \pi t + C$

23. $(0,\infty), \frac{1}{4}t^4 + \frac{1}{3}t^3 - 2t^{\frac{1}{2}} + 2t^{-\frac{1}{2}} + C$

25. $F(x) = e^x + \frac{1}{2}x^2 - x + 5$

27. $\int |x|dx = \frac{1}{2}x|x| + C$ on $\mathbb{R}$

29. $F(x) = \begin{cases} -\frac{1}{x} + \frac{1}{2} \text{ if } x > 0 \\ -\frac{1}{x} \text{ if } x < 0 \end{cases}$

31. a. acceleration $= \frac{1}{900}$ miles per second squared

  b. $d(5) = \frac{625}{9,000} \approx 0.069$ miles

33. a. Samantha hits the water about 0.2 second after Scott hits the water.

  b. Samantha and Scott hit the water with about the same speed: approximately 14 meters per second.

35. a. $y = \sqrt[3]{x + C}$

  b. $y = \frac{1}{\sqrt[5]{-15x+C}}$

  c. $y = \sqrt[4]{x^2 - x + C}$

  d. $y' = 2e^{2x} + 2x, y'' = 4e^{2x} + 2$

  e. $y' = 3e^x + 2e^{2x}, y'' = 3e^x + 4e^{2x}$

SECTION 5.2

1. $\int_a^b 0\,dx = 0$

3. $\int_{-1}^{1} 7x^6\,dx = 2$

5. $\int_0^{-1} x^4(2x - 1)\, dx = \frac{8}{15}$

7. $\int_1^3 \frac{1}{x^4}\, dx = \frac{26}{81}$

9. $\int_1^2 (x^4 + 1)/x^2\, dx = \frac{17}{6}$

11. $\int_1^2 x\sqrt{x}\, dx = \frac{2}{5}(4\sqrt{2} - 1)$

13. $\int_{-2}^5 |x|\, dx = \frac{29}{2}$

15. $\int_{-1}^0 |2x + 1|\, dx = \frac{1}{2}$

17. $\int_{-2}^0 e^{x+2}\, dx = e^2 - 1$

19. $\int (3x - x^2)\, dx$ is increasing on the interval $(0, 3)$, decreasing on the intervals $(-\infty, 0)$ and $(3, \infty)$, and it is maximized when $a = 0, b = 3$.

21. antiderivative, $\dfrac{1}{t^2 + t + 1}, 0, 0$

23. **a.** $H = F \circ h$, where $h(x) = x^2$.
 b. $H'(x) = 2x f(x^2)$
 c. $G = -F \circ g = m \circ F \circ g$, where $g(x) = x^3$ and $m(x) = -x$.
 d. $G'(x) = -3x^2 f(x^3)$
 e. $K = G + H, K'(x) = 2x f(x^2) - 3x^2 f(x^3)$
 f. $H = F \circ h, G = m \circ F \circ g, K = G + H$
 $H'(x) = f[h(x)]\, h'(x), G'(x) = -f[g(x)]\, g'(x),$
 $K'(x) = f[h(x)]\, h'(x) - f[g(x)]\, g'(x)$
 g. $4x^7 e^{x^4} - x e^x$
 h. $\dfrac{2}{4x^2 + 1} - \dfrac{e^x}{e^{2x} + 1}$

SECTION 5.3

1. $\frac{2}{3}(x + 2)^{\frac{3}{2}} + C$

3. $\frac{1}{3}(x^2 + 1)^3 + C$

5. $\frac{2}{9}(x^3 + 9)^{\frac{3}{2}} + C$

7. $\frac{10}{81}$

9. $\frac{2}{3}(x^3 + 21x + 8)^{\frac{1}{2}} + C$

11. $\frac{1}{3}$

13. $-e^{\frac{1}{x}} + C$

15. $-\dfrac{1}{2(e^x - 2)^2} + C$

17. **a.** $(uv)' = u'v + uv'$
 b. Integrate the equality in a.
 c. No
 d. $xe^x - e^x + C$
 e. $\dfrac{xe^{3x}}{3} - \dfrac{e^{3x}}{9} + C$
 f. $(x^2 - 2x + 2)e^x + C$
 g. $(x^2 - x + 2)e^x + C$
 h. $(x^3 - 3x^2 + 3x - 6)e^x + C$
 i. Step 1: $u = P(x), v' = e^x$, thus $\int P(x)\, e^x\, dx = P(x)\, e^x - \int P'(x)\, e^x\, dx$.
 Step 2: $u = P'(x), v' = e^x$, thus $\int P(x)\, e^x\, dx = P(x)\, e^x - P'(x)\, e^x + \int P''(x)\, e^x\, dx$.
 And so on, until reaching a derivative of P that is a constant. To evaluate the integral, one needs to use integration by parts a number of times equal to the degree of P.

j. Step 1: $u = P(x), v' = e^{kx}$, thus $\int P(x)\,e^{kx}\,dx = \frac{1}{k}P(x)\,e^{kx} - \int \frac{1}{k}P'(x)\,e^{kx}\,dx.$

Step 2: $u = P'(x), v' = \frac{e^{kx}}{k}$, thus $\int P(x)\,e^{kx}\,dx = \frac{1}{k}P(x)\,e^{kx} - \frac{1}{k^2}P'(x)\,e^{kx} + \int \frac{1}{k^2}P''(x)\,e^{kx}\,dx.$

And so on, until reaching a derivative of P that is a constant. To evaluate the integral, one needs to use integration by parts a number of times equal to the degree of P.

CHAPTER 5 REVIEW

1. $\mathbb{R}, \frac{9}{5}x^5 + C$

3. $[0, \infty), \frac{3}{55}x^{11/2} + C$

5. $(-\infty, 0) \cup (0, \infty), \frac{1}{x} + C$

7. $\mathbb{R}, \frac{15}{4}x^{4/3} + C$

9. $(-\infty, 0) \cup (0, \infty), e^x - \frac{3}{2}x^{2/3} + C$

11. $\mathbb{R}, \frac{1}{6}x^6 + \frac{1}{5}x^5 + \frac{9}{2}x^2 + 9x + C$

13. $(-\infty, 0) \cup (0, \infty), -\frac{3}{2}x^{2/3} + \frac{6}{5}x^{5/3} - x^2 + x + C$

15. $(-\infty, 0) \cup (0, \infty), -x^{-1} - x^{-2} + \frac{1}{3}x^{-3} + C$

17. a. $\left[(x - 1)e^x + C\right]' = xe^x$
b. $\left[(x^2 - 2x + 2)e^x + C\right]' = x^2 e^x$
c. $\left[(x^3 - 3x^2 + 6x - 6)e^x + C\right]' = x^3 e^x$
d. Check that the function $x^n e^x$ is the derivative of
$$[x^n - nx^{n-1} + n(n - 1)x^{n-2} - \cdots + (-1)^{n-1}n!\,x + (-1)^n n!]e^x + C.$$

19. The antiderivatives of f are of the form
$$F(x) = \begin{cases} -x^2 + x + C & \text{if } x \le 1/2, \\ x^2 - x + \frac{1}{2} + C & \text{if } x \ge 1/2. \end{cases}$$

21. $F(x) = \begin{cases} -x^2 - 9 & \text{if } x \le 0, \\ x^2 - 9 & \text{if } x \ge 0. \end{cases}$

23. $F(x) = \begin{cases} \dfrac{x^2}{2} - \dfrac{1}{x} & \text{if } x < 0, \\ \dfrac{x^2}{2} - \dfrac{1}{x} + 1 & \text{if } x > 0. \end{cases}$

25. The ball's initial velocity is -14 feet per second. Its velocity at impact is -46 feet per second.

27. $\frac{2}{101}$

29. -4

31. $\frac{256\sqrt{2} - 67}{36}$

33. $3e - 6$

35. 1

37. a. 6
  b. 0
  c. 12

39. $\frac{2}{3}(3\sqrt{3} - 2\sqrt{2})$

41. $\frac{1}{3}(2x + 3)^{3/2} + C$

43. $\frac{1}{297}(x^3 + 5)^{99} + C$

45. $e^{x^2 - x + 2} + C$

47. $\frac{1}{4}$

49. 0

CHAPTER 6

SECTION 6.1

1. For the first square, argue as in Case 2 (see the beginning of Section 6.1), with
$a = \frac{1}{10}$.
For the second square, argue as in Case 3, with $a = \pi = 3.14159\ldots$.

3. a. $\frac{1}{2}pa$

   b. $p = 2na \cdot \tan\left(\frac{\pi}{n}\right)$

     Area $= na^2 \tan\left(\frac{\pi}{n}\right)$

   c. Area $= 2\sqrt{3}\,a^2$

SECTION 6.2

1. a. Counting only whole squares: $\alpha = 38, \beta = 76$.
     Counting half-squares: $\alpha = 46.5, \beta = 65.5$.

   b. Counting only whole squares: $\gamma = \beta - \alpha = 38$.
     Counting half-squares: $\gamma = \beta - \alpha = 19$.

   c. Counting only whole squares: $\frac{\alpha + \beta}{2} = 57$.
     Counting half-squares: $\frac{\alpha + \beta}{2} = 56$.

   d. The answer here depends on the finer grid that is used.

   e. A finer grid leads to better approximations.

   f. Your approximations will approach the true area of the figure.

3. a.

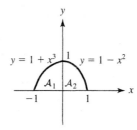

   b. $\frac{39}{64} < \mathcal{A}_1 < \frac{55}{64}$

   c. $\frac{17}{32} < \mathcal{A}_2 < \frac{25}{32}$

   d. $\frac{73}{64} < \mathcal{A} < \frac{105}{64}$

5. $M_5 = \frac{5 - 0}{5}\left(e^{-\frac{1}{2}} + e^{-\frac{3}{2}} + e^{-\frac{5}{2}} + e^{-\frac{7}{2}} + e^{-\frac{9}{2}}\right) \approx 0.953052$

7. a.

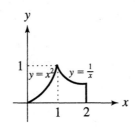

b. Compute T_n for $f(x)$ on $[0,1]$ and T_n for $g(x)$ on $[1,2]$. Add the results to get T_{2n} for the whole region.

c.

n	T_{2n}
20	1.02705
40	1.02662
60	1.02655
80	1.02652
100	1.02650
120	1.02650
140	1.02649
160	1.02649
180	1.02649
200	1.02649
220	1.02649
240	1.02648
260	1.02648

d. 1.026

SECTION 6.3

1. $A = 40$

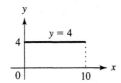

3. $A = \frac{9}{2}$

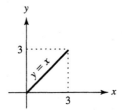

5. $A = 170$

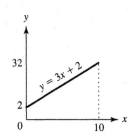

7. $A = 9$

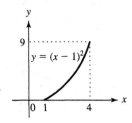

9. $A = e - 1$

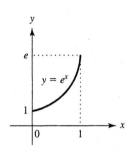

11. $A = \frac{27}{4}$

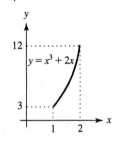

13. $A = \frac{\pi}{4}$

15. $A = \frac{1}{2}\ln(2\sqrt{6} + 5) + \sqrt{6} \approx 3.59$

17. $A = 1$

19. $A = \frac{1}{2}\ln\left(\frac{65}{16}\right) \approx 0.70$

21. $A = 6 - 2e \approx 0.56$

23.   **2.** $\frac{14}{3}$

   3. $\frac{17}{12}$

   5. $1 - e^{-5} \approx 0.99$

   6. $\frac{3}{4}$

   8. $\frac{5}{6} + \frac{4}{3}\sqrt{2} \approx 2.72$

SECTION 6.4

3. The Cartesian equation of an ellipse centered at point (x_0, y_0) is of the form $\left(\frac{x - x_0}{a}\right)^2 + \left(\frac{y - y_0}{b}\right)^2 = 1$, where a and b are the radii of the ellipse.

5. Consider the rhombus inscribed in the ellipse and the rectangle circumscribed about the ellipse. The area of the rhombus is $2ab$ and the area of the rectangle is $4ab$. It follows that the area of the ellipse can be approximated by $3ab$. (The exact value of the area is πab.)

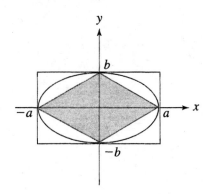

SECTION 6.5

1. $\mathcal{A} = 18$

3. $\mathcal{A} = \frac{8}{3}$

5. $\mathcal{A} = \frac{9}{2}$

7. $\mathcal{A} = \frac{1{,}741}{2{,}592}$

9. $\mathcal{A} = \frac{3}{2}$

11. a. $\mathcal{A} = \frac{1}{4}$ (area of circle with radius 1 − area of square with side length $\sqrt{2}$)

$\qquad = \frac{\pi}{4} - \frac{1}{2}$

  b. $\mathcal{A} = \int_0^1 (x - 1 + \sqrt{1 - x^2})\, dx = \frac{\pi}{4} - \frac{1}{2}$

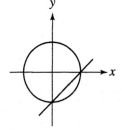

13. The two ribbons have the same area.

Argument 1. Let f be the function whose graph is the lower curve. Then, the upper curve is the graph of $f + b$. By the Fundamental Theorem of Calculus, the area of the curved ribbon is ab. This is exactly the area of the rectangular one.

Argument 2. Since f is continuous, the region between the two graphs can be, for n large, approximated by n parallelograms, each having base b and "height" $\frac{a}{n}$. The sum of the areas of all of these parallelograms is ab. As $n \to \infty$, the n parallelograms fit more and more the curved ribbon, and thus the area of the latter is ab.

Argument 3. Imagine, as Cavalieri did, that the curved ribbon is made up of a stack of line segments with the same length b. Then, rectify it into a rectangle.

CHAPTER 6 REVIEW

1. Hint: Divide a unit square into six congruent rectangles, as appropriate.

3. Area $= a^2 + 2ah$. Using a ruler, find approximate values for a and h and use these values to approximate the area $a^2 + 2ah$.

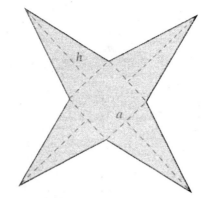

5. Area $\approx 2{,}940$ square meters.

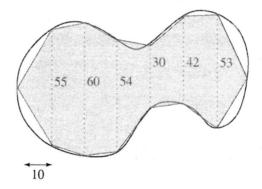

7. Area ≈ 1.207

9. Area $\approx \frac{496}{315}$

11. Lower bound $= f(-2) + f(-1) + f(0) + f(0) + f(1) + f(2) \approx 19.414$

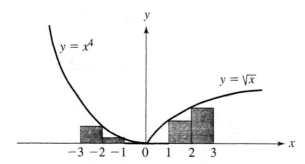

Upper bound $= f(-3) + f(-2) + f(-1) + f(1) + f(2) + f(3) \approx 102.146$

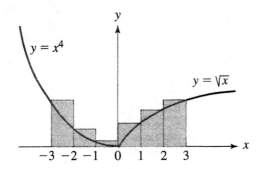

13. Using the left rectangular method, $\sum_{i=1}^{9} 1/i^2$ can be interpreted as the Riemann sum of the function $f(x) = 1/x^2$ over the interval $[1, 10]$ with nine tiles. (You may find other solutions.)

15. Using the trapezoidal method, $\frac{1}{20}\left(\frac{3}{2} + 2\sum_{i=1}^{9}\frac{1}{1 + \frac{i}{10}}\right)$ can be viewed as the Riemann sum of the function $f(x) = \frac{1}{1+x}$ over the interval $[0, 1]$ with ten tiles. (You may find other solutions.)

17. Area = 6

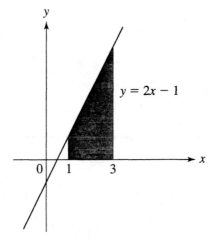

19. Area = $\frac{2}{5}$

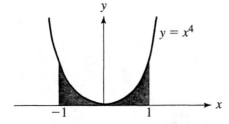

21. Area $= \frac{3}{8}$

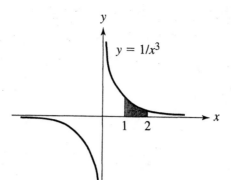

23. Area $= \frac{125}{6}$

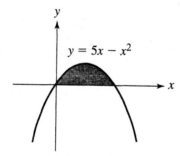

25. Area $= \frac{4}{15}$

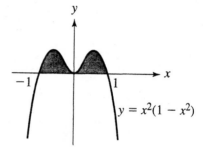

27. Area $= \frac{16}{3} - 2\sqrt{3}$

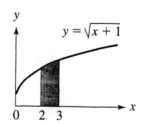

29. Area $= \frac{3}{10}$

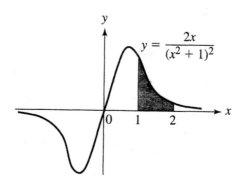

31. Area $= -\sqrt{3} + 2\sqrt{6}$

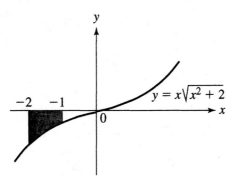

33. Area $= \frac{1}{2}(e - 1)$

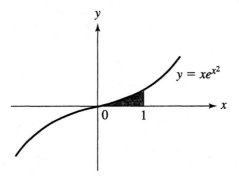

35. Sample answer: $\int_{-1}^{1}(7 - x^2)\,dx$ represents the area of the region in the plane that is above the x-axis and below the graph $y = 7 - x^2$, where $-1 \le x \le 1$.

37. Sample answer: $\int_{3}^{10}\left(\frac{1}{x+1} - \frac{1}{x^2+1}\right)dx$ represents the area of the region in the plane that is above the graph $y = \frac{1}{x^2+1}$ and below the graph $y = \frac{1}{x+1}$, where $3 \le x \le 10$. This region can also be described as the region in the plane that is above the x-axis and below the graph $y = \frac{1}{x+1} - \frac{1}{x^2+1}$, where $3 \le x \le 10$.

39. Area $= \frac{101}{4}$

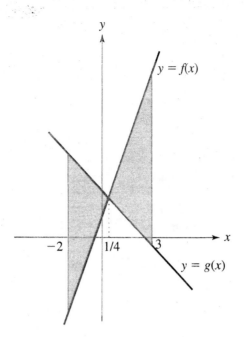

41. Area $= \frac{1}{3}$

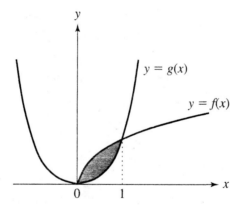

43. Area $= \frac{1}{15}$

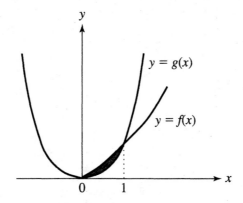

45. Area $= \frac{7}{12}$

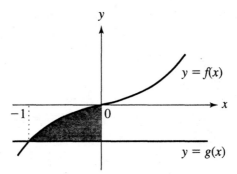

47. Area $= \frac{10 - 4\sqrt{2}}{3}$

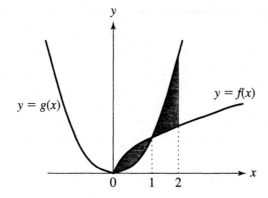

49. Area $= \frac{137 - 54\sqrt{3}}{15}$

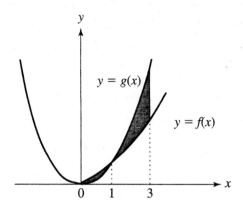

CHAPTER 7

SECTION 7.1

1. a.

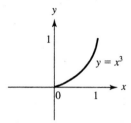

b.

n	$\mathcal{L}_n$
2	1.52317
4	1.54135
6	1.54496
8	1.54623
10	1.54682
12	1.54714
14	1.54733
16	1.54746
18	1.54754
20	1.54760

c. $\mathcal{L}_{20}$ is the best approximation, since every $\mathcal{L}_n$ is a lower bound for $\mathcal{L}$, and $\mathcal{L}_{20}$ is the largest value in the table.

d. $\mathcal{L} = \int_0^1 \sqrt{1 + 9x^4}\,dx \approx 1.54787$

3. a.

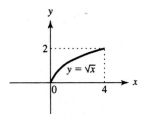

b.

n	L_n
2	4.53351
4	4.58118
6	4.60172
8	4.61310
10	4.62027
12	4.62515
14	4.62867
16	4.63131
18	4.63336
20	4.63498

c. L_{20} is the best approximation.

d. $L = \int_0^4 \sqrt{1 + \frac{1}{4x}}\,dx \approx 4.64678$

5. $L = \int_0^1 \sqrt{1 + 4x^2}\,dx$

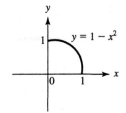

7. $L = \int_{-3}^3 \sqrt{16x^2 + 8x + 2}\,dx$

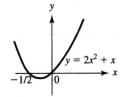

9. $L = \int_1^2 \sqrt{16x^6 - 32x^3 + 17}\,dx$

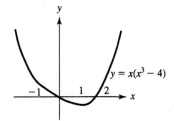

11. $f(x) = x^2/2, 0 \le x \le 2$

13. $f(x) = x^2/2 + x, 0 \le x \le 5$

15. The length of the graph $y = f(x) + \alpha$ is $\mathcal{L}$ since this graph is a vertical translation of $y = f(x)$; translating a graph does not change its length.

Alternatively, the formula for computing the length of the graph of a function depends only on the derivative of the function, and $f(x)$ and $f(x) + \alpha$ have the same derivative, $f'(x)$.

17. $\mathcal{L} = \frac{14}{3}$

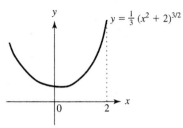

19. a. The Cartesian equation of an ellipse is $\left(\frac{x - x_0}{a}\right)^2 + \left(\frac{y - y_0}{b}\right)^2 = 1$, where (x_0, y_0) is the center of the ellipse and a, b are its radii.

  b. You can use objects that have the shape of an ellipse, measure their circumferences using a string and a ruler, measure their radii a and b, and try to see how the circumferences relate to the radii. Then, if possible, try to make a conjecture about the circumference of an ellipse in general.

  c. The circumference of the ellipse is given by $\frac{2}{a} \int_{-a}^{a} \sqrt{\frac{a^4 + (b^2 - a^2)x^2}{a^2 - x^2}} \, dx$.

  d. The integrand does not have an obvious antiderivative; it cannot be evaluated by hand using the techniques of integration learned so far. The answer provided by the calculator is too complicated.

  In the case of a circle, $a = b = r$, so the integrand becomes $\frac{2a}{\sqrt{a^2 - x^2}}$. The integral still cannot be computed using only the techniques of integration learned so far, but a calculator gives the result.

SECTION 7.2

1. $S = 56x^2$

3. $S = \frac{3\sqrt{3}}{2}x^2 + 3x\sqrt{h^2 + \frac{3}{4}x^2}$

5. $S = \int_0^2 2\pi \cdot e^x \sqrt{1 + e^{2x}}\, dx$

7. $S = \int_1^5 2\pi \cdot \frac{\sqrt{x^4+1}}{x^3}\, dx$

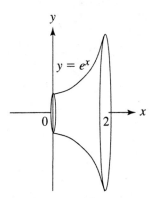

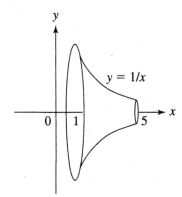

9. $S = 18\pi$

11. $S = 16\sqrt{5}\pi$

13. $S = \frac{2\sqrt{2} - 1}{9}\pi$

15. $S = 4\pi$

17. $S = \frac{\pi}{6}(17^{\frac{3}{2}} - 5^{\frac{3}{2}})$

19. a. Yes. You obtain the sphere by revolving about the x-axis the graph $y = \sqrt{R^2 - x^2}$, where $-R \le x \le R$. The surface area of the ith band is given by the integral $\int_{x_i}^{x_{i+1}} 2\pi R\, dx = 2\pi R(x_{i+1} - x_i)$. In each case, $x_{i+1} - x_i = \frac{R}{6}$. Thus, the area of each of the twelve bands is $\frac{\pi R^2}{3}$.

b. Yes. The integral can be computed as in the case $n = 12$, but now $x_{i+1} - x_i = \frac{2R}{n}$, so the area of each of the n bands is $\frac{4\pi R^2}{n}$.

SECTION 7.3

1. 12,800 cubic feet

3.

    12. a. Radius 1 cm, height 10 cm

    b. Volume of empty space $= 40 - 10\pi \approx 8.6$ cm³

    c. $\frac{\text{volume of can}}{\text{volume of box}} = \frac{\pi}{4}$

    d. The ratio $\frac{\text{volume of can}}{\text{volume of box}}$ is always $\frac{\pi}{4}$, independent of the size of the can and box.

    13. a. The shorter cylinder has the greater volume.

    b. The shorter cylinder has the greater surface area.

5. $V = \frac{2\pi}{5}$

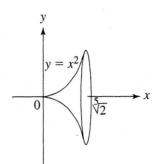

7. $V = \frac{2{,}186\pi}{7}$

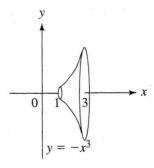

9. $V = \frac{15\pi}{32}$

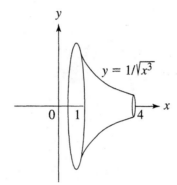

11. $V = 16\pi$

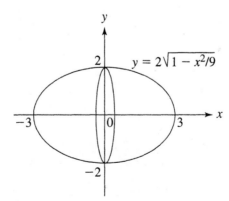

13. $V = \frac{\pi}{2}$

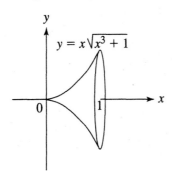

$y = x\sqrt{x^3 + 1}$

15. $V = 2\pi$

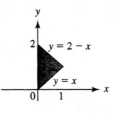

$y = 2 - x$

$y = x$

17. $V = 4\pi$

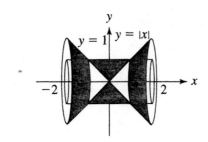

$y = 1$ $y = |x|$

19. I. a.

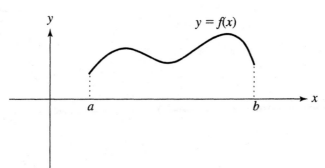

$y = f(x)$

b.

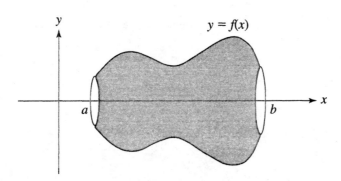

$y = f(x)$

c. Cylinders, cones, truncated cones, spheres, ellipsoids.

d. $V = \pi \int_a^b [f(x)]^2 \, dx$

II. a.

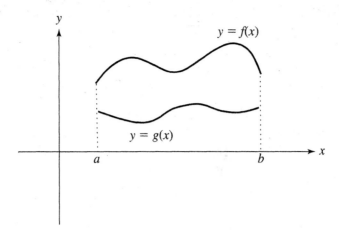

b.

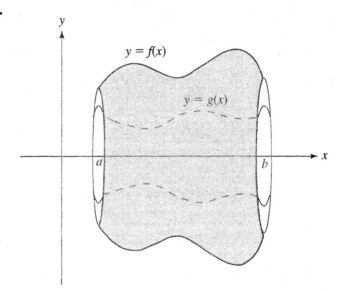

c.

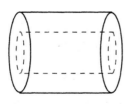

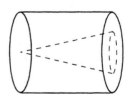

A cylinder with a
cylindrical hole

A cylinder with a
conical hole

d. Denote by V_f and V_g the volumes of the solids that are obtained by revolving about the x-axis the graphs of f and g, respectively. We have $V = V_f - V_g$.

e. $V = \pi \int_a^b \left[[f(x)]^2 - [g(x)]^2 \right] dx$

CHAPTER 7 REVIEW

1. a.

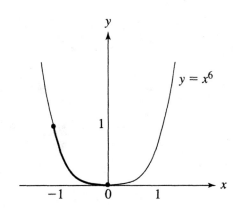

b. $x_i = a + i\dfrac{b-a}{n}$

$$L_n = \sum_{i=0}^{n-1} \sqrt{\frac{(b-a)^2}{n^2} + [f(x_{i+1}) - f(x_i)]^2}$$

n	L_n
5	1.663
10	1.67016
15	1.67149
20	1.67195
25	1.67217
30	1.67228
35	1.67235
40	1.6724
45	1.67243
50	1.67245

c. 1.67245

3. a.

b. $x_i = a + i\dfrac{b-a}{n}$

$$L_n = \sum_{i=0}^{n-1} \sqrt{\frac{(b-a)^2}{n^2} + [f(x_{i+1}) - f(x_i)]^2}$$

n	$\mathcal{L}_n$
10	9.06569
20	9.07082
30	9.07209
40	9.0726
50	9.07286
60	9.07301
70	9.07311
80	9.07317
90	9.07322
100	9.07325

c. 9.07325

5. a.

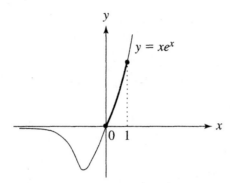

b. $x_i = a + i\dfrac{b - a}{n}$

$$\mathcal{L}_n = \sum_{i=0}^{n-1} \sqrt{\frac{(b - a)^2}{n^2} + [f(x_{i+1}) - f(x_i)]^2}$$

n	$\mathcal{L}_n$
2	2.92295
4	2.93002
6	2.93133
8	2.9318
10	2.93201
12	2.93213
14	2.9322
16	2.93224
18	2.93227
20	2.93229

c. 2.93229

7. $\int_0^3 \sqrt{2 + 2e^x + e^{2x}}\, dx$

9. $\int_{-1}^{0} \sqrt{\frac{4x + 9}{4x + 8}} \, dx$

11. Any graph $y = -\frac{1}{2x^2} + C$, where $3 \le x \le 4$ and C is a constant, works.

13. Any graph $y = \frac{2}{3}(x + 2)^{3/2} + C$, where $-1 \le x \le 1$ and C is a constant, works.

15. Any graph $y = \frac{1}{3}x^3 + \frac{1}{4x} + C$, where $1 \le x \le 5$ works.

17. Length $= \frac{80\sqrt{10}}{27} - \frac{8}{27}$

19. Length $= \frac{123}{32}$

21. Surface area $= 2\pi \int_0^1 x^2 \sqrt{1 + 4x^2} \, dx$

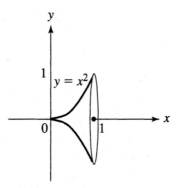

23. Surface area $= -2\pi \int_{-1}^{0} xe^x \sqrt{1 + (x^2 + 2x + 1)\, e^{2x}} \, dx$

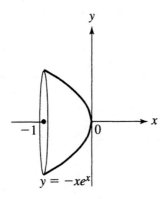

25. Surface area $= 2\pi \int_{-2}^{2} (4x^2 - x^4)\sqrt{1 + (8x - 4x^3)^2} \, dx$

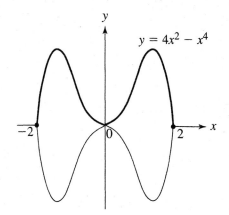

$$y = 4x^2 - x^4$$

27. Surface area $= 144\pi$

29. Surface area $= 6\sqrt{17}\,\pi$

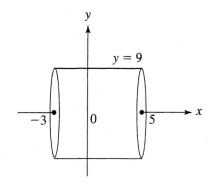

$$y = 9$$

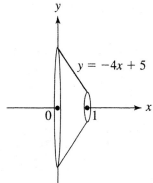

$$y = -4x + 5$$

31. Surface area $= \dfrac{5\sqrt{10}}{3}\,\pi$

33. Surface area $= \dfrac{27 - 5\sqrt{5}}{6}\,\pi$

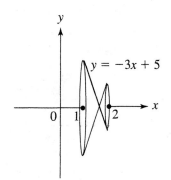

$$y = -3x + 5$$

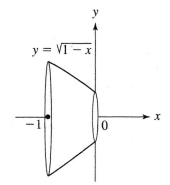

$$y = \sqrt{1 - x}$$

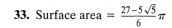

35. Volume $= 4\pi$

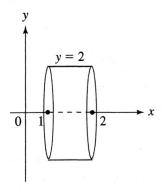

37. Volume $= \frac{13}{3}\pi$

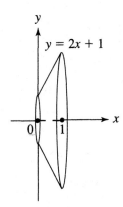

39. Volume $= \frac{2\pi}{35}$

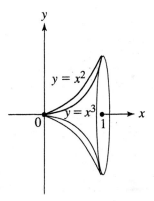

41. Volume $= \frac{5}{3}\pi$

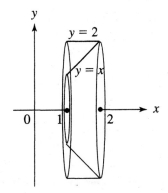

43. Volume $= \frac{596}{15}\pi$

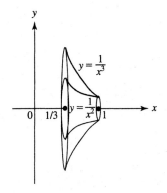

45. Volume $= 24\pi$

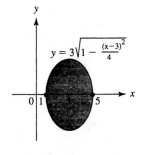

47. Volume $= \frac{5}{2}\pi$

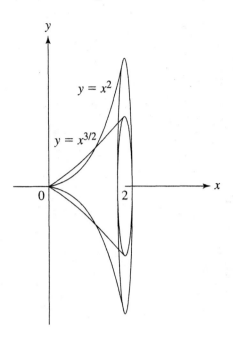

49. Volume $= \frac{2,048}{15}\pi$

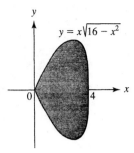

Photo Credits

Page 2   Reprinted with permission from *Mathematics in Context: Patterns and Figures.* © 2003 by Encyclopaedia Britannica, Inc.

Page 3   Reprinted with permission from *Mathematics in Context: Patterns and Figures.* © 2003 by Encyclopaedia Britannica, Inc.

Page 4   Reprinted with permission from *Mathematics in Context: Patterns and Figures.* © 2003 by Encyclopaedia Britannica, Inc.

Page 5   Reprinted with permission from *Mathematics in Context: Patterns and Figures.* © 2003 by Encyclopaedia Britannica, Inc.

Page 8   Used by permission of McDougal Littell, Inc., a division of Houghton Mifflin.

Page 9   Used by permission of McDougal Littell, Inc., a division of Houghton Mifflin.

Page 26   Used by permission of McDougal Littell, Inc., a division of Houghton Mifflin.

Page 27   Used by permission of McDougal Littell, Inc., a division of Houghton Mifflin.

Page 28   From *Connected Mathematics: Frogs, Fleas and Painted Cubes* by Glenda Lappan, James T. Fey, William M. Fitzgerald, Susan N. Friel, and Elizabeth Defanis Phillips. © 2004 by Michigan State University. Published by Pearson Education, Inc., publishing as Pearson Prentice Hall. Used by permission.

Page 29   From *Connected Mathematics: Frogs, Fleas and Painted Cubes* by Glenda Lappan, James T. Fey, William M. Fitzgerald, Susan N. Friel, and Elizabeth Defanis Phillips. © 2004 by Michigan State University. Published by Pearson Education, Inc., publishing as Pearson Prentice Hall. Used by permission.

Page 30   From *Connected Mathematics: Frogs, Fleas and Painted Cubes* by Glenda Lappan, James T. Fey, William M. Fitzgerald, Susan N. Friel, and Elizabeth Defanis Phillips. © 2004 by Michigan State University. Published by Pearson Education, Inc., publishing as Pearson Prentice Hall. Used by permission.

Page 31   From *Connected Mathematics: Frogs, Fleas and Painted Cubes* by Glenda Lappan, James T. Fey, William M. Fitzgerald, Susan N. Friel, and Elizabeth Defanis Phillips. © 2004 by Michigan State University. Published by Pearson Education, Inc., publishing as Pearson Prentice Hall. Used by permission.

Page 39   From *Connected Mathematics: Growing, Growing, Growing* by Glenda Lappan, James T. Fey, William M. Fitzgerald, Susan N. Friel, and Elizabeth Defanis Phillips. © 2004 by Michigan State University. Published by Pearson Education, Inc., publishing as Pearson Prentice Hall. Used by permission.

Page 40   From *Connected Mathematics: Growing, Growing, Growing* by Glenda Lappan, James T. Fey, William M. Fitzgerald, Susan N. Friel, and Elizabeth Defanis Phillips. © 2004 by Michigan State University. Published by Pearson Education, Inc., publishing as Pearson Prentice Hall. Used by permission.

Page 43   From *Connected Mathematics: Growing, Growing, Growing* by Glenda Lappan, James T. Fey, William M. Fitzgerald, Susan N. Friel, and Elizabeth Defanis Phillips. © 2004 by Michigan State University. Published by Pearson Education, Inc., publishing as Pearson Prentice Hall. Used by permission.

Page 46   Used by permission of McDougal Littell, Inc., a division of Houghton Mifflin.

Page 47   Used by permission of McDougal Littell, Inc., a division of Houghton Mifflin.

Page 48   Used by permission of McDougal Littell, Inc., a division of Houghton Mifflin.

Page 49   Used by permission of McDougal Littell, Inc., a division of Houghton Mifflin.

Page 61   Used by permission of McDougal Littell, Inc., a division of Houghton Mifflin.

Page 62   Used by permission of McDougal Littell, Inc., a division of Houghton Mifflin.

Page 66   Reprinted with permission from *Mathematics in Context: Tracking Graphs.* © 2003 by Encyclopaedia Britannica, Inc.

Page 67	Reprinted with permission from *Mathematics in Context: Tracking Graphs.* © 2003 by Encyclopaedia Britannica, Inc.
Page 68	Reprinted with permission from *Mathematics in Context: Tracking Graphs.* © 2003 by Encyclopaedia Britannica, Inc.
Page 72	Reprinted with permission from *Mathematics in Context: Tracking Graphs.* © 2003 by Encyclopaedia Britannica, Inc.
Page 73	Reprinted with permission from *Mathematics in Context: Tracking Graphs.* © 2003 by Encyclopaedia Britannica, Inc.
Page 74	Reprinted with permission from *Mathematics in Context: Tracking Graphs.* © 2003 by Encyclopaedia Britannica, Inc.
Page 75	Reprinted with permission from *Mathematics in Context: Tracking Graphs.* © 2003 by Encyclopaedia Britannica, Inc.
Page 78	Reprinted with permission from *Mathematics in Context: Tracking Graphs.* © 2003 by Encyclopaedia Britannica, Inc.
Page 79	Reprinted with permission from *Mathematics in Context: Tracking Graphs.* © 2003 by Encyclopaedia Britannica, Inc.
Page 80	Reprinted with permission from *Mathematics in Context: Tracking Graphs.* © 2003 by Encyclopaedia Britannica, Inc.
Page 81	Reprinted with permission from *Mathematics in Context: Tracking Graphs.* © 2003 by Encyclopaedia Britannica, Inc.
Page 82	Reprinted with permission from *Mathematics in Context: Tracking Graphs.* © 2003 by Encyclopaedia Britannica, Inc.
Page 154	Reprinted with permission from *Mathematics in Context: Get the Most Out of It.* © 1997 by Encyclopaedia Britannica, Inc.
Page 155	Reprinted with permission from *Mathematics in Context: Get the Most Out of It.* © 1997 by Encyclopaedia Britannica, Inc.
Page 158	Reprinted with permission from *Mathematics in Context: Get the Most Out of It.* © 1997 by Encyclopaedia Britannica, Inc.
Page 164	From *MathScape: Family Portraits.* © 1991, Glencoe McGraw-Hill. Reprinted with permission.
Page 176	Used by permission of McDougal Littell, Inc., a division of Houghton Mifflin.
Page 177	Used by permission of McDougal Littell, Inc., a division of Houghton Mifflin.
Page 183	Used by permission of McDougal Littell, Inc., a division of Houghton Mifflin.
Page 184	Used by permission of McDougal Littell, Inc., a division of Houghton Mifflin.
Page 217	Used by permission of McDougal Littell, Inc., a division of Houghton Mifflin.
Page 218	Used by permission of McDougal Littell, Inc., a division of Houghton Mifflin.
Page 232	From *Connected Mathematics: Covering and Surrounding* by Glenda Lappan, James T. Fey, William M. Fitzgerald, Susan N. Friel, and Elizabeth Defanis Phillips. © 2004 by Michigan State University. Published by Pearson Education, Inc., publishing as Pearson Prentice Hall. Used by permission.
Page 249	Used by permission of McDougal Littell, Inc., a division of Houghton Mifflin.
Page 250	Used by permission of McDougal Littell, Inc., a division of Houghton Mifflin.
Page 251	Used by permission of McDougal Littell, Inc., a division of Houghton Mifflin.
Page 252	Used by permission of McDougal Littell, Inc., a division of Houghton Mifflin.
Page 253	Used by permission of McDougal Littell, Inc., a division of Houghton Mifflin.
Page 254	Used by permission of McDougal Littell, Inc., a division of Houghton Mifflin.
Page 281	Used by permission of McDougal Littell, Inc., a division of Houghton Mifflin.
Page 282	Used by permission of McDougal Littell, Inc., a division of Houghton Mifflin.
Page 283	Used by permission of McDougal Littell, Inc., a division of Houghton Mifflin.
Page 288	Used by permission of McDougal Littell, Inc., a division of Houghton Mifflin.
Page 289	Used by permission of McDougal Littell, Inc., a division of Houghton Mifflin.
Page 290	Used by permission of McDougal Littell, Inc., a division of Houghton Mifflin.
Page 293	*Discovering Geometry,* 3rd Edition, Key Curriculum Press, 1150 65th Street, Emeryville, CA 94608, 1-800-995-MATH, www.keypress.com.

Index